江西水问题研究与实践丛书

水工安全与防灾减灾

SHUIGONG ANQUAN
YU FANGZAI JIANZAI

江西省水利科学研究院◎编

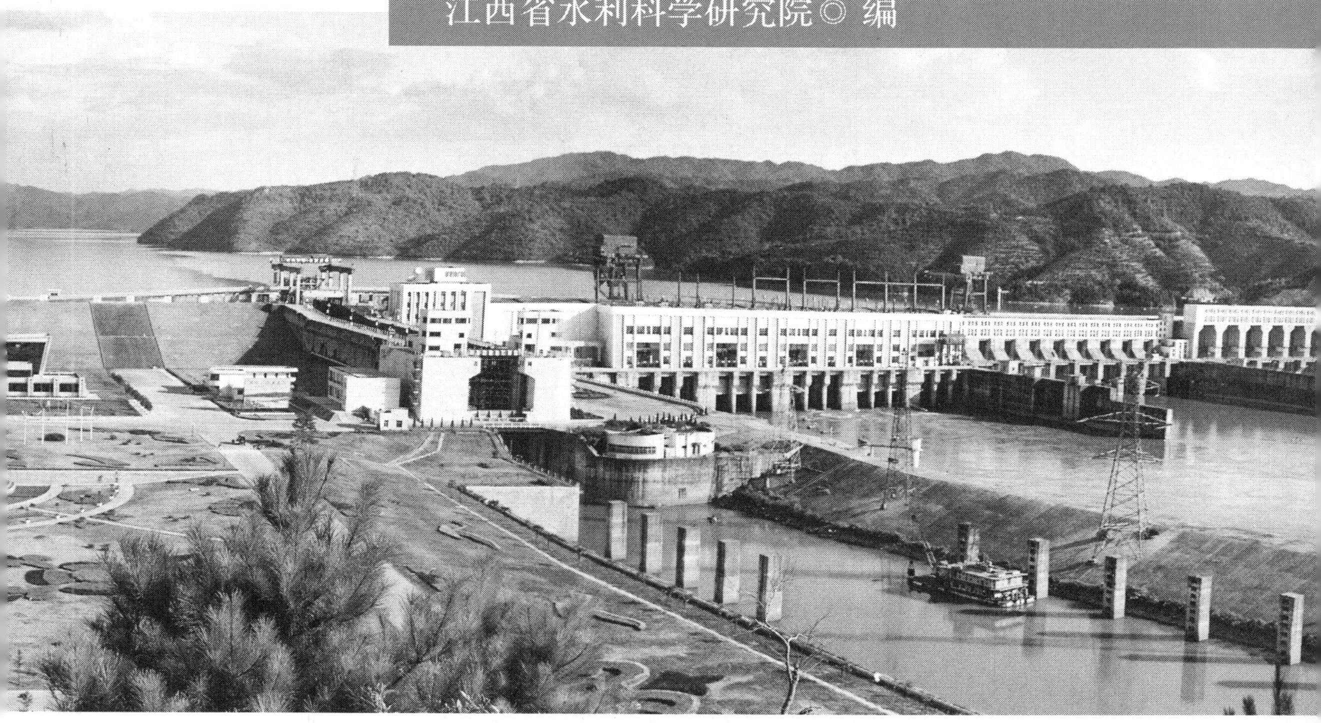

中国水利水电出版社
www.waterpub.com.cn

内 容 提 要

本书是《江西水问题研究与实践丛书》的《水工安全与防灾减灾》分册，收集和展现了江西省水利科学研究院水工程安全与灾害防治研究所、建材与岩土研究所、信息与自动化研究所、大坝安全管理科与新组建的水工程安全与防灾减灾科研创新团队所承担的公益科研、技术服务和技术咨询项目中代表性成果，包括工程安全分析与评价、工程风险评估、工程技术推广应用、工程管理等几个方面，旨在总结和提炼水工程安全与防灾减灾领域已有科研和技术成果，夯实学科发展基石，鼓励科研创新，促进学科发展和技术交流。

本书可供从事水工结构、岩土工程及大坝安全管理等领域的科研、规划、设计、施工、建设管理人员参考和借鉴，也可作为高校相关专业的参考用书。

图书在版编目（CIP）数据

水工安全与防灾减灾 / 江西省水利科学研究院编
. -- 北京 : 中国水利水电出版社, 2014.8
(江西水问题研究与实践丛书)
ISBN 978-7-5170-2428-6

Ⅰ. ①水… Ⅱ. ①江… Ⅲ. ①水利水电工程－工程施工－安全管理－江西省②水利水电工程－灾害防治－江西省 Ⅳ. ①TV513②TV698

中国版本图书馆CIP数据核字(2014)第204386号

书　　名	江西水问题研究与实践丛书 **水工安全与防灾减灾**
作　　者	江西省水利科学研究院　编
出版发行	中国水利水电出版社 (北京市海淀区玉渊潭南路1号D座　100038) 网址：www.waterpub.com.cn E-mail：sales@waterpub.com.cn 电话：(010) 68367658（发行部）
经　　售	北京科水图书销售中心（零售） 电话：(010) 88383994、63202643、68545874 全国各地新华书店和相关出版物销售网点
排　　版	中国水利水电出版社微机排版中心
印　　刷	北京纪元彩艺印刷有限公司
规　　格	184mm×260mm　16开本　22.5印张　534千字
版　　次	2014年8月第1版　2014年8月第1次印刷
印　　数	0001—1000册
定　　价	**88.00**元

《江西水问题研究与实践丛书》
编　委　会

《水工安全与防灾减灾》
编　委　会

序

水是生命之源、生产之要、生态之基，是人类社会发展中最不可或缺的因子。水具有利、害两重性，决定了人类发展须不断进行兴水利、除水害。水利作为公益性、基础性、战略性行业，水安全关系到防洪安全、供水安全、粮食安全、经济安全、生态安全、国家安全，水利的地位和作用在国家事业发展中日益突出。

鄱阳湖位于江西北部，长江中下游南岸，是江西的母亲湖，中国最大的淡水湖，列入国际重要湿地名录。流域总面积16.22万km^2，约占长江流域面积的9%，江西94%的国土面积属鄱阳湖流域，鄱阳湖对江西发展及长江下游水生态安全具有重要的影响。

科学技术是第一生产力，科技创新是提高社会生产力和综合国力的战略支撑。随着经济社会的快速发展，以水资源短缺、水生态退化、水环境恶化及水土流失加剧为标志的水危机，已成为制约我国经济社会发展的瓶颈。应对水危机的严重挑战，水利科技人员必须率先发声。

走着，走着，就60年了！作为个人，该喝满花甲酒，作为一个单位，尤其是科研单位是该回顾回顾、总结总结，因为还要往下走，还要走得更远！

江西省水利科学研究院建院已届60年，该院作为一家省级公益性水利科研单位，经过几代人的共同开拓，取得了丰硕的科研成果。60年来，几代水利科技工作者紧紧围绕江西经济社会发展战略目标及水利中心工作任务，针对江西存在的重大水问题，以保持鄱阳湖一湖清水为至高点，积极开展鄱阳湖流域综合治理技术研究，有力地支撑了江西水利事业的发展。江西省水利科学研究院在多年水利技术研究及管理的基础上，总结水利试验与研究、水工安全与防灾减灾、农业水利技术与应用、水资源综合调控与管理、水生态环境保护与综合治理等方面的技术成果，编辑出版了《江西水问题研究与实践》丛书。该丛书既是该院多年来的科技创新成果，又可作为今后水问题研究与实践的基础与借鉴，对促进江西水利事业发展具有积极作用。

习近平总书记从战略的高度提出“节水优先、空间均衡、系统治理、两手发力”的16字治水新思路。水利一方面迎来难得的改革发展机遇，更迎来了巨大挑战，这就要求江西水利科技工作者，坚持以科学发展观为指导，准确把握水利发展的形势与要求，找准制约江西经济社会发展的水利关键技术问题，充分发挥高端技术人才优势，积极开展水利技术研究，加强水利科技创新，以技术创新引领江西水利事业快速发展。

回首60年的创新发展历程，令人欣慰，眺望水利科技发展前景，激人奋进。我有幸在江西省水利科学研究院工作了6年，借作序之际，祝愿江西省水利科学研究院始终发扬“献身、负责、求实”的水利行业精神和“团结、求实、开拓、创新”的团队精神，保持“大江东去，浪淘尽，千古风流人物”的豪迈勇气，“问苍茫大地，谁主沉浮”的责任担当，“水利万物而不争”的敬业奉献，为实现江西水利改革发展宏伟目标，建设富裕和谐秀美江西做出新的更大贡献。

孙晓山

2014年7月24日

前　言

水资源作为基础性自然资源和战略性经济资源，对经济社会发展具有决定性作用。江西省位于我国南方，地处长江中下游南岸，降雨充沛，水系发达。全省多年平均降水量约1638.4mm，列全国第四位；多年平均水资源总量1565亿m^3，列全国第七位。全省地表水多年平均年径流量为1545.5亿m^3，平均径流深为925.7 mm，多年平均地下水资源量为379.0亿m^3，水资源可利用量约为423亿m^3。但由于降雨时空分布不均，加上工业化、城镇化、农业现代化进程的加快，我省局部地区水资源保障程度还不高、水生态退化还未遏制、水环境污染还很严重、水工程调控能力还不足，水管理手段还很落后、水土流失还在加剧等水问题日益突出。因此，充分发挥水利科技作用，以水利科技创新引领水利事业及支撑经济社会发展具有重要战略意义。

总结过去，成绩斐然。江西省水利科学研究院成立于1954年，是全国较早成立的专业门类齐全，主要从事水工安全与防灾减灾、水资源、水生态环境、农村水利、水利信息化、工程质量检测、河湖治理、水利发展战略等领域科学研究、技术支撑和成果推广为一体的公益性省级综合水利水电科研机构。全院现有职工194人，其中享受国务院特殊津贴专家2人、水利部5151工程部级人选1人，江西省赣鄱英才555工程人选1人，江西省新世纪百千万人才工程人选4人。各类专业技术人员159人，其中教授级高级工程师10人，高级工程师35人，工程师47人。本科及以上学历142人，其中博士12人，硕士83人。建有鄱阳湖模型试验研究基地、江西省鄱阳湖水资源与环境重点实验室、江西省水工安全工程技术研究中心、院士工作站、博士后科研工作站等5大科研平台，配备有一批先进的试验仪器设备。拥有大坝安全鉴定、病险水库除险加固蓄水安全鉴定、水闸安全鉴定、水资源论证、岩土工程类检测、混凝土工程类检测、量测类检测、金属结构类检测、机械电气类检测等9项甲级资质；水土保持监测、水文水资源调查与评价、水利水电工程施工总承包等3项乙级资质（2级）；地质灾害治理工程设计、工程咨询、水利工程设计综合等3项丙级资质。多年来，以水利及当地经济发展需求为导向，共承担各级各类科研项目300余项，发表论文600余篇，出版专著10多部，获各级奖励100余项，获专利及软件著作权20余项。在防洪抗旱、水资源调控与管理、水生态环境治理与保护、水工程安全鉴定与监测、农田水利、节水灌溉、水利信息化、工程质量检测、河湖治理、水工模型试验、水政策研究等方面取得了一大批先进实用成果，有力地支撑区域经济社会发展。

展望未来，任重道远。为适应江西水利改革发展需求，更好助推区域经济社会发展，江西省水利科学研究院在“节水优先，空间均衡，系统治理，两手发力”新时期治水思路指导下，按照“立足江西、面向全国、放眼世界”的发展定位；“团队—项目—基地—人才”四位一体的发展模式；“科研立院、技术兴院、人才强院、管理固院、文化铸院”的发展文化；着力打造水资源综合调控与管理、农村水利、水工安全与防灾减灾、鄱阳湖模型试验、水生态环境保护与综合治理等五大科研创新团队，实现全国省级一流水科院的目标。出版《江西水问题研究与实践》丛书，一是对多年来所取得的优秀成果进行梳理与总结，二是可为深入开展相关领域技术研究、技术集成及技术推广提供借鉴，三是可起到与水利界同仁进行技术交流作用。

本丛书的出版得到许多曾在江西省水利科学研究院工作过的领导、专家的帮助和支持，在此表示衷心的感谢。鉴于水平有限，本书难免存在疏漏和不足，恳请读者批评指正。

编者

2014 年 7 月

目　录

二、工程风险评估

三、工程技术推广应用

四、工程管理

一、工程安全分析评价

SHUIGONGANQUANYUFANGZAIJIANZAI

Research on the Key Technology & Its Application to Reinforcement Project Post-evaluation for Dangerously Weak Dams

Wu Haizhen, *Hong Wenhao*, *Hu Qiang*

JiangXi Provincial Institute of Water Sciences

Abstract: According to the Fuzzy Analytic Hierarchy Process (FAHP) and the project success degree comprehensive evaluation theory, this paper established the judgment matrix with the aid of expert opinions, and put forward the fuzzy comprehensive evaluation theory for reinforcement project post-evaluation of dangerously weak dams. Based on the characteristics of the reinforcement project for dangerously weak dams, as well as on the research of the hierarchical relationships and relative importance among various factors, the paper proposed the main influence factor of reinforcement project which may have impact on the effectiveness and efficacy in each reinforcement stage, derived the weighted value of each index, and gave the scoring methods and scoring details for each index so as to establish the reinforcement project post-evaluation index system for dangerously weak dams. Depending on the total scores of dangerously weak dam reinforcement project, targeting the evaluation of the total effect of reinforcement project and with the aid of expert scores, the paper also studied and determined the fuzzy mapping matrix of each index, judgment matrix and index weight, and put forward a fuzzy comprehensive evaluation method with strong operability and high practical value for dangerously weak dam reinforcement project post-evaluation. The practical engineering application proves that the above-mentioned method is both feasible and rational, and is of high academic value and popularization and application value.

Key words: Dangerously weak dam; Reinforcement project; Fuzzy comprehensive evaluation; Post-evaluation

1 Introduction

Safety of reservoir (and hydropower station) dams concerns the national welfare and the people's livelihood. There are more than 30,000 dangerously weak dams throughout

本文发表于 2013 年。

the country, accounting for 36% of the total dams. If any dam failure occurs, it will not only cause damage to the dam structure, but also bring personal injury to the downstream public and lead to serious economic, social and environmental losses. Since 2000, China has been intensifying investment on the reinforcement of dangerously weak reservoirs (hereinafter referred to reinforcement) and launching a massive reinforcement project construction. By 2012, the national total investment on reinforcement exceeded RMB 70 billion Yuan which has been completed basically. After reinforcement, with the significant reduction of the ratio of dangerously weak dams, a large number of reinforced reservoirs begin to produce better social and economic benefits. However, dangerously weak dams have the characteristics of discontinuity and periodicity and they occur and develop in dynamic process. Therefore, the reinforcement should be a long-term, regular and systematic work. Correspondingly, the reinforcement project need huge investment, long construction period and have wide influence, which involve many work links and processes. So reinforcement project quality control, afterwards running condition and performance evaluation, economic benefits and social impact estimation are common concerns to the decision-making departments, engineering construction and operation management units. To solve these problems, it is urgent to establish a scientific, reasonable and feasible project post-evaluation method. In the present study, some post-evaluation methods both at home and abroad for railway engineering, highway engineering and other construction projects have been discussed. Also, some positive explorations have taken place in the field of construction of water conservancy project, and some research results have been achieved. For example, Wen Hui[1] made some exploratory research on the definition and significance of the post evaluation of water conservancy construction project as well as the characteristics, content, principles and others of the post evaluation, Zhang Rentian and Zhang Jinmiao[2] studied the post evaluation system for water conservancy construction projects. Han Dong[3] made pilot study on the application of the post-evaluation methods in the water conservancy construction project post evaluation. Sun Xiangqin and Qian Shanyang[5] discussed the main contents involved in the the water conservancy construction project post evaluation, including revaluation of fixed assets, socio-economy, finance, engineering, environment, operating management, society and immigrants. However, none of these studies involved reinforcement project. Few research results related to reinforcement project post-evaluation index system, methods and other fields can be found.

Based on the Fuzzy Analytic Hierarchy Process (FAHP) and the Comprehensive Evaluation Theory of Project Success Degree, this paper studies and puts forward the fuzzy comprehensive evaluation of reinforcement project post-evaluation. The paper builds an index system for reinforcement project post-evaluation according to the reinforcement project characteristics. Also, with the aid of expert scores method, the paper determines the fuzzy mapping

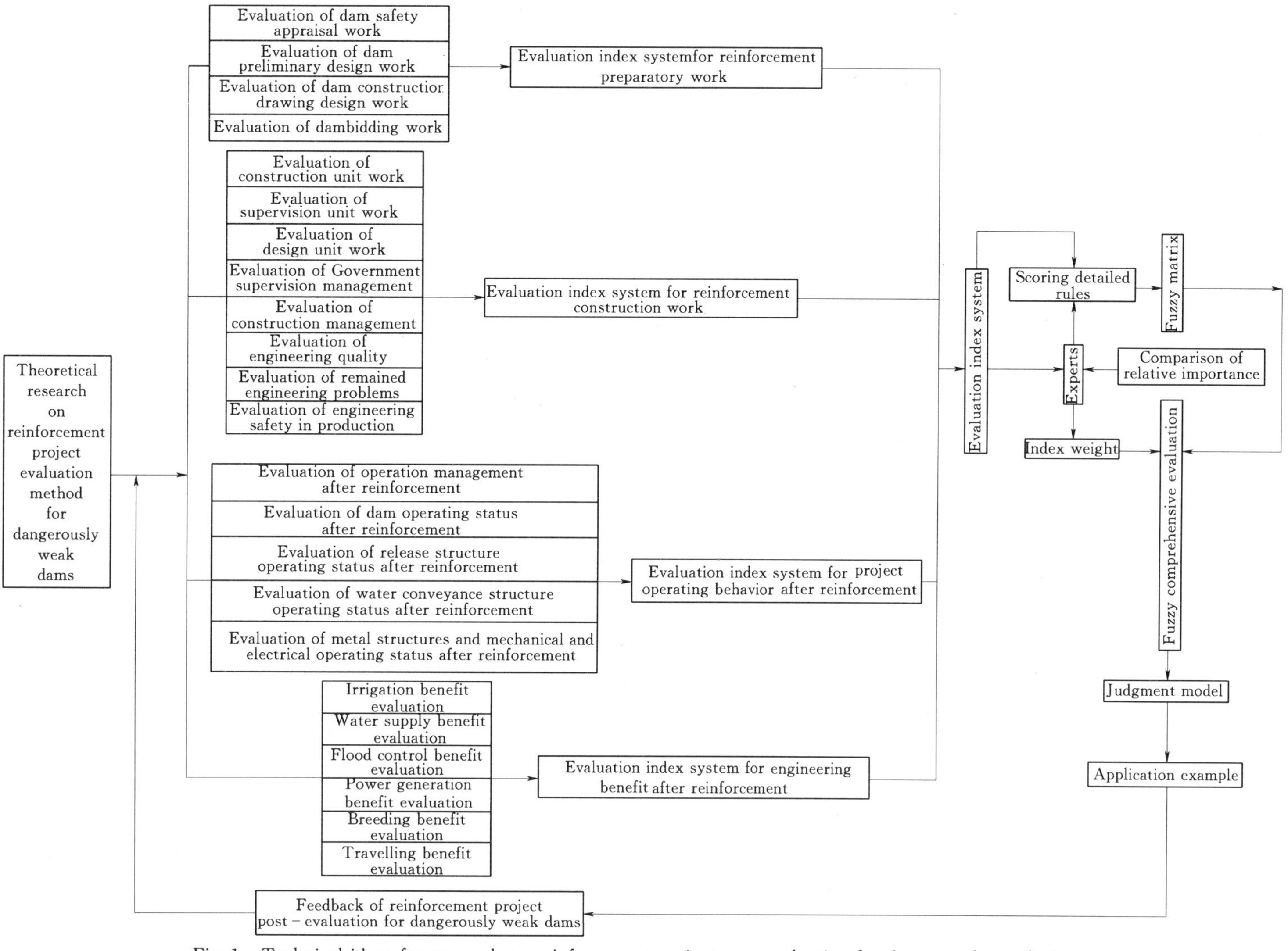

Fig. 1 Technical ideas for research on reinforcement project post-evaluation for dangerously weak dams

matrix of each index, judgment matrix and index weight, and puts forward a fuzzy comprehensive evaluation method with strong operability and high practical value to apply it to practical engineering. The main technical route of this paper is shown in Fig. 1.

2 Post-evaluation methods for dangerously weak dams

2.1 *General idea*

The effect of reinforcement project is of extensive ambiguity. To achieve the objectives[6], the membership rule of fuzzy theory is often applied to such uncertainty for quantitative analysis. Also, the method of project success degree can be applied to measure the reinforcement effect. On these bases, with the idea of AHP method and the comprehensive evaluation theory of project success degree, this paper studies and proposes the fuzzy comprehensive evaluation method for proposed reinforcement projects. Its overall outline and implementation steps are described as follows:

(1) Division of project implementation phase. In combination with the features and the main contents of reinforcement project, the project implementation process can be divided into reinforcement preparatory work, reinforcement construction process, engineering operating status after reinforcement and project benefits after reinforcement.

(2) Build-up of evaluation structure model. With the idea of AHP method, the factors affecting reinforcement effect can be classified by properties to establish the hierarchical relationships of its influence, and ultimately to form a four-level analytical structure model.

(3) Establishment of index system. Based on the basic theory of fuzzy comprehensive evaluation method, an evaluation index system can be established for various stages of reinforcement project.

(4) Setup of weight matrix at various levels. Based on the evaluation index system, the judgment matrix will be set up at the appropriate level. With the expert opinions, the weight corresponding to each index at various levels will be derived to build the weight matrix at various levels.

(5) Establishment of scoring details for the bottom indexes. With the expert opinions, based on the scoring details, the comment of reinforcement effect will be converted to the corresponding scores. The fuzzy mapping matrix corresponding to the index will be calculated through the matrix.

(4) The matrix obtained in the previous step will be multiplied by the weight matrix corresponding to its previous level to get the new fuzzy mapping matrix. Through such repeated calculations, the target matrix of this reinforcement project can be obtained. Finally, the target matrix will be multiplied by the score matrix, and this product is just the total scores of reinforcement project.

(7) The overall satisfaction of the reinforcement project will be determined based on

Evaluation Level.

Set and Their Corresponding Scores for Overall Satisfaction of Reinforcement Project (Table 1) .

Table 1　Evaluation level set and their corresponding meanings for overall satisfaction of reinforcement project

Degree of satisfaction	Scores	Explanation
Excellent	90～100	In the whole process of reinforcement of dangerously weak dam, each work link has completely reached or exceeded the relevant requirements. None shortcoming exists in each work link, and all work links are closely linked up. All work links comply with the relevant statutory and regulatory processes and their technical specifications
Good	80～89	In the whole process of reinforcement of dangerously weak dam, the vast majority of work links have completely reached or exceeded the relevant requirements. None serious problem exists in each work link basically, and all work links are closely linked up. Most work links comply with the relevant statutory and regulatory processes and their technical specifications
Critical	70～79	In the whole process of reinforcement of dangerously weak dam, most work links have reached or exceeded the relevant requirements except shortcomings of a few links. Most work links comply with the relevant statutory and regulatory processes and their technical specifications
Poor	60～69	In the whole process of reinforcement of dangerously weak dam, only part work links have reached the relevant requirements. Some work links cannot comply with the relevant statutory and regulatory processes and their technical specifications
Very poor	0～59	In the whole process of reinforcement of dangerously weak dam, most work links cannot meet the relevant provisions. Basically, none work links can meet the relevant requirements. Most work links have shortcomings, or does not comply with the relevant statutory, regulatory processes and their technical specifications

2.2　*Establishment of evaluation structural model*

When the hierarchical structure model for evaluation at the reinforcement stage is established with the AHP method, the evaluation object will be methodized and layered. Then, a four-level analytic structure model (Table 2) will be formed in accordance with aggregate combinations at different levels. Among which, the bottom level of the hierarchical structure is the expert score index level.

2.3　*Composition of index system*

Taking into account of the characteristics of reinforcement project, the post-evaluation index system will be mainly established from four aspects, i. e. , reinforcement preparatory work, reinforcement construction process, operating status after reinforcement and engineering benefits after reinforcement.

Meanwhile, the reinforcement project evaluation index system is divided into four levels, that is, first-level index (target level), second-level index (middle level), third-level index (middle level) and forth-level index (bottom) .

Each level in the index system consists of different indexes. In various levels, determination of indexes depends on the characteristics of the engineering project itself. According to statistics, in a certain reinforcement project post-evaluation index system (see Table 2), there are 179 influence indexes at different levels, including 4 first-level indexes, 23 second-level indexes, 41 third-level indexes and 111 forth-level indexes.

Table 2 Reinforcement project post-evaluation index system for dangerously weak dams

Target level	First-level index	Second-level index	Third-level index	Forth-level index name code
Reinforcement project post-evaluation index system for dangerously weak dams	Reinforcement preparatory work Q	Safety appraisal stageQ1	On-site inspection (testing) Q11	Inspector Q111
				⋮
			⋮	⋮
				⋮
		⋮	⋮	⋮
	Reinforcement construction process G	Construction unit G1	Institution G11	Qualification G111
				⋮
			⋮	⋮
		⋮	⋮	⋮
				⋮
		Engineering Safety production G8	Safety production G81	Project legal person G811
				⋮
	Operating status after reinforcement Y	⋮	⋮	⋮
				⋮
		⋮	⋮	⋮
				⋮
	Project benefits X	⋮	⋮	⋮
		Water supply benefits X2	Gross tonnage of water supply X21	
		⋮	⋮	

Note Each index in the table represents a unique letter, such as Q for reinforcement preparatory work, Q1 for safety appraisal, Q2 for preliminary design, and so on.

2.4 *Establishment of scoring details for bottom index*

In formulating the scoring details of the bottom index (score level), it is necessary to follow the principles of detailed, comprehensive, scientific, reasonable and practical. By reference to the relevant technical regulations, specifications, and considering engineering practice, the scoring details should be determined after repeated discussions and comprehensive analysis, and improving continuously in the process of application. For example, the scoring details of the index of "Qualification of Geological Survey Unit (Q121)" in

"Geological Survey Work at Safety Appraisal Stage (Q12)", which belongs to the first index of "Reinforcement preparatory work of dangerously weak dam", can be graded from the following two aspects (the total score of this index is 100): ① If the geological survey of large reservoirs undertaken by Class A geological survey units, then medium-sized reservoirs undertaken by Class B ones and small reservoirs undertaken by Class C ones, 90% of the total score can be granted to this index. ② If the geological survey of medium-sized reservoirs or small reservoirs undertaken by Class A geological survey units, then small reservoirs undertaken by Class B ones, the total score can be granted to this index.

2.5 *Build-up of judgment matrix*

The pairwise comparisons of the importance among various factors at the same level under a single norm will be carried out. The matrix formed by their coefficients of comparisons is called the judgment matrix, which indicates the relative importance of various factors at this level relative to the previous level[6].

Assuming there are three second-level indexes C_1, C_2, C_3 under first-level index B_1, the experts will be invited to conduct the pairwise comparisons to determine the relative importance. According to the assigned values by the experts based on the "0.1~0.9" metrization in Table 3, the relative importance factors can be obtained, as shown in Table 4.

Table 3　Metrization of judgment matrix elements

Metrization (quantized value)	Explanation
0.9	One element is extremely important compared to the other one.
0.8	One element is more important compared to the other one.
0.7	One element is strongly important compared to the other one.
0.6	One element is a little important compared to the other one.
0.5	One element is same important compared to the other one.
0.1~0.4	Converse comparison

Table 4　Relative importance coefficient table of the pairwise comparisons relative to B_1

B_1	C_1	C_2	C_3
C_1	c_{11}	c_{12}	c_{13}
C_2	c_{21}	c_{22}	c_{23}
C_3	c_{31}	c_{32}	c_{33}

From the above table, the judgment matrix M_{B1} relative to B_1 can be obtained as:

$$M_{B1}=\begin{pmatrix} c_{11} & c_{12} & c_{13} \\ c_{21} & c_{22} & c_{23} \\ c_{31} & c_{32} & c_{33} \end{pmatrix} \tag{1}$$

2.6 *Calculation of weight related to each index*

1. *Unification of judgment matrix*

The above-mentioned judgment matrix M_{B1} will be unified to derive the Fuzzy consistent judgment matrix W:

$$W=\begin{pmatrix} w_{11} & w_{12} & w_{13} \\ w_{21} & w_{22} & w_{23} \\ w_{31} & w_{32} & w_{33} \end{pmatrix} \tag{2}$$

2. *Solution of weight*

From W, determine the weight set $A=(A_1, A_2, \cdots, A_n)$ to indicate weight distribution of various factors:

$$\begin{cases} w_i = \sum_{j=1}^{n} C_{ij} \\ w_{ij} = (w_i - w_j)/2n + 0.5 \\ i = 1,2,\cdots,j = 1,2,\cdots,n \end{cases} \tag{3}$$

$$\begin{cases} l_i = \sum_{j=1}^{n} w_{ij} - 0.5 \\ \sum_{i=1}^{n} l_i = \dfrac{n(n-1)}{2} \\ i = 1,2,\cdots,n \\ A_i = l_i / \sum_{i=1}^{n} l_i = 2l_i/[n(n-1)] \end{cases} \tag{4}$$

in formula, n represents of order of matrix; l_i represents the importance of index i relative to the previous-level objects. l_i will be normalized to obtain the weight of each index.

2.7 *Formation of fuzzy mapping matrix*

According to the expert assessment results, the proportion of each index level will be derived statistically, namely, the degree of membership of each index. The matrix formed by this membership index constitutes the judgment matrix of the fuzzy mapping relationship evaluation.

It is assumed to invite several experts to pass judgment on C_1, C_2 and C_3 at D_1 level, resulting in the results shown in Table 5. On this basis, the judgment matrix R_1 of fuzzy mapping relationship corresponding to D_1 is obtained as follows:

Table 5 Solution of degree of membership

D_1	Excellent	Good	Critical	Poor	Very poor
C_1	r_{11}	r_{12}	r_{13}	r_{14}	r_{15}
C_2	r_{21}	r_{22}	r_{23}	r_{24}	r_{25}
C_3	r_{31}	r_{32}	r_{33}	r_{34}	r_{35}

Note r_{ij} represents the degree of membership of the i Index Element C_i relative to Index Element j of the comment set. For example, r_{12} represents proportion of Index C_1 to be rated to be GOOD accounting for all ratings of Index C_1.

$$R_1=\begin{bmatrix} r_{11} & r_{12} & r_{13} & r_{14} & r_{15} \\ r_{21} & r_{22} & r_{23} & r_{24} & r_{25} \\ r_{31} & r_{32} & r_{33} & r_{34} & r_{35} \end{bmatrix} \tag{5}$$

According to the above method, the matrix R_n of fuzzy mapping relationship corresponding to other indices can be obtained.

2.8 *Realization of fuzzy comprehensive evaluation*

According to the above method, the total fuzzy mapping matrix $r_{Q\sim\hat{X}}$ *of* various indexes for reinforcement project post-evaluation will be established as:

$$r_{Q\sim\hat{X}}=\begin{bmatrix} f_{Qa1} & f_{Qa2} & f_{Qa3} & f_{Qa4} & f_{Qa5} \\ f_{Ga1} & f_{Ga2} & f_{Ga3} & f_{Ga4} & f_{Ga5} \\ f_{Ya1} & f_{Ya2} & f_{Ya3} & f_{Ya4} & f_{Ya5} \\ f_{Xa1} & f_{Xa2} & f_{Xa3} & f_{Xa4} & f_{Xa5} \end{bmatrix} \tag{6}$$

Total weight matrix of various indexes: $A_{Q\sim X}=[q_Q,\ q_G,\ q_Y,\ q_X]$

Total fuzzy mapping matrix:

$$\hat{R}=A_{Q\sim X}\times r_{Q\sim\hat{X}}=[F_{a1},F_{a2},F_{a3},F_{a4},F_{a5}] \tag{7}$$

Score matrix P:

$$\begin{aligned} P &=[(100+90)/2,(89+80)/2,(70+79)/2,(69+60)/2,(59+0)/2] \\ &=[95,84.5,74.5,64.5,29.5] \end{aligned} \tag{8}$$

Synthesis score of reinforcement project post-evaluation (Overall effect):

$$\begin{aligned} F_Z &=\hat{R}\times P^{\mathrm{T}}=[F_{a1},F_{a2},F_{a3},F_{a4},F_{a5}]\times\begin{bmatrix} 95.0 \\ 84.5 \\ 74.5 \\ 64.5 \\ 29.5 \end{bmatrix} \\ &=\theta(0<\theta<100) \end{aligned} \tag{9}$$

According to the scores, based on the meanings listed in Table 1, the overall effect after reinforcement and the reinforcement effect at various stages can be judged.

3 Engineering Applications

3.1 *Introduction to engineering*

The Hongqi Reservoir[7] is located in the Wotou Village, Sidu Town, Xiushui County in Jiangxi Province, it is also in the upstream of the Penggu River (tributary of the Xiushui River) . The reservoir can control a drainage area of 10.9km^2 above the dam site, with a normal pool level of 152.68m (the Yellow Sea elevation, the same below), with a total capacity of 1231$\times10^4$m^3 and design irrigation area of 11400 mu (1mu≈666.67m^2). Built in 1969, the reservoir dam mainly consists of the dam, spillway, dam culvert pipe, outlet tunnel and top grade power station, etc. , which is a medium-sized reservoir giving priority to irrigation, and combining with flood control, power generation, aquaculture and other comprehensive utilization efficiency. The reservoir was included in the second batch of large and medium-sized dangerously weak dam reinforcement planning projects of Jiangxi Province. Its safety appraisal work was completed in September 2005, and its reinforcement design work was completed in February 2006. The reinforcement construction was started in August 2007, and completed and put into operation in November 2010. The situation after reinforcement is shown in Picture 1 and Picture 2.

Picture 1 Downstream revetment situations of Hongqi Reservoir after reinforcement

Picture 2 Upstream revetment situations of Hongqi Reservoir after reinforcement

3.2 *Implementation process of post-evaluation*

According to the actual situation of this project, the bottom indexes will be scored by the expert group. In combination with the index system of the four stages of reinforcement project, each evaluation matrix, index weights and fuzzy mapping matrix will be established to calculate the final scores.

1. *Establishment of evaluation matrix and solution of weight sets*

With the aid of expert experience, together with the actual situation of this project, the judgment matrix F at the relevant level will be established to derive the corresponding weight A.

$$F=\begin{bmatrix}0.5 & 0.5 & 0.5 & 0.5\\0.5 & 0.5 & 0.5 & 0.5\\0.5 & 0.5 & 0.5 & 0.5\\0.5 & 0.5 & 0.5 & 0.5\end{bmatrix},A=[0.25,0.25,0.25,0.25]$$

2. *Setup of fuzzy mapping matrix*

The bottom index matrix will be calculated step by step to derive the global fuzzy mapping matrix R of reinforcement project post-evaluation:

$$R=\begin{bmatrix}0.577 & 0.173 & 0.250 & 0 & 0\\0.657 & 0.217 & 0.126 & 0 & 0\\0.125 & 0.504 & 0.249 & 0 & 0.122\\0.50 & 0.50 & 0 & 0 & 0\end{bmatrix}$$

3. *Correlative matrix operation and computed results*

The actual scores of the four stages of reinforcement project post-evaluation of this dam can be computed as 92. 40, 90. 36, 90. 40 and 80. 54 points, with a final synthesis score of 87. 10.

3. 3 *Summing-up*

From the computed score results of the four stages, the comprehensive effect of this reservoir dam is rated to be GOOD. Among which, the remaining three stages are in the EXCELLENT level except that the project benefit is totally in a GOOD level.

Through field visits, investigation, survey and interview, as well as access to information of the project, the above-mentioned evaluation results basically comply with the actual conditions of the project, which validates the feasibility and rationality of the relevant theoretical method of reinforcement project post-evaluation of dangerously weak dams proposed in this paper.

4 Conclusion

This paper deeply analyzed the characteristics of reinforcement project process and the contents of post evaluation. On these bases, the paper built the main influence factor set to measure the effectiveness and efficacy of the works at various stages of reinforcement project based on the FAHP and the theory as well as method of comprehensive judgment of project success degree, gave the scoring methods and scoring details for various indexes, established the index system of the four stages of reinforcement project post-evaluation and put forward the fuzzy comprehensive evaluation method for reinforcement project post-evaluation. Practical engineering application has proven such method was feasible and rational.

The research results further enrich, and develop the FAHP and the comprehensive evaluation method of project success degree. They are of important practical significance to: ①summerize the experiences and lessons learned at various stages of reinforcement project construction and management; ②improve and perfect the reinforcement project approval and decision-making, planning and design, quality control and others of dangerously weak dams; ③improve project investment efficiency, fund utilization rate and management level of decision-makers and to promote the virtuous cycle and healthy development of investment in water conservancy projects.

References

[1] Wen Hui. Discussion on post-evaluation of water conservancy construction project [J]. Sci-Tech Information Development & Economy, 2006 (4): 218 - 219.

[2] Zhang Rentian, Zhang Jinmiao. Study of post-evaluation of water conservancy project [J]. Jiangsu Water Resources, 2004 (9): 7 - 11.

[3] Han Dong. Comparison of two AHP approximation algorithms in water conservancy project post-evaluation [J]. China Rural Water and Hydropower, 2006 (4): 88 - 89.

[4] Zhang Rui, Yang Gaosheng, et. al. Taihu Lake Basin's Wangyu River project post-evaluation [J]. Journal of Economics of Water Resources, 2005 (1): 23 - 24.

[5] Sun Xiangqin, Qian Shanyang. Post-evaluation: Important process of water conservancy construction projects [J]. Academic Journal of Zhejiang Water Conservancy and Hydropower College, 2002 (1): 37 - 40.

[6] Weng Yue-jiao, Tang Deshan, Wang Yinyin. Post-evaluation of water conservancy reinforcement and reconstruction engineering effect based on F-AHP method [J]. Water Resources and Power, 2009, 27 (5): 145 - 148.

[7] Jiujiang Hydropower Planning and Design Institute. Preliminary Design Report of Reinforcement for Xiushui County Hongqi Reservoir [R]. 2008, June.

Discussion of method for computing temperature load during operation period on high arch dam

Wang Zhiqiang[a,b], *Wu Xiaobin*[b]

[a]*Jiangxi Provincial Institute of Water Sciences*

[b]*College of Hydrodynamic and Ecology Engineering, Nanchang Institute of Technology*

Abstract: According to the existing arch dam design specification, the temperature load is obtained based on analytic solution of infinite long free plate, it only considers the average temperature and the equivalent linear temperature difference of dam section in operation period, the influence of non-linear temperature difference has not been considered. This kind of simplification has some approximation; especially it can not reflect spatial effect of larger temperature gradient changes in the upstream and downstream water level change area. This paper first takes the free plate as an example, has calculated the temperature field corresponding to analytical method, finite element method, standard method, explains the rationality of using written program to carry on finite element analysis. On this basis, take Xiluodu arch dam as an example, the dam temperature load and temperature stress has been analyzed by using finite element method and standard method, the difference between both has been discussed.

Keywords: high arch dam; temperature load; free plate; finite element method

1 Introduction

The temperature load is one of main design loads on arch dam. According to the existing arch dam design specification in our country, it only considers average temperature and equivalent linear temperature difference of dam section in operation period, the influence of non-linear temperature difference has not been considered [1]. How much of the error caused by not considering non-linear temperature difference and how to consider non-linear temperature difference have been people's concern. After hydration heat of dam concrete and impact of initial temperature difference disappeared, the dam enters into quasi-steady temperature field; temperature load is the difference between quasi-steady tempera-

本文发表于2013年。

ture field and joint closure temperature field. This method can consider the influence of non-linear temperature difference. Based on actual air temperature and water temperature, etc., dam quasi-steady temperature field can be obtained using finite element simulation [2].

Based on introducing the relevant calculation principles, this paper first takes free plate as an example, has calculated the temperature field corresponding to analytical method, finite element method, standard method, explains the rationality of using written program to carry on finite element analysis. Then takes Xiluodu arch dam as an example, the dam temperature load and temperature stress has been calculated by using finite element method and standard method, the difference between both has been discussed.

2 Calculation Principles

2.1 *Temperature load corresponding to analytical method*

Temperature load corresponding to analytical method is obtained for an infinite free plate; the infinite long free plate sees reference [3].

Among them, the temperature of plate surface point on upstream and downstream is equal to boundary temperature.

When calculating plate internal point temperature, multi-annual average temperature on upstream and downstream side is reflected by linear interpolation, multi-annual average temperature amplitude on upstream and downstream side is reflected by the following formula.

For the plate satisfying heat conduction equation and boundary conditions, the theoretical solution sees reference [3].

After calculating the temperature of various points, the temperature load of various points is obtained by subtracting initial temperature.

2.2 *Temperature load corresponding to finite element method*

Based on heat conduction equation, considering initial conditions and outside air temperature, water temperature, ground temperature and other boundary conditions, annual change quasi-steady temperature field is calculated by finite element method, temperature load is the difference between quasi-steady temperature field and joint closure temperature field [3]. Here the quasi-steady temperature field is non-heat-source non-steady temperature field, it is different from non-steady temperature field that considering concrete hydration heat of construction process.

2.3 *Temperature load corresponding to standard method*

After T_m and T_d of cross-section has been computed using standard method, the temperature load at various points of cross-section is obtained through the following method. Equivalent linear temperature is $T_e(x) = T_m + T_d \cdot x/L$, in dam upstream surface, $x=-L/2$, equivalent temperature is $T_e(x)=T_m-T_d/2$, in dam downstream surface, x

$=L/2$, equivalent temperature is $T_e(x)=T_m+T_d/2$, in dam interior, equivalent temperature is obtained by linear interpolation.

2.4 *Temperature stress corresponding to finite element method*

The actual project is often complex, its temperature stress is very difficult to be obtained by analytical method, at present it is mainly obtained by finite element method. The principle of using finite element method to calculate temperature stress sees reference [3].

3 Example confirmations

In order to explains the rationality of using written program to carry on finite element analysis, this paper takes free plate as an example, has calculated the temperature load by analytical method, finite element method, standard method.

The plate height is 400m, thickness is 40m, annual average temperature is 18℃, annual amplitude of air temperature is 10℃, initial temperature is 16℃, upstream water level is 400m, downstream is no water, water depth of constant temperature strata is 60m, concretes thermal diffusivity is $3m^2$/month.

Finite element mesh is divided into 20 along horizontal direction and vertical direction. Temperature load of temperature rise in mid-July on upper 380m and lower 20m elevation is shown in Fig. 2. In which the time of calculating temperature field by finite element method is 50 years, the temperature tends to be stable 20 years later, and this paper takes the temperature in mid-July of 49 years as the temperature corresponding to temperature-rise.

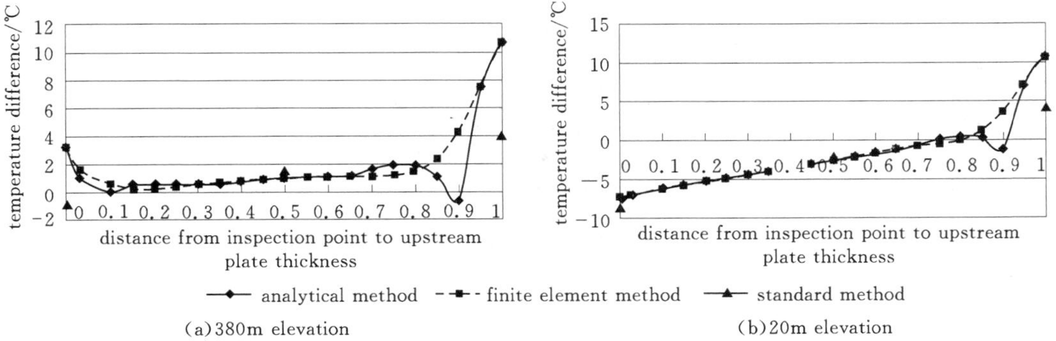

Fig. 2 Temperature load in mid-July at different elevation of plate

From the above results we can see that: ① the results obtained by finite element method and analytic method are basically same, explaining the finite element analysis program in this paper is reasonable; ②in the place not far away from the downstream surface, the result obtained by analytic method has catastrophic phenomenon, it is not the real case, it may be numerical singularity occurred in the area, needing further study.

4 Example analysis

The height of Xiluodu arch dam is 278m, crest elevation is 610m, annual average temperature is 21.7℃, annual amplitude of air temperature is 9.25℃, joint closure temperature of elevation above 560m is 16℃, joint closure temperature of elevation below 560m is 13℃, upstream normal water level is 600m, corresponding downstream water level is 378m, water depth of constant temperature strata is 140m, concretes thermal diffusivity is 1.68m^2/month.

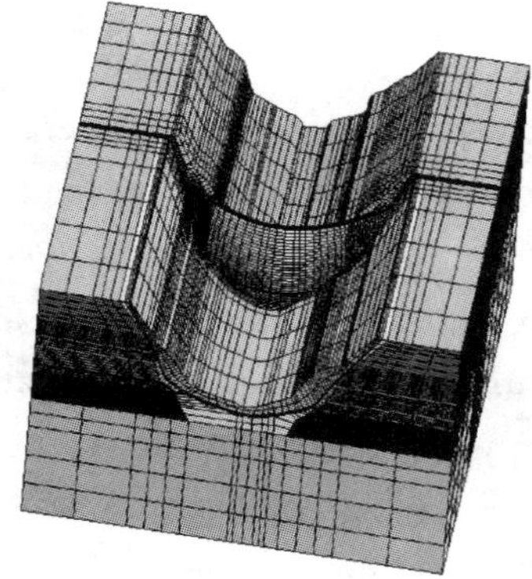

Fig. 3 Finite element mesh of Xiluodu arch dam

The temperature load and temperature stress of temperature drop in mid-January are calculated by standard methods and finite element method. In which the time of calculating temperature field by finite element method is 50 years, the temperature tends to be stable 20 years later, and this paper takes the temperature in mid-January of 49 years as the temperature corresponding to temperature-drop.

The finite element mesh of Xiluodu arch is shown in Fig. 3, the temperature load of operation period in mid-January at 590m elevation and 400m elevation are shown in Fig. 4.

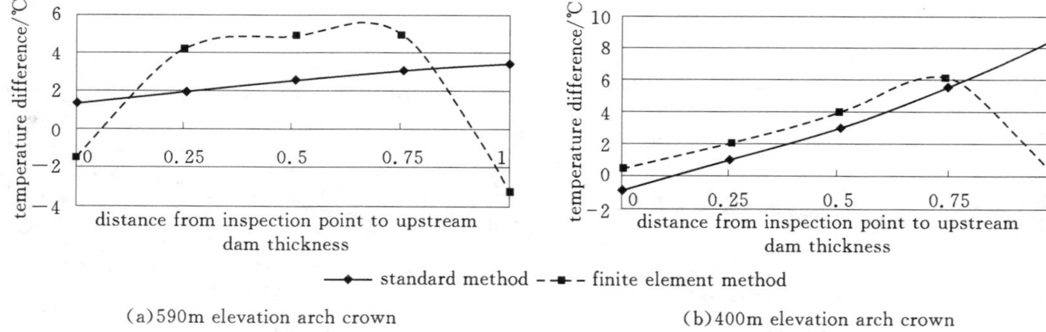

Fig. 4 Temperature load in mid-January at different elevation

From the above results we can see that: ①in the upper dam, because dam thickness is small and temperature gradient change is big at upstream and downstream surface, the difference between the results obtained by standard method and finite element method is big in the whole cross-section, the influence of non-linear temperature difference almost affects entire cross-section; ②In the lower dam, because dam thickness is big, temperature amplitude of upstream dam surface is small or even zero, temperature amplitude of downstream dam surface is big, the difference between the results obtained by standard method and finite element method is small near upstream surface and in dam interior but big near downstream surface. The influence of non-linear temperature difference only affects the region near downstream surface.

5 Conclusion

(1) This paper takes free plate as an example, has calculated the temperature field of temperature rise in mid-July by analytical method, finite element method, standard method, the results obtained by finite element method and analytic method are basically same, explains the rationality of using written program to carry on finite element analysis.

(2) This paper takes Xiluodu arch dam as an example, the dam temperature load and temperature stress of temperature drop in mid-January has been calculated by using finite element method and standard method. The influence of non-linear temperature difference is big in the whole cross-section of upper dam and near downstream surface of lower dam.

(3) Using finite element method to calculate temperature load of arch dam, can consider the influence of non-linear temperature difference, the result matches practical case. Along with the computer level's development, in the practical application, finite element method can be used to calculate temperature load of arch dam.

References

[1] SL 282—2003. *Design specification for concrete arch dam*. Beijing: China Water & Power Press, 2003.

[2] Yang Bo, Zhu Bofang. *Analysis of autogenous thermal stress in operating period in Arch dams*. Water Resources and Hydropower Technology, 2003, 34: 24 - 26.

[3] Zhu Bofang. *Thermal stresses and temperature control of mass concrete*. Beijing: China Electric Power Press, 1999.

浅析小型水库除险加固工程设计中水文水利计算

虞　慧[1,2]，张丽玲[3]，胡久伟[1]

1. 江西省水利科学研究院；2. 江西省水工程安全工程技术中心；3. 福建省三明市水利局

摘　要： 阐述小型水库除险加固工程设计中水文水利计算的基本方法，总结了在水文水利计算工作中应注意的问题和经验。

关键词： 小型水库；除险加固；水文水利计算；基本方法

水文水利计算是水库除险加固设计的重要组成部分。水文水利计算是在调查水库基本情况的基础上，复核水库的基本参数及洪峰流量，计算出相应频率下的洪水过程。经调洪计算得特征水位后，分析水库大坝在不同的工况下能否满足防洪要求。其工作的主要特点时间节点性强，工作量大，收集资料多，与各种专业相关性强，同时需要遵循“多种方法结合、综合分析数据、合理选用成果”的工作原则。因此，针对小型水库水文水利计算工作而言，其往往时间紧任务重，如何找到切合实际、简便且行之有效的水文水利计算方法，是小型水库除险加固设计的主要工作之一。

1　主要工作内容

水文水利计算的主要工作内容为由设计暴雨而求出的各频率下的洪水过程线、由实测量算得出的水位—库容曲线与泄流建筑物的水位—泄量关系曲线进行洪水调节计算，从而得出相应频率下的水位，再根据坝顶高程复核计算得出其最终的校核水位及设计水位。

1.1　计算基本流程

小型水库水文水利计算的基本流程如图 1 所示。

1.2　设计暴雨

江西省小型水库多处于小流域，故基本没有适合的水文站进行水文比拟法计算，一般做法是根据坝址以上流域中心位置查阅 2010 年编制的《江西省暴雨洪水查算手册》（以下简称《手册》）得出年最大 24h、年最大 6h、年最大 1h 的暴雨均值和变差系数。如雨量站有降雨量资料，可将其与《手册》法结果对比，综合选取适合本水库的暴雨均值和变差系数。

选定暴雨均值（P_{24}）后，将暴雨均值（P_{24}）、模比系数（K_p）与点面系数（a）三

本文发表于 2013 年。

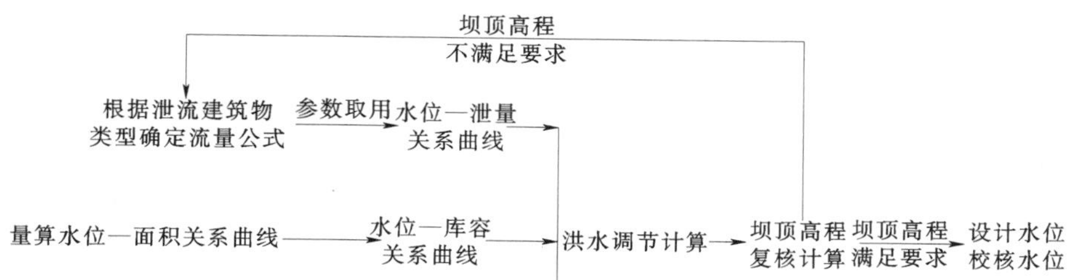

图1　小型水库水文水利计算流程图

者相乘之积得出各个时段的面雨量 $H_{24面}$。再根据《手册》中相应时段的暴雨时程分布，按各时段所占百分数计算各时段雨量，得到各频率下的时段暴雨分配量。

1.3　设计洪水

设计洪水计算分为产流计算和汇流计算，其最终目的是为得到洪水过程线。

1.3.1　产流计算

查阅《手册》中产流分区，相应流域最大蓄水量 I_M 及前期土壤含水量 P_a。由 24h 平均暴雨强度 $I=H_{24}/24$ 查阅相应附表，根据经验公式得到各频率下的下渗率，后得到净雨过程表。

1.3.2　汇流计算

确定水库在《手册》中推理公式法分区。根据下列公式计算地面洪峰流量：

$$Q_t=0.278\frac{h}{t\times F} \tag{1}$$

$$\tau=\frac{0.278L}{mJ^{1/3}Q_\tau^{1/4}} \tag{2}$$

设 $\theta=L/J^{1/3}$，即

$$Q_t=\left(\frac{0.278\theta}{m\tau}\right)^4 \tag{3}$$

式中：h 为各时段对应的总净雨量，mm；τ、t 为时间，h；F 为流域面积，km^2；Q_t 为流量，m^3/s；L 为主河道长，km；J 为河道平均坡降；m 为汇流参数。

通过点绘 Q_t—t 曲线和 Q_τ—τ 曲线，两曲线的交点即为所求的地面设计洪峰流量（$Q_{M地下}$）和汇流时间（τ）。地面流量过程线由概化五点折腰多边形过程线推求，各转折点的值由相应的百分数确定。地下径流峰值按以下公式计算

$$Q_{M地下}=\frac{R_下\ F}{3.6T} \tag{4}$$

$$T=\frac{9.6W}{Q_{M地面}} \tag{5}$$

$$W=0.1hF \tag{6}$$

式中：$Q_{M地下}$ 为地面径流终止点，地下径流过程采用以 $Q_{M地下}$ 为顶点的等腰三角形，底宽为 2 倍的地面径流过程时间；$R_下$ 为地下径流深，mm；T 为地面径流过程底宽，h；W 为

洪水总量，$10^4\mathrm{m}^3$。

自 $Q_{M地下}$ 开始向后每增加一个时段，流量均随之减少一个 $\Delta Q_{M地下}$，（$\Delta Q_{M地下}=\Delta t/TQ_{M地下}$）即得地下径流过程。地面流量过程和地下径流过程相加得设计洪水过程。

1.4 调洪计算

在洪水调节计算中，有3个基本要素，分别为洪水过程线、泄流曲线和库容曲线。洪水过程线由设计洪水得出。库容曲线通过对地形图进行复核计算得出水位—容积关系。泄流曲线为小型水库的水位—流量关系曲线，小型水库溢洪道一般为宽顶堰或明渠，根据相应流量计算公式得到溢洪道泄流曲线。

根据以上的3项基本资料，利用水量平衡公式，采用试算法对各频率设计洪水进行洪水调节计算。

1.5 坝顶高程复核

通过洪水调节计算出相应频率下对应的最高洪水位，需要复核其坝顶高程能否满足规范要求。

江西省大部分小型水库大坝为土石坝，根据《碾压式土石坝设计规范》（SL 274—2001）要求，坝顶高程＝最大波浪在坝顶的爬高（R）＋最大风壅水面高度（e）＋安全加高（A）。在规范中有几种求坝顶爬高（R）的方法，常用为采用莆田试验站公式和鹤地水库公式，在一般情况下两种公式都能适用，实际采用时需根据水库所处的地理条件及流域条件而定。

2 应注意的问题

2.1 资料收集

水库除险加固设计中要求基本资料全面、充分及准确，这样复核得出的水文成果，才能保障水库除险加固设计合理、可行且发挥水库最大效益。因而收集水库基本资料显得尤为重要。应收集的资料有：

（1）水库初步设计资料、三查三定资料（即对1981年年底前对已建成水利工程查安全、定标准；查效益、定措施；查综合经营、定发展计划的资料）、水库安全鉴定书、水库登记注册资料及水库历年除险加固等资料；

（2）水库万分之一地形图和库区地形图，地勘测量图。通过万分之一地形图确定水库的位置、集雨面积、主河道长、主河道比降，同时确保高程系统的一致性。而就目前所采用的方法而言，在量算流域面积、河流长度、河道比降上仍沿用老版地形图，存在一定的偏差；如采用已矢量化后的新版地形图同时结合地理信息软件提取流域三大要素更为精确。在设计洪水中考虑后期人为对产、汇流条件改变或新建水库的影响程度。

（3）原设计的水位—面积—容积关系曲线。在小型水库除险加固设计中，往往会沿用原设计的水位—面积—容积关系，但由于长期运行中，水库会存在一定程度的淤积，此时的水位—面积关系会发生一定的变化，如运用DEM与遥感技术相结合对水库的水位—面积关系进行测定计算，可弥补因水库在丰水期等原因无法对其进行实测的缺陷。

2.2 现场检查

在以往的设计中，水文设计人员往往忽略现场检查，而对于小型水库而言，由于其地形条件复杂，来水情况不一，现场情况多变，进行现场检查是十分必要的。在现场检查中需要对以下资料进行收集：

（1）水库在本地区的流域条件、洪水成因情况等。现场检查时详细了解洪水的来水情况，是否存在梯级水库调度问题，是否为库中库，是否存在引水情况等。如来上游引水为水库泄水，需查阅上游水库的设计相关资料；如来水有渠道或涵洞引流，应首先了解其是否为闸门控制及控制的方式；再了解渠道坡降，长度、高度及净宽，衬砌形式等；

（2）水库历年运行管理情况及对历史洪水进行调查。如历年险情，大坝是否曾进行加高加固，历年最高水位等；

（3）溢洪道历年过水深度，溢洪道控制段高程是否曾提高或降低及其原因；溢洪道类型、溢流净宽、衬砌形式；有闸门控制的溢洪道，闸顶高程、闸孔数、闸孔净宽以及闸门开启度；如有其他建筑物参与泄洪，其断面尺寸、进出口底板高程、纵坡降及衬砌形式。

2.3 洪水标准

依据《水利水电工程等级及洪水标准》（SL 252—2000）及针对小型水库的基本坝型为土石坝进行分析统计，小（1）型水库设计洪水标准为30年一遇，校核洪水标准为300年一遇；小（2）型水库设计洪水标准为20年一遇，校核洪水标准为200年一遇。但在实际应用中，常综合考虑水库下游交通设施重要程度、水库总库容的变化、坝高及上下游水头差等条件，适当调整洪水标准。

2.4 合理性分析

水文水利计算结果的合理性分析，一般采用将设计洪水成果与附近流域的水库的设计成果进行比较分析的方法。如差异较大，需要重新复核所使用的参数的准确性。在洪水调洪计算中，水位的合理性也能反映设计洪水的合理性，两者相辅相成。水位计算成果与历年水位进行比较分析；在坝顶高程复核中，出现坝顶高程不满足要求时，调查其是否存在漫顶情况，根据历年情况进行分析判断。水文水利计算结果需结合实际情况分析水位的合理性。

2.5 正常蓄水位及死水位的确定

通常在设计中，水库的正常水位及死水位等水位往往维持不变，但在设计中需综合考虑。如水库原库容较小或水库承担较为重要的责任（如水源地），可与水工人员协调调整水位的可行性，同时上报上级主管部门；如水库因其他原因曾降低正常水位，此时应和当地水利部门协调重新确定正常水位。死水位与坝顶高程复核息息相关，如当地水利部门需要将涵管的进口底高程与渠道相接，或是通过抬高该水位使得改善其当地的农业条件。这些水位的调整都需要在反复论证中确定该方案的可行性。

2.6 与当地水利部门协调

在小型水库设计中，水文水利计算工作还需要和水利工程相关专业进行配合。此外，更为重要的是及时、有效地与当地的水利部门进行沟通，掌握水库更详细的信息，实现工程的最优化设计。

3 结语

水文水利计算在小型水库除险加固设计工程中占基础地位，但对整个设计工作影响重大，其准确性直接与水库的安全、投资及效益挂钩。水文水利计算成果的准确性与基础资料的可靠性、准确性息息相关。在计算过程中，应着重于基础资料分析复核，精细于流域三要素的提取与暴雨取值，对水位—面积—容积关系曲线、水位—泄量关系曲线进行详细的复核及调整，同时将计算的结果进行合理性分析。通过细致、准确、多方法论证、合理性分析及多角度沟通的基础上为水库除险加固工程设计提供切合实际及可靠的计算成果。

在今后的小型水库除险加固工程设计的水文水利计算中，应将重点研究小型水库在小流域及特小流域上计算的参数及方法和利用 GIS 技术、建立数据库的分析方法对水文计算中的基础资料进行提取和分析。在设计洪水中，有些小型水库集雨面积不到 $1km^2$，其下垫面和下渗的情况都有一定的差异，同时在设计洪水合理性分析上没有类似的水库进行比较，故这类水库的实际情况与最终结果存在一定的偏差。而对于小型水库而言，其最为缺乏的是实测洪水资料，如在重点的小流域上建立雨量站或水文站，将设计结果与监测资料进行比较分析，同时建设小型水库监测分析数据库，对比不同的流域上设计洪水结果的差异，寻求出适合小型水库的方法。

参考文献：

[1] 中华人民共和国水利部．SL 252—2000 水利水电工程等级划分及洪水标准［S］．北京：中国水利水电出版社，2000.

[2] 江西省暴雨洪水查算手册［K］．江西：江西省水文局，2010.

[3] 中华人民共和国水利部．SL 274—2001 碾压式土石坝设计规范［S］．北京：中国水利水电出版社，2002.

基于渗流与应力耦合的封闭式坝体防渗墙的应力研究

高江林[1,2]

1. 天津大学建筑工程学院；2. 江西省水利科学研究院

摘　要： 为探讨土石坝加固工程中封闭式坝体防渗墙的应力状态及其适用性，利用ABAQUS有限元软件建立渗流与应力耦合数值模型，实现了同时考虑墙体与坝体相互接触、应力与渗流耦合共同作用的模拟。通过对典型土石坝工程的计算，在比较流固耦合作用对计算结果影响的基础上，对不同工程条件下的封闭式坝体防渗墙的应力进行了分析。分析结果表明，在蓄水工况下，是否考虑耦合作用对防渗墙应力的计算结果影响明显；不同墙体弹性模量、坝体土变形模量及坝基透水率等条件下的墙体应力分布有明显区别。本文分析结论对于土石坝除险加固工程中混凝土防渗墙的应用及设计计算方法具有一定的参考意义。

关键词： 水工建筑物；防渗墙应力；流固耦合；土石坝；有限元法；除险加固

1　引言

在近几年国内大规模开展的病险水库除险加固工程建设中，混凝土防渗墙得到了大量应用。对于坝基覆盖层不厚的土石坝，一般采用直接嵌入基岩的封闭式坝体防渗墙进行防渗处理。在江西省近年来的土石坝加固工程中，绝大多数采用了这种形式的防渗墙。

目前，关于新建土石围堰堰体防渗墙和深厚覆盖层中坝基防渗墙的应力变形研究较多[1-4]。针对土石坝除险加固工程中增建混凝土防渗墙的应力研究相对较少。由于封闭式坝体防渗墙主要承受水平荷载，且建设防渗墙时一般在大坝运行几十年之后，土体固结对墙体应力的影响基本可以忽略不计[5]。因而，土石坝加固工程中增建的封闭式坝体防渗墙的承载形式及土体固结的影响，与一般新建土石围堰中的堰体防渗墙及深厚覆盖层中的坝基防渗墙均有明显区别。因此，针对目前土石坝加固工程中的封闭式坝体防渗墙，有必要探讨墙体在不同工程条件下的应力状况及混凝土防渗墙的适用性。

研究方法方面，早期一般采用简化方法来计算墙体应力，如采用简化地基梁进行计算[6]。简化方法应用相对简单，但由于假设条件的偏差，往往导致结果与实际情况相差较大。随着有限元分析方法在岩土工程领域的发展，国内学者采用数值计算方法对防渗墙应力变形开展了一系列研究，取得很多有益成果。但这些研究中，较少考虑渗流与应力耦合

本文发表于2013年。

作用。而土石坝属于典型的流固耦合问题，为反映渗流对防渗墙应力的影响，往往通过水压力和湿容重（或干容重）来实现[7]，或将渗流计算出的渗透力作为荷载加到墙体上再进行应力计算[8]。间接方法需先假定或计算渗流场，分析过程繁琐，且无法从本质上反映坝体渗流与墙体应力的直接作用关系及相互影响。因此，考虑渗流与应力耦合作用是防渗墙应力变形研究方法进一步发展的必然要求。

为分析土石坝加固工程中增建的坝体防渗墙的受力性状，笔者利用 ABAQUS 有限元软件建立二维有限元耦合数值模型，同时考虑坝体土与防渗墙接触、渗流与应力耦合的共同作用。在对比耦合作用对墙体应力计算结果影响的基础上，重点分析不同条件下的墙体应力分布及各条件对墙体应力的影响规律，为土石坝加固工程中的防渗墙应用和设计计算提供一些有益参考。

2 防渗墙与坝体相互作用的数值模拟

土石坝中防渗墙应力的有限元分析，关键是处理好防渗墙与坝体土相互作用的模拟。主要包括三个方面的计算：初始应力状态、渗流与应力耦合、墙体与坝体接触（包括渗流的计算方法）。

2.1 初始应力状态

土石坝除险加固工程中增建混凝土防渗墙一般是在大坝运行几十年以后进行，此时坝体土大多已固结完成，土体固结对墙体应力的影响很小。另一方面，防渗墙施工时一般空库状态，即大坝上游无水压力作用。因此，进行防渗墙与坝体相互作用分析的初始状态应该是加固完后水库蓄水前的天然状态，坝体和墙体存在初始应力但初始位移为零。

为获得初始应力状态，须进行初始应力平衡。在 ABAQUS 中实现的方法为：首先将重力荷载施加于土体，并施加符合工程实际情况的边界条件，计算出在重力荷载下的应力场；根据得到的初始应力场，计算出各单元节点应力，并将其和重力荷载一起施加到原始有限元模型中计算，即可得出初始状态的应力，同时实现初始位移为零。需要注意的是，由于混凝土防渗墙施工多为水下浇筑混凝土，且施工时一般为空库状态，初始应力状态计算时不应考虑墙体与坝体的接触，否则会导致墙体初始应力偏大。

2.2 渗流与应力的耦合分析

土石坝中的渗流与应力耦合分析，需统一考虑饱和与非饱和渗流计算。本文假定非饱和与饱和土体均服从 Darcy，两者的区别在于孔隙水压力和渗透系数的不同[9]。因此，非饱和渗流与饱和渗流可以用统一的方程形式，此时可将非饱和与饱和区域作为统一区域进行渗流计算。渗流与应力耦合计算需同时满足应力平衡方程和渗流连续方程。

某体积域内的平衡方程可用某一时刻的虚功原理表述：

$$\int_V \sigma\delta_\varepsilon \mathrm{d}V = \int_S f_s\delta_v \mathrm{d}S + \int_V f\delta_v \mathrm{d}V + \int_V sn\rho_w g\delta_v \mathrm{d}V \tag{1}$$

式中：σ 为柯西应力；δ_ε 为虚应变；δ_v 为虚位移；f_s 为单位面积上的面力；f 为单位体积上流体重力之外的体积力；s 为土体饱和度；n 为土体孔隙率；ρ_w 为流体的密度；g 为重力加速度。

计算域用有限元网格离散后，单元网格中流体运动满足连续性，即单位时间内流入土体的流体体积等于流体体积的增加速率，其连续方程可表示为：

$$\frac{\mathrm{d}}{\mathrm{d}t}\left(\int_V \frac{\rho_w}{\rho_w^o} s\boldsymbol{n}\,\mathrm{d}V\right) = -\int_S \frac{\rho_w}{\rho_w^o} s\boldsymbol{n}\boldsymbol{n} \cdot v_w \mathrm{d}S \tag{2}$$

式中：$\boldsymbol{n}$ 为渗流面 S 的法线方向；v_w 为流体的平均流速；ρ_w^o 为流体参照密度。

在上述应力平衡方程和渗流连续方程的基础上，采用 Galerkin 有限元格式，将节点位移和孔隙水压力作为节点自由度进行空间离散，形成应力平衡方程和渗流连续方程的矩阵形式。在同时满足位移边界和渗流边界的条件下，直接对耦合控制方程进行求解。

2.3 墙体-坝体的接触单元

2.3.1 接触模拟

防渗墙与坝体土的接触采用 ABAQUS 提供的接触单元进行模拟，接触行为包含接触面的法向作用和切向作用。接触面的法向模拟采用硬接触[10]，即两物体只有在压紧状态时才能传递法向压力 p，若两物体之间有间隙时则不能传递法向应力。

接触面的切向作用采用库仑摩擦模型，用摩擦系数 μ 来模拟在两个表面之间的摩擦行为。当接触面处于闭合状态（即有法向接触压力）时，接触面可以传递切向应力（摩擦力）。设定极限剪应力：$\tau_{\mathrm{crit}}=\mu p$。若摩擦力小于某一极限值 τ_{crit} 时，认为接触面处于黏结状态；若摩擦力大于 τ_{crit} 后，接触面开始出现相对滑移变形，称为滑移状态。

2.3.2 渗流模拟

为实现整个计算域的渗流与应力耦合计算，必须采用能够考虑渗流与应力耦合作用的接触单元。本文采用 ABAQUS 提供的能够考虑孔压-位移耦合的接触单元，除满足上述传力机理外，还满足以下渗流条件：

(1) 孔压在接触对应的实体上完全连续，用公式表示为 $p_A-p_B=0$，其中 p_A、p_B 为接触面两侧节点上的孔隙水压力。

(2) 接触面的切线方向无流体流动。

(3) 接触面压力为有效压力，不包含孔隙水压力的作用。

接触面单元渗流模式如图 1 所示。

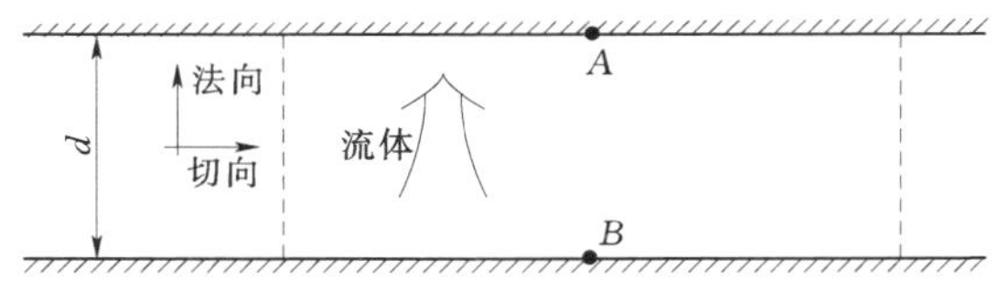

图 1 接触面单元渗流模式示意图

3 计算模型

本文以一般地基上的均质土坝作为分析对象。坝高 30.0m，坝基覆盖层厚 5.0m，防渗墙自坝顶嵌入至基岩 0.5m，墙深 35.5m，墙厚 0.6m，强风化和弱风化基岩厚度均取为 10.0m，正常蓄水位 25.0m，下游水位为坝脚位置。具体计算条件如图 2 所示。

有限元计算网格如图 3 所示。

土体采用 Mohr - Coulomb 弹塑性模型，采用变形模量进行计算。根据本文计算结果，考虑到混凝土压应力水平较低及抗拉状态时的应力应变曲线接近线性关系[5,11]，且整个基岩应力水平较低，本文计算中防渗墙和基岩材料采用弹性模型。对于封闭式防渗墙而言，坝基岩层一般为渗透系数较小或进行过灌浆处理的基岩，可作为相对不透水层看待。

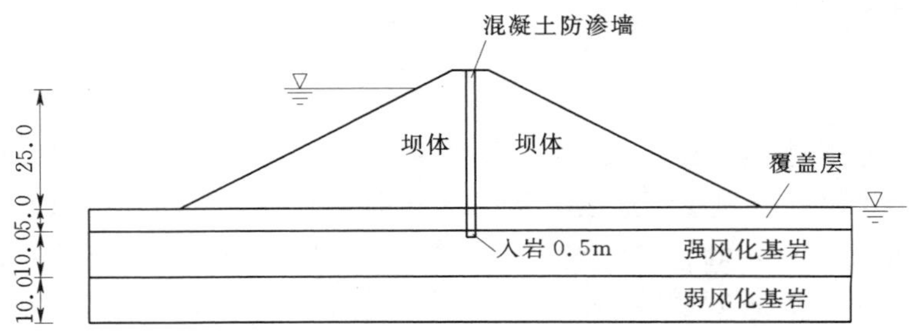

图 2 计算模型图（单位：m）

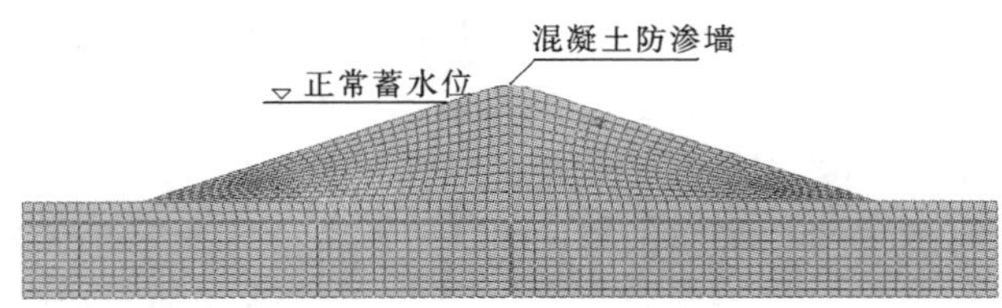

图 3 有限元网格图

部分有限元计算参数见表 1 所示。

表 1 **主要计算参数**

材料	干密度 ρ_d /(kg·m^{-3})	变形（弹性）模量 E /MPa	泊松比 μ	黏聚力 c /kPa	内摩擦角 φ /(°)	渗透系数 k /(cm·s^{-1})
坝体土	1600	100	0.3	20.0	20.0	5.0×10^{-5}
覆盖层	1700	200	0.3	5.0	30.0	1.0×10^{-4}
强风化基岩	2400	2.0×10^{4}	0.25	—	—	1.0×10^{-6}
弱风化基岩	2600	2.5×10^{4}	0.2	—	—	1.0×10^{-6}
刚性混凝土	2400	2.0×10^{4}	0.25	—	—	1.0×10^{-7}
塑性混凝土	2400	500	0.25	—	—	1.0×10^{-7}

4 耦合作用对防渗墙应力结果的影响

为分析耦合作用对墙体应力计算结果的影响，分别对是否考虑耦合时墙体的最大、最小主应力进行比较。不考虑耦合时，仅考虑重力荷载和迎水面水压力荷载作用，无坝体渗流作用。由于 ABAQUS 中应力以拉为正，因此，最大主应力反映的是墙体最大拉应力，而最小主应力则反映墙体的最大压应力。以刚性混凝土防渗墙为例进行分析。

4.1 墙体应力

是否考虑耦合时墙体的最大、最小主应力分布如图 4 所示。

由图 4 可知，考虑耦合作用时防渗墙最大拉应力和压应力分布为 4.51MPa 和 3.56MPa，均出现基岩面位置附近。最大拉应力大于压应力，相对混凝土的极限承载能力而言，墙体破坏由拉应力控制。拉应力和压应力呈对称分布，说明墙体承受典型的弯矩作用，墙侧土摩阻力引起的墙体压应力贡献相对较小。

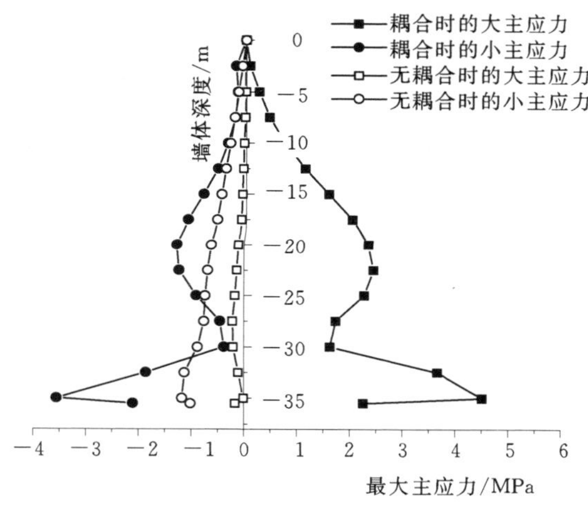

图 4 防渗墙的大、小主应力

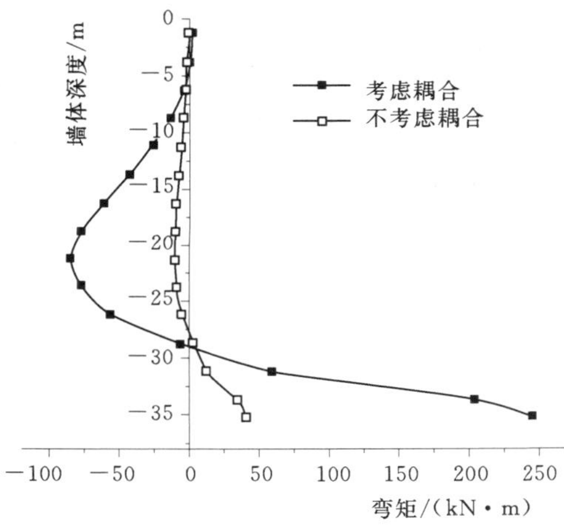

图 5 刚性墙墙身弯矩分布

不考虑耦合作用时墙体的最大主应力和最小主应力均为负值，说明此时防渗墙整体处于受压状态，此时墙侧摩阻力引起的压力作用更加突出。最大压应力为 1.18MPa，远小于墙体材料的允许值，说明不考虑渗流与应力耦合会导致墙体应力计算结果偏小，尤其是控制墙体破坏的拉应力。

4.2 墙身弯矩分布

为探讨正常蓄水工况下防渗墙的受力机理，进一步分析墙身弯矩分布。利用墙体各单元节点上的应力结果，分别计算出防渗墙迎水面和背水面的竖向应力，进一步可推算出墙体弯矩。墙体弯矩沿深度分布如图 5 所示。弯矩方向以顺时针为正。

根据图 5 可以看出，考虑耦合作用时墙体弯矩在中部（21.0m 深左右）和底部出现两处极值，分别为−84.6kN·m 和 245.8kN·m，之间弯矩发生反向，说明弯矩主要有中下部水平荷载引起。最大弯矩值出现在墙体底部。弯矩的大小与分布规律与应力相对应。可以推断，对于高坝或更高水位工况下墙体会出现更加不利的应力状态。

不考虑渗流与应力耦合时，最大弯矩值为 26.7kN·m，出现在墙体底部，分布规律类似，但弯矩值远小于考虑耦合作用的计算结果。

4.3 坝体应力场和渗流场的比较

为说明耦合作用对墙体应力影响，进一步分析耦合作用下的坝体渗流场和应力场。

4.3.1 坝体孔压分布

渗流与应力耦合作用下坝体孔压分布如图 6 所示。

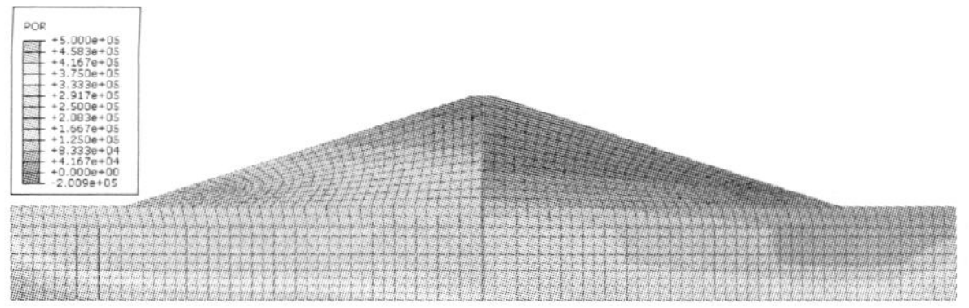

图 6 坝体孔压分布

图 6 所示的灰色区域为非饱和土，分界线为坝体浸润线位置。由于防渗墙的截渗作用，墙体前后水头有明显突变，水头差约 12.5m。

4.3.2 坝体应力场

考虑耦合时和不考虑耦合时的坝体竖向有效应力分布如图 7 和图 8 所示。为方便比较，仅列出坝体部分的土体有效应力分布。

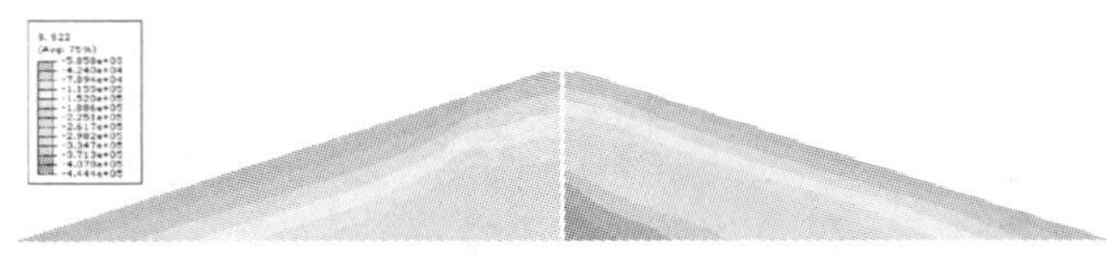

图 7 坝体竖向有效应力分布（考虑耦合）

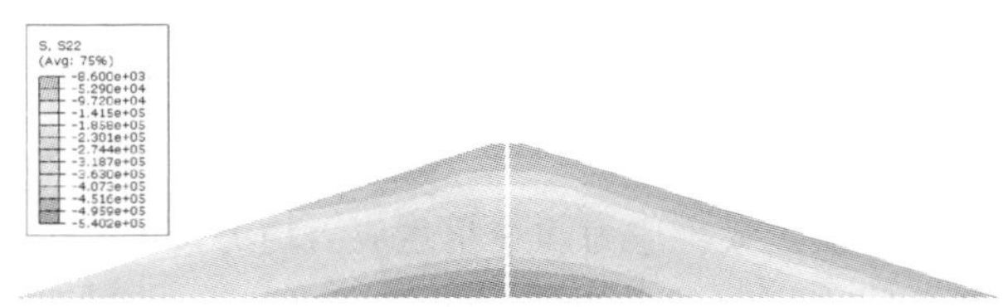

图 8 坝体竖向有效应力分布（不考虑耦合）

由图 7 可知，考虑耦合作用时的最大竖向有效应力为 0.44MPa，相对集中出现在防渗墙底部附近，且防渗墙前后土体的有效应力相差明显，且整体上墙后土体的有效应力大于墙前土体，说明防渗墙形成向下游挤压的趋势。根据墙体前后土体有效应力差和墙体弯矩分布，可以进一步推断，引起墙体向下游侧挤压的主要作用力来自于上游土体中的孔隙水压力。

由图 8 可知，不考虑耦合作用时的最大坝体竖向有效应力为 0.54MPa，由于无孔隙水压力作用，最大有效应力值大于考虑耦合时的计算结果。此时防渗墙前后坝体的有效应力基本相当。由此看出，不考虑渗流与应力耦合时，无法反映墙体前后水头差对墙体应力的影响。

5 不同条件下的墙体应力分析

结合实际工程情况，为探讨不同条件下的混凝土防渗墙应力情况及其在土石坝中的适用性，本文重点分析不同混凝土弹模、坝体填筑材料、坝基透水率等条件下的墙体应力及变化规律。根据前面分析可知，封闭式坝体防渗墙主要承受水平荷载作用，墙体破坏由拉应力控制，下文仅对不同条件下的最大主应力（拉应力）进行分析。

5.1 不同混凝土弹模条件下的墙体应力

在近年来江西省已建成的土石坝坝体防渗墙中，墙体混凝土的抗压强度从几兆帕的塑性混凝土到二十几兆帕刚性混凝土都有，相应弹性模量差别也很大。在工程质量控制上，一般仅把混凝土抗压强度和渗透系数作为墙体材料的质量控制指标，很少对墙体材料的弹性模量进行控制。

为分析混凝土弹性模量对墙体应力的影响，分别对刚性和塑性混凝土防渗墙的应力进行计算。刚性混凝土弹性模量 E_c 分别取为 2.5×10^4MPa 和 2.0×10^4MPa，塑性混凝土弹性模量取为 1500MPa 和 500MPa。不同弹模对应的最大主应力如图 9 所示。

根据图 9 可知，墙体应力随着墙体材料弹性模量的增大而增大，且刚性混凝土防渗墙的拉应力的远大于塑性墙，并超出材料本身的极限值。而塑性混凝土防渗墙拉应力均在允许范围内，当 $E_c=500$MPa 时，墙体基本无拉应力存在。由此看出，对于承受水平荷载为主的封闭式坝体防渗墙，墙体弹模对应力影响显著。墙体弹性模量越小（与坝体土弹模越接近），越有利于墙体受力，即对墙体受力而言，混凝土抗压强度不是越大越好。因此，在实际工程中应增加对墙体材料弹性模量的上限控制。

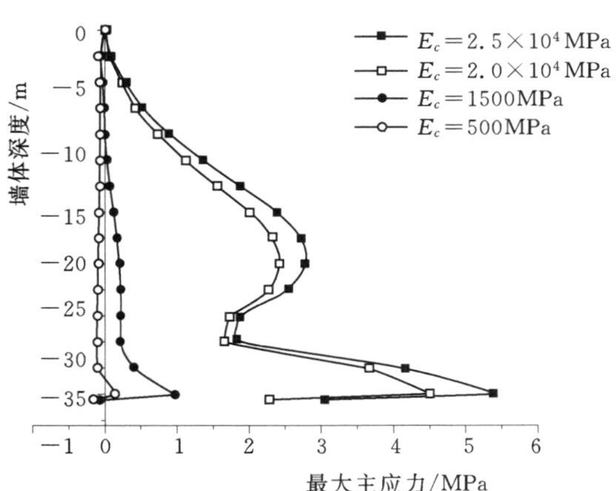

图 9　不同混凝土弹模时的墙体最大主应力

5.2　不同土体模量条件下的墙体应力

由上述分析可知，墙体和坝体土弹性模量相差越大，对于墙体受力越不利。在实际工程中，由于坝体填筑材料种类的不同，坝体土本身的变形模量同样存在一定的差异。为探讨混凝土防渗墙在不同坝体填筑材料中的应力情况及其适用性，进一步分析不同坝体土变形模量对刚性和塑性墙体应力的影响。坝体土变形模量 E_s 分别为 50MPa、100MPa、300MPa、600MPa。

不同土体变形模量时刚性混凝土防渗墙的最大主应力分布如图 10 所示。

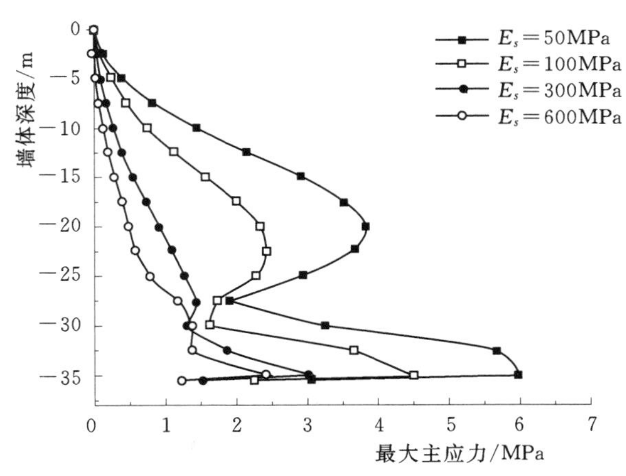

图 10　不同土体模量时刚性墙的最大主应力

由图 10 可以看出，坝体土的变形模量对刚性混凝土墙体应力影响明显，土体模量越小，墙体拉应力越大。当坝体填筑材料为软黏土时（$E_s<100$MPa），墙体应力急剧增加，最大拉应力达到 6.0MPa，大大超出混凝土材料的极限值。当坝体填筑材料为模量较大的硬质黏土或砂性土时（$E_s>300$MPa），对墙体受力更为有利。因此，进行刚性混凝土防渗墙设计时，尤其是对于坝体填筑材料为变形模量较小的软黏土时，应充分考虑墙体材料弹模过大对墙体应力的不利影响。

进一步分析不同坝体土变形模量时塑性墙体的应力情况，如图 11 所示。

由图 11 可知，增建塑性混凝土防渗墙时，墙体最大拉应力同样随坝体土变形模量的减小而增大，但塑性混凝土墙的最大拉应力远远小于刚性混凝土墙。一般情况下，塑性墙

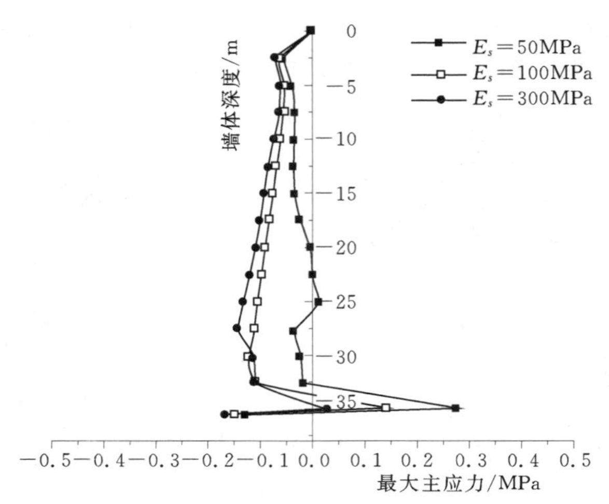

图 11　不同坝体土模量时塑性墙的最大主应力

体的最大拉应力均在材料允许值范围内。从墙体受力角度考虑，对于不同的坝体填筑材料，塑性墙较刚性墙具有更好的适用性。

5.3　不同坝基透水率时的墙体应力

由于地质条件和所采取工程措施的不同，实际土石坝工程中的坝基透水率存在一定差异。风化程度较低或进行了灌浆处理的坝基透水率往往较小，而对于一般强风化的基岩透水率则较大。一般情况下，坝基渗透系数在 1.0×10^{-4}～1.0×10^{-6}cm/s 之间。为分析不同坝基透水率时混凝土防渗墙的应力情况，坝基渗透系数 K_r 分别取为 1.0×10^{-4}cm/s、1.0×10^{-5}cm/s、1.0×10^{-6}cm/s，此时模型中的强风化和弱风化基岩采用相同的渗透系数。不同坝基渗透系数对应的刚性混凝土防渗墙的最大主应力分布如图 12 所示。

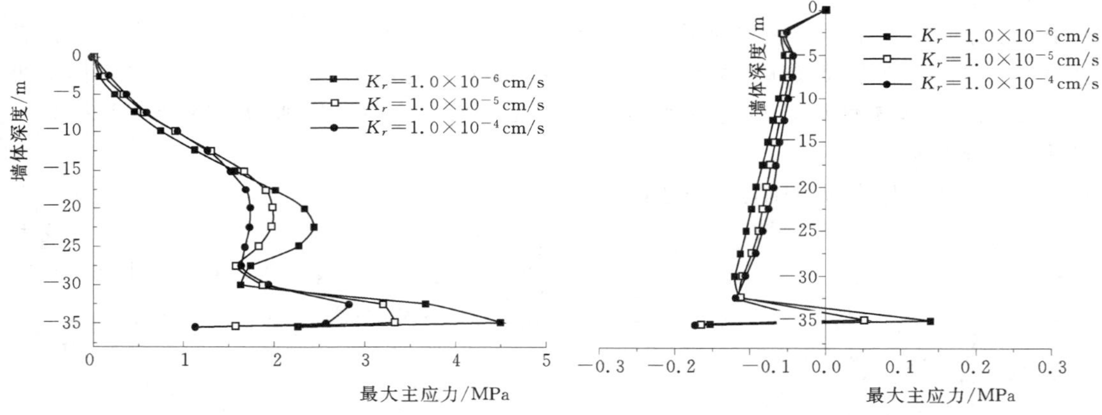

图 12　不同坝基透水率时刚性墙的最大主应力

图 13　不同坝基透水率时塑性墙的最大主应力

根据图 13 可以看出，坝基渗透系数越小，墙体最拉应力越大。这主要是由于坝基透水率越小，墙体前后形成的水头差越大（三个渗透系数对应的水头差分别为 7.5m、9.5m、12.5m），墙体所承受的水平荷载也越大，相应引起的拉应力也更大。当坝基渗透系数 $K_r=1.0\times10^{-6}$cm/s 时，最大拉应力为 4.51MPa，已超出混凝土极限值。当 $K_r=1.0\times10^{-4}$cm/s 时，最大拉应力为 2.84MPa，不同地基透水率的最大拉应力值相差近 2 倍，说明坝基透水性能对刚性墙体应力的影响不能忽略。

不同坝基渗透系数对应的塑性混凝土防渗墙的最大主应力分布如图 13 所示。

由图 13 可知，塑性混凝土防渗墙的最大拉应力同样随坝基渗透系数减小而增大，当 $K_r=1.0\times10^{-6}$cm/s 时，最大拉应力为 0.15MPa；当 $K_r=1.0\times10^{-4}$cm/s 时，最大主应力为−0.17MPa（整体为受压状态）。不同坝基透水率时的塑性混凝土防渗墙最大拉应力相差很小。说明不同坝基透水率对塑性混凝土防渗墙应力的影响较小。

可以看出，墙体应力与坝基透水性本身是一对矛盾，从基础防渗角度，坝基透水率越小越好，而从墙体受力而言，透水率越大，对墙体受力越有利。而建设防渗墙的根本目的就是防渗，针对坝基透水率小的情况，应首先考虑建设塑性混凝土防渗墙。

6 结语

（1）本文基于渗流与应力耦合分析，实现了防渗墙与坝体之间相互接触、渗流与应力耦合的共同作用的模拟。分析结果表明，所建立的耦合数值模型能较好地反映防渗墙与坝体的相互作用，可为土石坝中混凝土防渗墙的应力变形计算提供方法参考。

（2）是否考虑渗流与应力的耦合作用，对于防渗墙的应力结果影响明显。正常蓄水工况下，封闭式坝体防渗墙主要承受墙体前后水头差引起的水平荷载作用。仅考虑上游坝坡水压力作用（不考虑渗流）进行墙体应力计算是偏于不安全的。

（3）防渗墙墙体材料的弹模对于墙体应力影响明显，墙体材料弹模越大，墙体应力越大。水平荷载作用下，塑性混凝土应力远小于刚性混凝土墙，此时塑性混凝土防渗墙具有明显的承载优势。为保证封闭式坝体防渗墙的承载能力和工程安全，建议在防渗墙设计时明确墙体材料弹性模量的上限值。

（4）坝体填筑材料的变形模量对刚性混凝土墙体应力影响明显，坝体材料模量越小，墙体应力越大。当坝体材料为模量很小的软黏土时，对于刚性混凝土防渗墙受力极为不利。不同坝体填筑材料变形模量对塑性混凝土墙应力的影响相对较小，对墙体受力而言，塑性墙具有更广泛的适应性。因此，在土体模量相对较小的土石坝中增建混凝土防渗墙时，一定要充分考虑墙体材料弹模过大对墙体应力的不利影响。

（5）坝基透水率对刚性混凝土防渗墙的应力影响明显，坝基透水率越小，水平荷载越大，刚性混凝土防渗墙的墙体应力越大。相对而言，坝基透水率对塑性混凝土防渗墙的应力影响较小，不同渗透系数时的墙体应力均在材料允许范围内，且相差不大。因此，针对坝基透水率小的情况，应优先考虑建设塑性混凝土防渗墙。

参考文献：

[1] 沈新慧．防渗墙及周围土体的应力探讨［J］．水利学报．1995（11）：39-45.

[2] 卢廷浩，汪荣大．瀑布沟土石坝防渗墙应力变形分析［J］．河海大学学报．1998，26（2）：41-44.

[3] 王刚，张建民，濮家骝．坝基混凝土防渗墙应力位移影响因素分析［J］．土木工程学报．2006，39（4）：73-77.

[4] 熊欢，王清友，高希章，等．沙湾水电站一期围堰塑性混凝土防渗墙应力变形分析［J］．水力发电学报，2010，29（2）：197-203.

[5] 王清友，孙万功，熊欢．塑性混凝土防渗墙［M］．北京：中国水利水电出版社，2008.

[6] 郑秀培．土石坝地基混凝土防渗墙设计与计算［M］．北京：水利电力出版社，1979.

[7] 沈珠江，左元明．三峡工程二期混凝土防渗墙围堰初步设计方案的应力应变计算［J］．长江科学院院报．1987（3）：31-37.

[8] 段伟，骆原，陈珂，等．黄壁庄水库副坝渗流及防渗墙应力分析 [J]. 人民黄河．2011，33 (5)：124-127.
[9] 费康，张建伟．ABAQUS 在岩土工程中的应用 [M]. 北京：中国水利水电出版社，2010.
[10] ABAQUS Documentation [CP/DK]. ABAQUS Theory Manual (Version 6.10)，2010.
[11] 高丹盈，王四巍，宋帅奇，等．塑性混凝土单向受压应力—应变关系的试验研究 [J]. 水利学报. 2009，40 (1)：82-87.

预应力锚索内锚段有限元精细化模拟

郑　勇[1]，张　雄[2]

1. 江西省水利科学研究院；2. 武汉大学水资源与水电工程科学国家重点实验室

摘　要： 预应力锚索加固作为岩土工程的一种主要加固技术已得到日益广泛的应用，但现有的预应力锚索数值模拟方法却不能很好地反映其实际加固效果。针对预应力锚索内锚段模拟存在的难点问题，提出用精细化有限单元法进行模拟。该算法最突出的特点是详细地描述了钢绞线、砂浆和接触面等细部结构。算例表明，这种方法是合理和有效的，内锚段轴向位移和应力沿钢绞线方向由外到里呈指数性衰减。

关键词： 岩石力学；内锚段；有限单元法；数值模拟

预应力锚索加固技术因经济合理、安全可靠、对岩土体扰动小、施工快捷等优点而被广泛应用于岩土体的加固工程中，是其他传统方法不可替代的[1]。岩体预应力锚索普遍采用的胶结式内锚固段，是预应力锚固体系的重要部分，直接决定着锚固作用的效果。目前有关内锚固段作用机理的研究尚不成熟，工程应用中仍然较多地依赖经验或工程类比[2]，现有的计算方法很难充分体现这种加固作用。因此，进一步研究预应力锚固的机理，并在此基础上建立合理的分析方法，在理论上和实践中都具有重要意义。

内锚段的模拟是预应力锚索模拟的关键，目前主要有两类基本方法：一类是锚索和岩体的离散模拟；另一类则是锚索和岩体的等效连续模拟。在离散模拟方面，陈卫忠等[3]建立了二维预应力锚索模型，假定内锚段与围岩固结在一起用杆单元来模拟，而自由段则以两端的锚固力来模拟；丁秀丽等[4]将外锚头和内锚段由锚单位模拟，在自由段两端施加一对集中力；黄福德等[5]分别采用三维杆单元和梁单元来模拟预应力锚索自由段和锚固段；徐前卫等[6]对锚固段采用空间杆单元，而对自由段采用一种三维杆单元来揭示预应力锚索锚固的效果和机理。在加锚岩体等效模拟方面，王敏强等[7]将预应力锚索的外锚头、锚索体及内锚段均采用隐式方法隐含在普通单元中，锚索体单元的刚度按拉伸刚度产生附加刚度，经坐标系转换后计入整体刚度矩阵中；杨延毅等[8]则采用等效抹平的处理方法，从损伤加筋体的自一致理论出发，给出了能反映锚索加固作用的岩体本构关系。

上述数值分析方法多侧重于研究加锚后锚固体的力学效应变化，关于预应力锚索中的组成部分——钢绞线、砂浆和周边岩土体的共同作用的研究则很少[9]。为了深入分析预应力锚索内锚固段的作用机理，笔者建立了钢绞线、砂浆及其相互接触面的有限元精细化模

本文发表于2012年。

型，探索了内锚段应力分布及其传递规律。

1 基本原理

1.1 力学模型

内锚段的钢绞线通过砂浆与岩体相连，预应力主要以钢绞线一砂浆接触面和砂浆一岩体接触面剪力的形成分散到深度岩体中（见图 1）。砂浆、钢绞线、岩体均用实体单元模拟，接触面的模拟是一个难点。

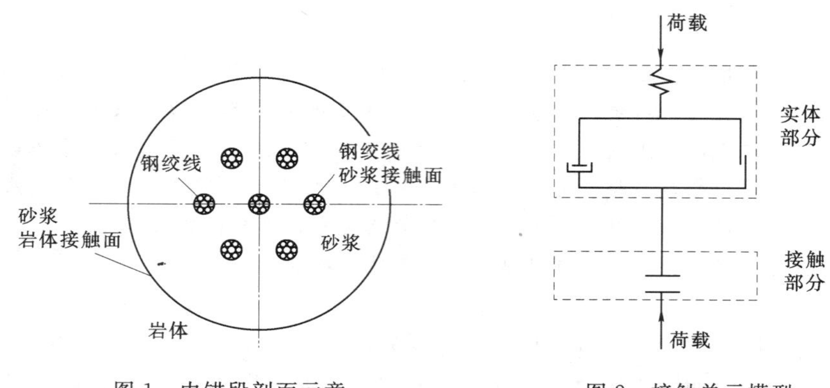

图 1 内锚段剖面示意　　图 2 接触单元模型

接触单元由薄层实体元件和接触元件串联而成（见图 2），具有以下两条基本原则：应变叠加原则，单元的应变增量等于实体与接触面的应变增量之和；应力一致原则，单元、实体块和接触面的应力增量相等。

在有限元计算中，这两条原则可分别表示为

$$\Delta\varepsilon = \Delta\varepsilon_R + \Delta\varepsilon_C \tag{1}$$

$$\Delta\sigma = \Delta\sigma_R = \Delta\sigma_C \tag{2}$$

式中：$\Delta\varepsilon_R$、$\Delta\sigma_R$ 分别为实体元件应变增量、应力增量；$\Delta\varepsilon_C$、$\Delta\sigma_C$ 分别为接触元件应变增量、应力增量。

定义整体坐标和局部坐标下的应力和应变相互转换关系如下：

$$\begin{cases} \Delta\varepsilon_j = \boldsymbol{T}\Delta\varepsilon_J \\ \Delta\sigma_J = \boldsymbol{T}^{\mathrm{T}}\Delta\sigma_j \end{cases} \tag{3}$$

式中：$\Delta\varepsilon_j$、$\Delta\varepsilon_J$ 分别为局部坐标下和整体坐标下的应变增量；$\Delta\sigma_j$、$\Delta\sigma_J$ 分别为局部坐标下和整体坐标下的应力增量；$\boldsymbol{T}$ 为整体与局部坐标转换矩阵。

1.2 本构关系

对薄层实体元件和接触元件分别进行具体化。

（1）实体元件的本构关系。弹性矩阵：

$$D_R=\begin{bmatrix}\lambda_R+2G_R & \lambda_R & \lambda_R & 0 & 0 & 0\\ \lambda_R & \lambda_R+2G_R & \lambda_R & 0 & 0 & 0\\ \lambda_R & \lambda_R & \lambda_R+2G_R & 0 & 0 & 0\\ 0 & 0 & 0 & G_R & 0 & 0\\ 0 & 0 & 0 & 0 & G_R & 0\\ 0 & 0 & 0 & 0 & 0 & G_R\end{bmatrix} \tag{4}$$

式中：λ_R、G_R 为岩块的 Lame 常数。

（2）接触元件的本构关系。弹性矩阵：

$$D_C=\begin{bmatrix}0 & 0 & 0 & 0 & 0 & 0\\ 0 & 0 & 0 & 0 & 0 & 0\\ 0 & 0 & k_n & 0 & 0 & 0\\ 0 & 0 & 0 & k_s & 0 & 0\\ 0 & 0 & 0 & 0 & k_s & 0\\ 0 & 0 & 0 & 0 & 0 & 0\end{bmatrix} \tag{5}$$

式中：k_n、k_s 分别为法向和切向刚度系数。

（3）单元的本构关系。将上述公式中的接触元件本构方程进行坐标变换，然后与实体混凝土元件的本构方程一起代入式（2），即可得到接触单元的本构方程：

$$\Delta\varepsilon=\boldsymbol{S}\Delta\sigma \tag{6}$$

其中

$$\boldsymbol{S}=\boldsymbol{D}_R^{-1}+\boldsymbol{T}^{-1}\boldsymbol{D}_C^{-1}(\boldsymbol{T}^{\mathrm{T}})^{-1} \tag{7}$$

2 算例分析

2.1 计算条件

取一 5m×5m×5m 的立方体试件进行分析，在立方体试件中心有一半径为 0.05m 的砂浆，砂浆中含有 7 根长 3m 的钢绞线，其半径为 0.005m，每根施加 9kN 的拉拔力，共计 63kN。接触面用 1mm 厚的接触单元模拟。钢绞线布置见图 3，有限元网格见图 4、图 5，共有 17784 个单元、19180 个节点，实体材料力学参数见表 1，接触面力学参数见表 2。

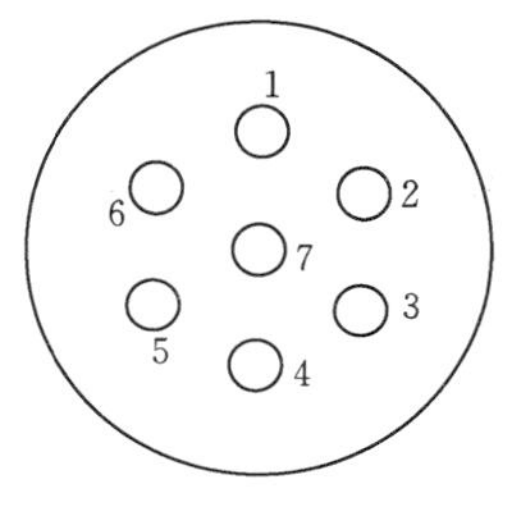

图 3　钢绞线分布

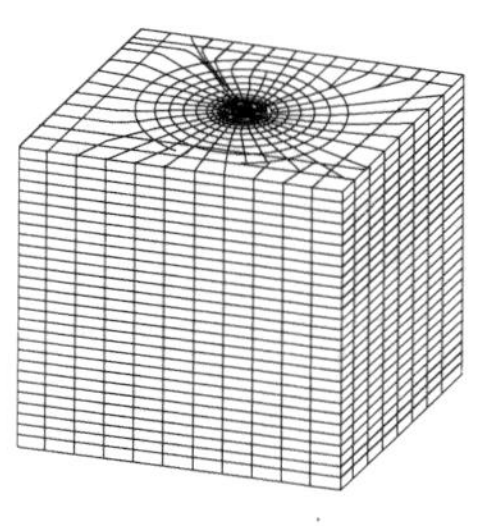

图 4　有限元网格

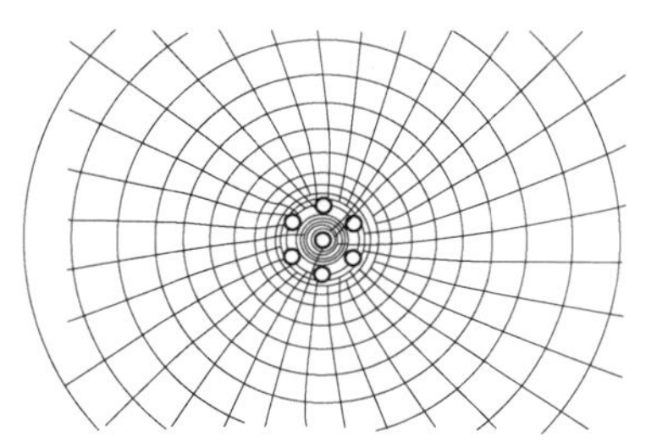

图 5　钢绞线局部网格

表 1　　材料的力学参数

材　　料	弹性模量 E/GPa	泊松比 μ	黏聚力 c/MPa	摩擦角 φ/(°)
钢绞线	200	0.25		
砂浆	26	0.18	2.5	45
岩石	25	0.20	1.7	45

表 2　　接触面的力学参数

材　　料	法向刚度/GPa	切向刚度/GPa	黏聚力 c/MPa	摩擦角 φ/(°)
岩石—砂浆	100	5	1.0	50
砂浆—钢绞线	100	8	2.0	58

2.2　计算结果及分析

内锚段中 7 根钢绞线的位移和应力基本上一致，因此只给出了 7# 钢绞线的计算结果，见图 6～图 8，岩体—砂浆接触面剪应力沿钢绞线长度分布见图 9。可以看出轴向位移和应力沿钢绞线方向由外到里呈指数性衰减，表面处轴向应力最大，外端点处达到 83MPa，在 0～0.5m 深度急剧衰减，0.5～1.5m 深度衰减趋缓，1.5～3m 深度慢慢减为 0，因此一般情况下内锚段并不是越长越好，3～5m 较为合适。这与文献［10］中介绍的现场监测结果相吻合。

通过分析平切位移矢量和纵切位移矢量可知：在拉拔力的作用下岩体朝上变位，上部岩体以锚索为中心向外发散，下部则向中间收缩。

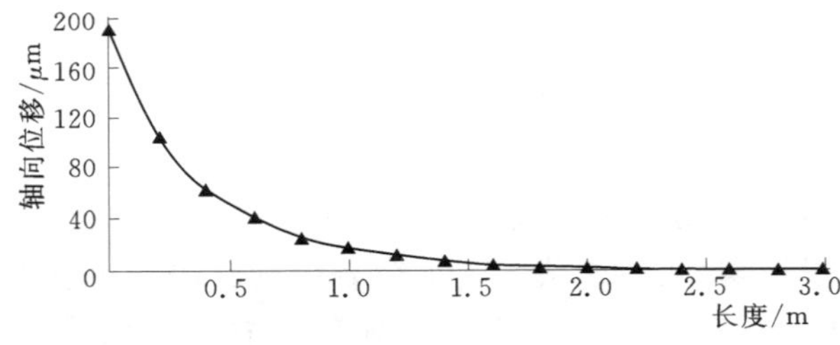

图 6　7# 钢绞线轴向位移沿钢绞线长度分布

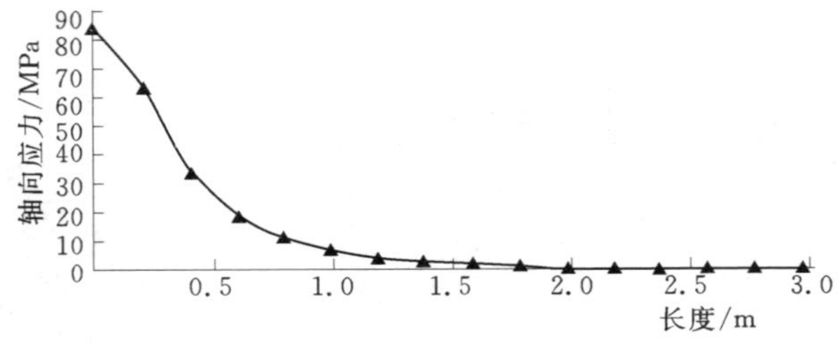

图 7　7# 钢绞线轴向应力沿钢绞线长度分布

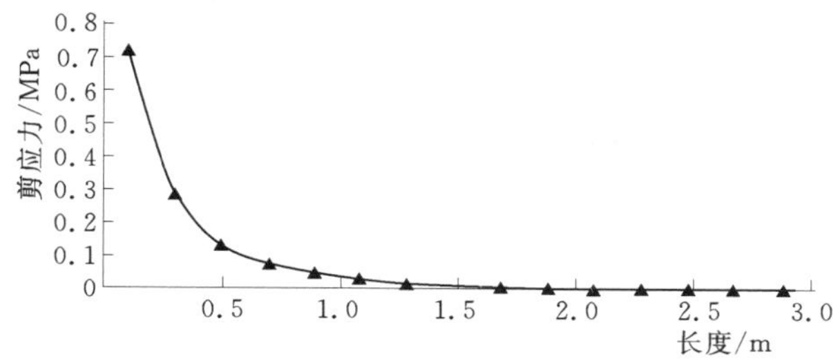

图 8　7# 钢绞线—砂浆接触面剪应力沿钢绞线长度分布

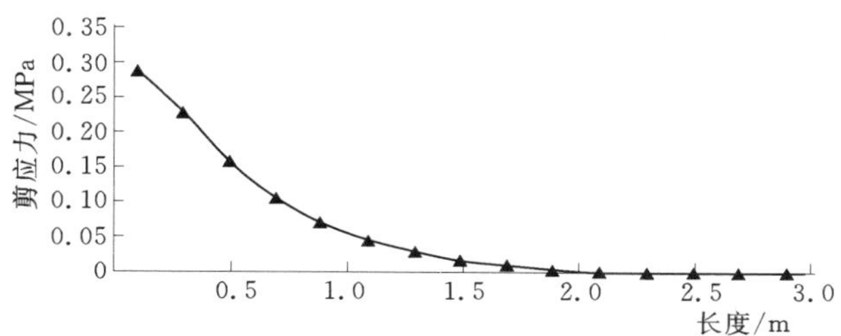

图 9　岩体—砂浆接触面剪应力沿钢绞线长度分布

3　结语

基于有限单元法提出了一种计算预应力锚索内锚段的精细化模型。其最显著的特征是精细地描述了每根钢绞线、砂浆等细部结构，所反映的规律符合工程实际，满足工程要求。内锚段轴向位移和应力沿钢绞线方向由外到里呈指数性衰减，与现场监测结果相吻合。

参考文献：

[1] 何君弼．“八五”攻关预应力群锚加固边坡机理研究［J］．建筑技术开发，1997，24（2）：9-12.

[2] 徐年丰，牟春霞，王利．预应力岩锚内锚段作用机理与计算方法探讨［J］．长江科学院院报，2002，19（3）：47-63.

[3] 陈卫忠，朱维申．节理岩体加固效果及其在边坡工程中的应用［J］．勘察科学技术，2001，19（1）：3-6.

[4] 丁秀丽，盛谦，韩军，等．预应力锚索锚固机理的数值模拟试验研究［J］．岩石力学与工程学报，2002，21（7）：980-988.

[5] 黄福德，赵彦辉，李宁．预应力锚固机理数值仿真分析研究［J］．西北水电，1996，55（1）：8-17.

[6] 徐前卫，尤春安，朱合华．预应力锚索的三维数值模拟及其应用研究［J］．岩石力学与工程学报，2004，23（S2）：4941-4945.

[7] 王敏强，刘晓刚，艾建申．某厂房边坡施工过程仿真及稳定分析［J］．岩石力学与工程学报，2002，21（S2）：2506-2510.

[8] 杨延毅，王慎跃．加锚节理岩体的损伤增韧止裂模型研究［J］．岩土工程学报，1995，17（1）：

9-17.

[9] 章青，陈爱玖，李珍，等．基于界面元法的改进压力集中型锚索作用分析［J］．工程力学，2008，25（4）：45-49.

[10] 张电吉，汤平，白世伟．节理裂隙岩质边坡预应力锚索锚固监测与机理研究［J］．岩石力学与工程学报，2003，22（8）：1276-1280.

封拱温度对砌石拱坝应力的影响研究

龚羊庆，吴海真，吴晓彬

江西省水利科学研究院

摘　要：温度荷载是砌石拱坝应力控制性边界条件之一，其中年平均温度场和变化温度场为自然环境边界条件，较难人为控制，而封拱温度场可采取工程措施加以控制。对于不设横缝、整体上升砌筑的拱坝，各拱层封拱温度随施工期气温、建筑材料及砌体温度变化而变化，不是一个常数，合理选择封拱温度区间是控制砌石拱坝应力的有效手段。通过对砌石拱坝进行温度场和应力场仿真计算，分析了不同封拱温度场对砌石拱坝应力的影响规律，建立了安全封拱温度计算模型，并给出了相应的表达式。算例表明，采用该计算模型确定安全封拱温度区间，可有效控制坝体应力。

关键词：砌石拱坝；拱坝应力；安全封拱温度；温度荷载

砌石拱坝具有经济安全、施工期短等优点，广泛应用于中小型水利水电工程中。拱坝为超静定结构，受外界温度的剧烈变化影响，坝体常出现显著的拉应力，对于未设横缝、整体上升砌筑的砌石拱坝（以下简称“砌石拱坝”）尤其明显。在运行初期遭遇寒潮的冲击后，由于坝高体薄，砌石拱坝常或多或少出现温度裂缝，一些为表面裂缝，一些甚至为贯穿性裂缝，这势必影响大坝的耐久性和整体性。

温度荷载一般占拱坝总荷载的30%左右[1]，是砌石拱坝应力控制性边界条件之一。目前规范给定的封拱温度范围较大[2-3]，可操作性不强，因此合理确定安全封拱温度是砌石拱坝设计的重要内容。虽然文献［4］提出了一种基于可靠性理论的分层确定拱坝封拱温度的方法，但由于该方法将拱坝视为一组互不相连的独立拱圈进行分析，因此其精度还有待研究。文献［5］给出了在不允许坝面开裂及允许坝面出现微小裂纹两种情况下，确定混凝土拱坝安全封拱温度的方法，但其作者认为坝体上下游面最大主应力与封拱温度呈线性关系，不甚合理。

本文通过分析不同封拱温度场对砌石拱坝应力的影响规律，提出了安全封拱温度计算模型，并给出了相应的表达式。该方法为砌石拱坝设计与施工提供了一种温度应力控制方法，相对于传统设计而言，更具有一定的实用价值和推广价值。

1　温度荷载和封拱温度

拱坝拱圈某一截面上的温度变化见图1。图1中阴影部分的面积就是温度变化值，即

本文发表于2011年。

温度荷载[6]。该温度荷载可以分解为均匀温度变化 T_m、等效线性温差 T_d 及非线性温差 T_n 三部分，其中，T_n 将产生自生温度应力，表现为坝体的表面应力，在进行拱坝整体应力分析时可不予考虑，而 T_m 和 T_d 则应按热传导理论进行计算，并求其最不利组合，它们的计算公式分别为：

$$
\begin{aligned}
T_m &= \frac{1}{L}\int_{-L/2}^{L/2} T\mathrm{d}x \\
T_d &= \frac{12}{L^2}\int_{-L/2}^{L/2} Tx\,\mathrm{d}x \\
T_n &= T - T_m - \frac{T_d x}{L}
\end{aligned}
\tag{1}
$$

式中：L 为坝体厚度；T 为温度变化值，并为坐标 x 的函数。

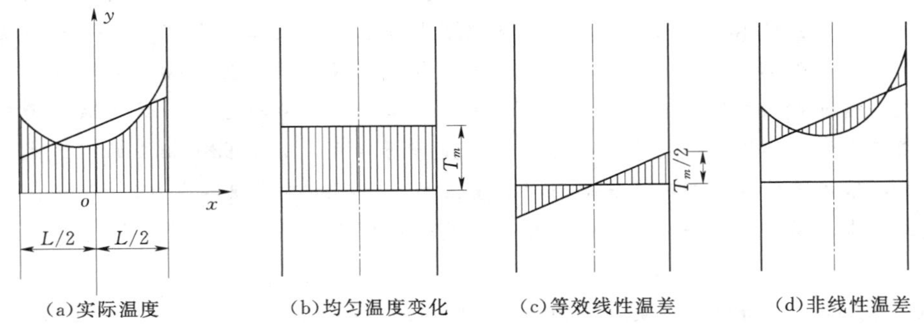

图 1　拱圈任一截面的温度荷载

拱坝的温度荷载可由封拱温度场、年平均温度场和变化温度场的相应值求得，其中，均匀温度变化 T_m 和等效线性温差 T_d 可分别按下式计算：

$$
T_m = T_{m1} - T_{m0} + T_{m2}, T_d = T_{d1} - T_{d0} + T_{d2} \tag{2}
$$

式中：T_m 和 T_d 的下标 0、1、2 分别表示与封拱、年平均和变化 3 个特征温度场所相应的值，具体计算可参考文献 [7]。

拱坝的温度荷载中的年平均温度场和变化温度场是自然环境边界条件，随环境温度变化而变化，难以人为控制，而封拱温度场可在施工过程中采取相应工程措施加以控制。

表 1 为江西省双门石砌石拱坝的温度荷载。该拱坝所在地区年平均气温为 15.0℃，气温年变幅为 11℃，日照影响为 2℃。该拱坝温度荷载的计算结果表明（表 1），大坝上部拱圈封拱温度 T_{m0} 均大于年平均气温，温降时 T_m 的绝对值较大，对于温降组合时的应力极为不利。

表 1　　双门石砌石拱坝的温度荷载（拱冠梁）

拱圈号	高程/m	T_{m0}/℃	T_{d0}/℃	正常+温降/℃		死水位+温降/℃		校核+温升/℃	
				T_m	T_d	T_m	T_d	T_m	T_d
1	197.5	22.0	0	−15.92	0	−15.23	0	9.92	0
2	194.0	26.0	0	−19.69	−0.10	−18.00	−0.10	5.69	0.21
3	190.0	20.0	0	−15.26	−1.20	−14.12	−1.10	9.75	2.32

续表

拱圈号	高程/m	T_{m0}/℃	T_{d0}/℃	正常+温降/℃		死水位+温降/℃		校核+温升/℃	
				T_m	T_d	T_m	T_d	T_m	T_d
4	185.0	16.0	0	−10.54	−2.10	−9.18	−2.06	9.56	4.71
5	180.0	13.3	0	−4.62	−3.65	−5.09	−1.00	9.93	8.02
6	175.0	13.3	0	−3.57	−4.58	−1.72	−5.61	8.20	10.49
7	170.0	13.3	0	−2.90	−5.10	−0.92	−6.44	6.90	12.33
8	165.0	13.3	0	−1.97	−4.42	−0.46	−6.74	5.85	13.56
9	159.0	13.3	0	−0.18	−0.77	0.15	0.20	−0.48	−0.78

在砌石拱坝设计中常以年平均气温以下的温度值作为封拱温度，其基本思想是充分利用砌石体抗压强度较高、抗拉强度较低的特点，以牺牲抗压安全度来换取拱圈不致发生受拉破坏为目的。同时，拱坝受力特点与重力坝不同，降低或提高封拱温度均可能导致拱圈弯矩增大，出现较大拉应力[8]。

笔者通过大量研究发现，坝体最大主拉应力与封拱温度存在一定的关系，如图2所示。即当封拱温度较低时，坝体大部分区域常处于温升状态且升幅较大，使坝体的整体性大为增强，但在夏季低水位时温升很高，拱圈就急剧向上游变形，特别是上层拱圈变幅更大，因而在下游面可能产生相当大的梁向拉应力（如死水位+温升）；当封拱温度较高时，在冬季低水位运行时温降很大，坝体表面温度急剧降低，砌石体内外温差较大，导致坝体拉应力过大而开裂（如死水位+温降）。

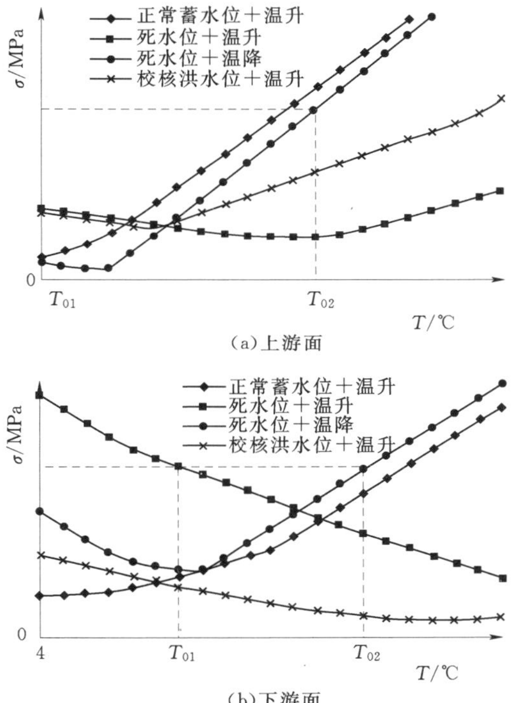

图2 最大主拉应力与封拱温度的关系曲线

2 安全封拱温度计算模型

笔者将坝体应力控制在规范中应力控制标准内的封拱温度称为安全封拱温度。安全封拱温度具有一定限度范围，其范围将随安全度指标的提高而缩小、降低而增大。

通常砌石拱坝各砌筑层没有明确的封拱部位、时间及温度，但其砌筑温度仍应按封拱温度来要求。受施工期气温影响，建筑材料及砌体的温度的不确定性影响，各砌筑层内的封拱温度不是一个常数，一般可根据温控措施和施工期日平均气温综合选取。目前规范未明确砌石拱坝封拱温度的确定方法，通常在设计过程中将封拱温度定为常量，这对砌石拱坝施工不甚合理，因为要控制砌筑温度为某一常量是极为困难的，且极不经济。因此合理选择封拱温度的安全限度对砌石拱坝设计具有重要意义。

砌石拱坝应力影响因素主要为大坝体型、砌体力学性质、工程地质特性、水荷载以及温度荷载[9]。当其他因素确定以后，坝体应力可表述为自变量封拱温度的函数，即：

$$\sigma = f(T_0) \leqslant [\sigma] \tag{3}$$

式中：σ 为坝体应力；T_0 为封拱温度；$[\sigma]$ 为应力控制标准；$f(T_0)$ 为以封拱温度为自变量，坝体应力为因变量的函数。通常可由拱梁分载法和有限元法进行砌石拱坝应力反演，由式（3）可得出安全封拱温度限度范围，即 $T_0 = [T_{0下限}, T_{0上限}]$。

3　算例

双门石水电站位于江西省资溪县马头山镇境内，工程于 2003 年 8 月动工兴建，2005 年 9 月基本完建。大坝为细石混凝土砌石双曲拱坝，拱圈为抛物线形，最大坝高 38.5m，坝顶拱弧中心角 90.0°，弧长 143.33m。本文采用拱梁分载法对封拱温度计算模型进行了验算。结果表明，大坝上游面拉应力控制工况为死水位＋温降组合，下游面拉应力控制工况为死水位＋温降组合和死水位＋温升组合。砌石拱坝最大主拉应力与封拱温度的关系见图 3。

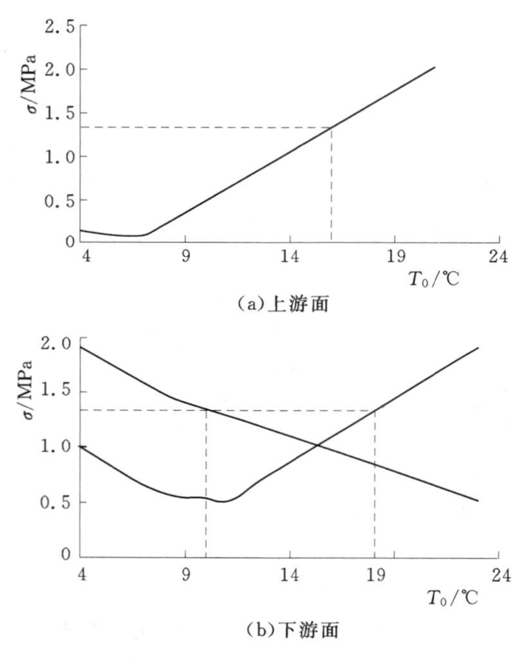

图 3　安全封拱温度区间

结合式（3）可知，双门石水电站大坝安全封拱温度 $T_0 = [10℃, 16℃]$，而设计封拱温度为 13.3℃，位于安全封拱温度区间内。但从坝体砌筑时间可知，中下部坝体施工期日平均气温总体与安全封拱温度相当，上部坝体砌筑时间主要为夏秋季节，虽然大坝浇筑从 6 月初开始为避开高温时段施工，但当地日平均气温较高，高于安全封拱温度上限约 10℃。笔者进行了封拱温度提高后的大坝应力分析，结果表明封拱温度对大坝应力有显著影响，尤其拉应力明显增大，导致局部应力超过规范要求。该工程在运行初期，已出现了最大宽度约 5mm 的贯穿性裂缝。

4　结论

（1）砌石拱坝各砌筑层通常没有明确的封拱部位、时间及温度，但其砌筑温度仍应按封拱温度来要求。受施工期气温影响，建筑材料及砌体的温度都有变化，各砌筑层内的封拱温度不是一个常数，一般可根据温控措施和施工期日平均气温综合选取。

（2）研究表明，当封拱温度较低时，砌石拱坝大部分区域常处于温升状态且升幅较大，大坝整体性大为增强，但在夏季低水位时温升很高，拱圈就急剧向上游变形，特别是上层拱圈变幅更大，因而在下游面可能产生相当大的梁向拉应力；当封拱温度较高时，在冬季低水位运行时温降很大，坝体表面温度急剧降低，砌石体内外温差较大，导致坝体拉应力过大而开裂。

(3) 安全封拱温度具有一定限度范围，其范围将随安全度指标的提高而缩小、降低而增大。采用安全封拱温度计算模型对砌石拱坝应力进行反演分析，可确定安全封拱温度限度范围，这对砌石拱坝设计和施工具有重要实用价值。

参考文献：

[1] 范福平．大花水水电站碾压混凝土拱坝封拱温度分析 [J]. 水力发电，2008，34 (7)：68 - 70.

[2] 上海勘测设计研究院，长江水利委员会长江勘测规划设计研究院．SL 282—2003 混凝土拱坝设计规范 [S]. 北京：中国水利水电出版社，2002.

[3] 中华人民共和国水利部．SL 25—2006 砌石坝设计规范 [S]. 北京：中国水利水电出版社，2002.

[4] 周筑宝，卢楚芬．拱坝封拱温度的分层确定法 [J]. 水力发电学报，1990，(4)：71 - 78.

[5] 周筑宝，卢楚芬．拉西瓦拱坝封拱温度研究 [J]. 水利学报，1995，(2)：52 - 57.

[6] 黎展眉．拱坝温度荷载与温度应力 [J]. 水利水运工程学报，2001，(2)：31 - 36.

[7] 黎展眉．砌石拱坝温度荷载影响的分析 [J]. 砌石坝技术，1989，(2)：9 - 13.

[8] 黎展眉．乌儿巢砌石拱坝应力分析 [J]. 贵州水力发电，2008，(3)：1 - 3.

[9] 刘德富．拱坝封拱温度场多目标非线性规划方法研究 [J]. 人民长江，1997，(5)：29 - 31.

基于有限元法的 $E-K$ 模型参数敏感性分析

龚羊庆[1]，吴海真[1]，黄　英[2]

1. 江西省水利科学研究院；2. 昆明理工大学

摘　要： 模型参数的确定是数值分析过程中的重要环节，合理把握参数的选取原则，可大大提高结构分析的可靠度以及参数反分析的计算效率。本文借助有限单元法考察了 $E-K$ 模型参数变化对一个均质堤坝的位移和应力值的影响程度。通过分析明确了 $E-K$ 模型中各参数对计算结果的敏感程度，同时提出了模型参数的选取原则。

关键词： $E-K$ 模型；参数；均质堤坝；位移；应力

1　引言

土体力学参数的取值因其复杂的力学特性而存在较大差异。它不仅随荷载大小而异，而且还与加载应力路径有关。此外，土体还会发生屈服，产生不可恢复的塑性变形[1]。诸多因素致使用土体试验参数作为计算输入参数进行数值分析时，所得结果往往与实际情况会有一定的误差，且不同程度地阻碍了数值计算在岩土工程中的应用。针对这一情况，国内外研究工作者开始对计算理论的反演问题开展研究，力图依据在工程现场获取的信息，尤其是根据位移量测信息反演，确定各类未知参数（如弹性模量和体积模量等）。因此，在土工计算中最棘手的问题是如何合理并切合实际地选取介质材料的物理力学参数。

邓肯-张模型在国内外广泛使用近 30 年，大量的试验成果表明，由于取样制样、试验仪器、试验方法、试验人员操作熟练程度、整理分析资料等诸多因素，使 $E-K$ 模型中的 7 个参数（c、φ、k、n、R_f、k_b、m）变化较大（其中 k 值可成倍甚至数量级的差别），用于计算所得结果差别也较大。所以有必要对这些参数进行敏感性分析，以正确掌握这些参数对计算结果的影响，这将有助于把握参数的选取原则，提高结构分析的可靠度以及参数反演分析的计算效率等。

2　敏感性分析方法

邓肯等人曾对 $E-K$ 模型中的 7 个参数作了初步讨论，并给出了几种不同类型土参数的取值范围（如表 1 所示）[2]。由于表 1 中的数值变化范围较大，不同取值对计算结果的影响没作进一步讨论。本文将借助分析土工结构中应力和变形问题的有限元软件 Geo-Studio[3] 来考察参数变化对一个均质堤坝（见图 1）的位移和应力的影响程度。考察某一

本文发表于 2011 年。

参数时，该参数值上下波动，其余 6 个参数保持试验值。

表 1　　邓肯-张模型参数取值范围

模型参数	软黏土	硬黏土	砂	砂卵石	石料
c/kPa	1～100	100～500	0	0	0
φ/(°)	20～30	20～30	30～40	30～40	40～50
k	50～200	200～500	300～1000	500～2000	300～1000
k_{ur}	3.0k	—	—	1.5～2.0k	—
n	0.5～0.8	0.3～0.6	0.3～0.6	0.4～0.7	0.1～0.5
R_f	0.7～0.9	0.7～0.9	0.6～0.85	0.65～0.85	0.6～1.0
k_b	20～100	100～500	50～1000	100～2000	50～1000
m	0.4～0.7	0.2～0.5	0～0.5	0～0.5	−0.2～0.4

注　k_{ur} 为土体回弹模量的试验常数。

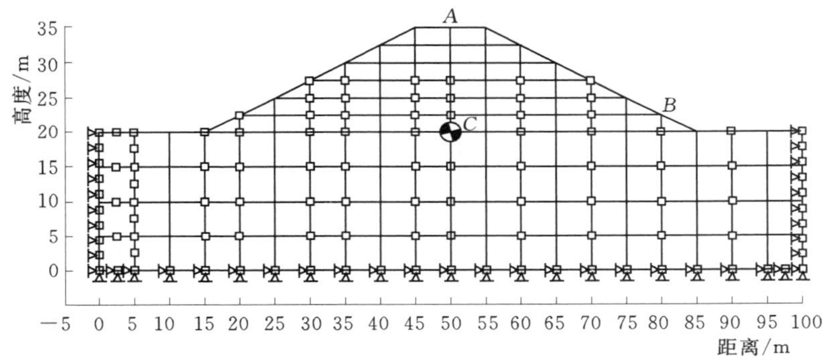

图 1　堤坝有限单元网格

3　结果分析及讨论

图 1 为模拟一座堤坝在填筑过程中的应力和变形分析的有限单元网格图。堤坝填筑土为黏性土，高 15m，上下游坡度均为 1∶2.0。堤坝地基约为 20m 的砂质黏土层。有关该实例的有限单元分析信息见表 2 和表 3。按表 3 中堤坝填筑土的试验参数进行有限元分析，得出堤坝的水平位移、垂直位移、竖向总应力、水平方向总应力以及坝轴线处竖向位移、竖向总应力和水平向总应力随高度的变化情况如图 2～图 5 所示。

表 2　　堤坝有限单元分析的有关信息

项　目	内　　容
分析对象	堤坝填筑过程的应力和变形分析
分析方法	总应力法（模型参数采用 CU 试验值）
分析模型	堤坝地基部分采用线弹性模型；堤坝填筑体采用非线性弹性模型（双曲线模型）
单元类型	堤坝上下游坝面采用 6 节点的三角形有限单元，其余部分采用 8 节点的四边形有限单元； 地基部分采用 4 节点的四边形有限单元，地基两侧采用 8 节点的四边形有限单元

续表

<table>
<tr><th colspan="2">项　目</th><th>内　容</th></tr>
<tr><td rowspan="2">边界条件</td><td>荷载</td><td>考虑到地基在自重作用下的固结过程已完成，故不计其自重荷载，只计施工过程中堤坝的自重荷载，堤坝自重荷载分 6 步分层施加</td></tr>
<tr><td>约束</td><td>地基底部铰接，无位移发生，地基两侧无侧向位移，其他边界无约束</td></tr>
</table>

表 3　堤坝填筑土的物理力学参数

c/kPa	φ/(°)	k	n	R_f	k_b	m	γ_W/(kN·m^{-3})
145	25.5	256	0.33	0.78	138	0.25	19.5

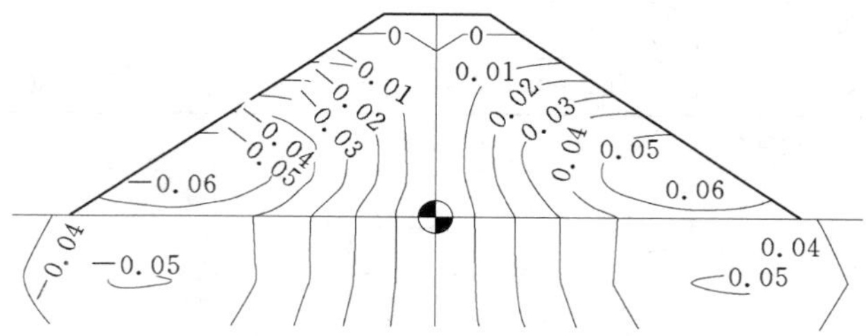

图 2　水平位移等值线图（单位：m）

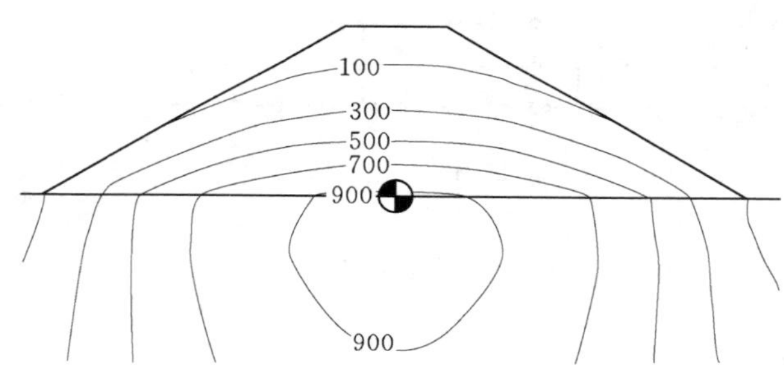

图 3　竖向总应力等值线图（单位：kPa）

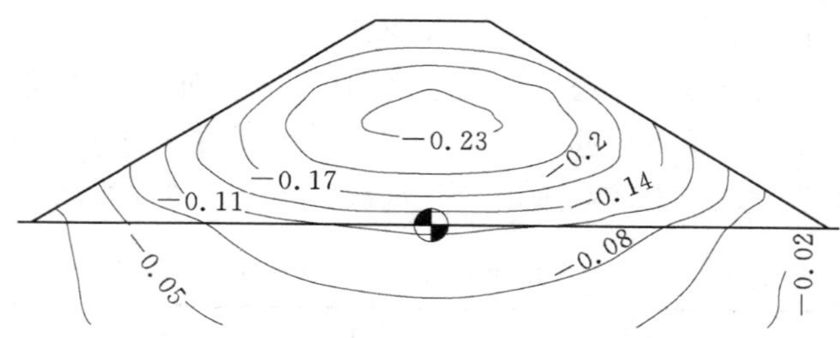

图 4　垂直位移等值线图（单位：m）

由图 2～图 5 可见，坝体最大竖向位移位于坝体中部区域；水平位移发生在上下游坝脚附近区域，这就是工程中常见的上下游坝脚附近隆起或滑动现象；最大竖向应力和水平应力均发生在坝底部区域。

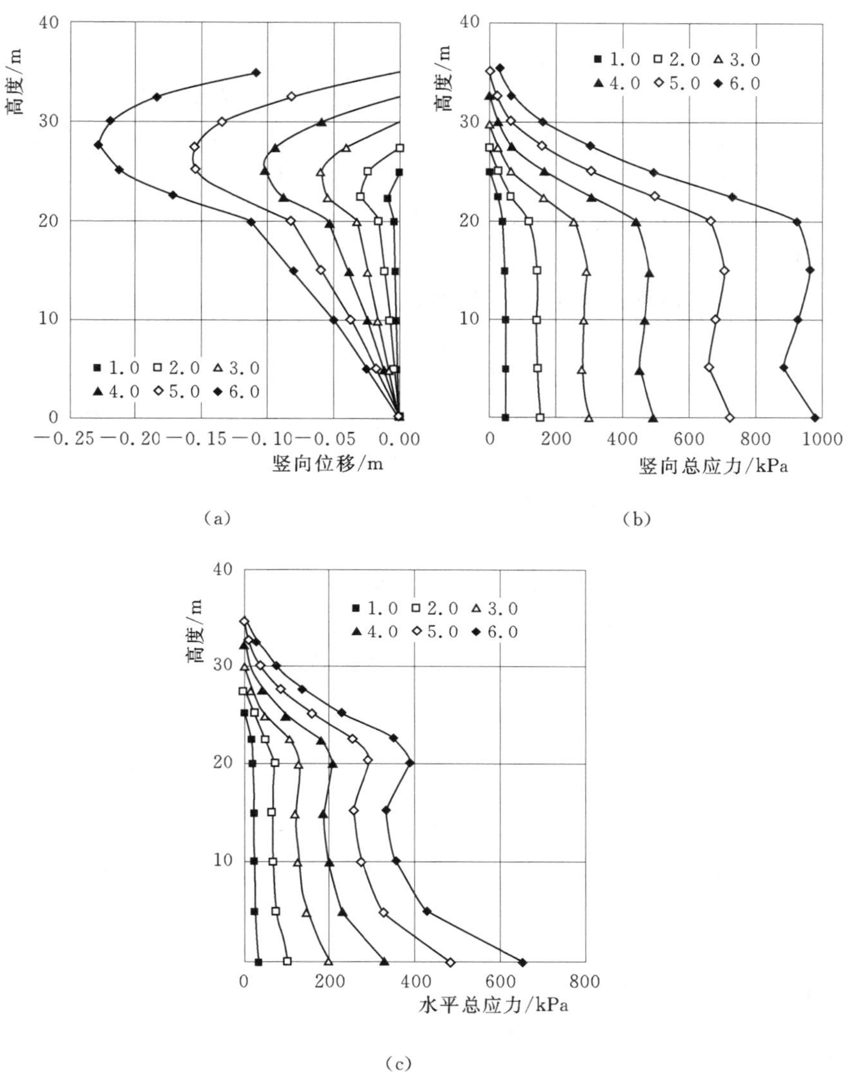

图 5　坝轴线处竖向位、竖向总应力和水平总应力随高度的变化

3.1　E_t 中各参数的影响

邓肯-张模型 E_t 中有 5 个须经试验确定的参数：摩尔-库仑准则的黏聚力 c、内摩擦角 φ、试验参数 k（相当于为 100kPa 时对应的 E_i 值）、初始模量 E_i 与围压力 σ_3 曲线的斜率 n 以及破坏比 R_f。下面以各参数的变化对坝顶 A 点的垂直位移 S_{VA}、坝脚 B 点的水平位移 S_{HB} 和 C 点的总应力的影响进行敏感性分析。

3.1.1　黏聚力 c

通过分析可知，黏聚力 c 对坝顶 A 点垂直位移 S_{VA}、坝脚 B 点水平位移 S_{HB} 和 C 点竖向总应力的影响如图 6 所示。

坝顶 A 点垂直位移 S_{VA} 和坝脚 B 点水平位移 S_{HB} 均随 c 值的增加而减小，其变化率在

c 值减小时大于 c 值增大时，如 c 值减小 50%时，S_{VA} 增大 5.53%，S_{HB} 增大 15.83%，c 值增大 50%时，S_{VA} 减少 2.93%，S_{HB} 减少 6.59%。

C 点竖向总应力与 c 值增减成正比，其变化率明显小于位移的变化率（以下各参数均有类似情况），这是因为应力的大小主要由填筑土容重的大小决定，而与各参数变化的关系不大。同时，在计算过程中还发现，c 值增减以后，堤坝的位移和应力等值线图与图2～图5所示的堤坝在基本试验值情况下得出的等值线图大致相同。

3.1.2 内摩擦角 φ

内摩擦角 φ 的增减对坝顶 A 点垂直位移 S_{VA}、坝脚 B 点水平位移 S_{HB} 和 C 点总应力的影响见图6～图8。

坝顶 A 点垂直位移 S_{VA}、坝脚 B 点水平位移 S_{HB} 随 φ 值的增加而减小，φ 值减小 25%时，A 点垂直位移 S_{VA} 和 B 点水平位移分别增加 3.95%和 6.27%；φ 值增加 25%时，A 点垂直位移 S_{VA} 和 B 点水平位移 S_{HB} 分别减小 2.90%和 4.77%。

C 点竖向总应力与 φ 值增减成反比，其变化率明显小于位移的变化率。

3.1.3 参数 n

初始模量 E_i 与围压力 σ_3 成指数关系，n 为指数，其影响见图6～图8。n 值增减对位移和应力水平的影响较小。当 n 增加 25%时，A 点垂直位移 S_{AB} 和 B 点水平位移 S_{HB} 分别减小 0.61%和增大 3.29%；n 值减小 25%时，分别增大 0.66%和减小 3.08%。同样 C 点竖向总应力与 n 值增减成反比。因此，n 值在整理分析时的局限性要小一些。

3.1.4 破坏比 R_f

邓肯-张模型定义破坏比为：

$$R_f=(\sigma_1-\sigma_3)_f/(\sigma_1-\sigma_3)_u$$

式中：$(\sigma_1-\sigma_3)_f$ 为土的抗剪强度取值；$(\sigma_1-\sigma_3)_u$ 是土的极限抗剪强度理论值；多数取值 $R_f=0.70\sim0.90$。R_f 的影响见图6～图8。A 点垂直位移 S_{VA}、B 点水平位移 S_{HB} 和 C 点总应力均随 R_f 的增加而增大，但变化率均不大。依据 R_f 的物理意义，R_f 值越大，计算使用的抗剪强度 $(\sigma_1-\sigma_3)$ 越接近土的理论极限抗剪强度 $(\sigma_1-\sigma_3)_u$。土体出现 $(\sigma_1-\sigma_3)_f$ 状态时，已经产生了相应的较大变形，濒于破坏。

3.1.5 参数 k

k 为 $\sigma_3=100\text{kPa}$ 时的初始模量 E_i，但无量纲。文献［1］及大量试验资料表明，由于各种原因，k 值可以相差数倍到几十倍。常规三轴试验过程中使用百分表量测轴向应变，试验点绘在普通坐标中求取 k 值。表面上看是 $\varepsilon_a\rightarrow0$ 时的纵坐标倒数，实际上多数情况下仅求取到 $\varepsilon_a=10^{-3}$ 级的相应值[1]，与原本物理意义相去甚远，也是众多资料相差较大的原因之一。

本文算例中堤坝填筑土的 k 值是由常规三轴试验法求取的，k 值对位移和应力的影响见图6～图8。

A 点垂直位移 S_{VA} 和 B 点水平位移 S_{HB} 位移均与 k 值的变化率相反，当 k 变化较大时，位移与应力也相应地变化较大。

k 从 256 减小至 64，S_{VA} 增加 119.05%，S_{HB} 增加 204.88%，σ_V 增加 5.81%；k 从 256 增至 460.8 时，S_{VA} 减小 15.67%，S_{HB} 也减小 35.83%，σ_V 减小 0.67%。

当 k 值变化率较大时，C 点的竖向应力变化较大，其变化率大于其他参数引起的变化，例如当 k 值减少 75%时，C 点的竖向应力 σ_V 增加 5.81%。

3.2 K_t 中各参数的影响

3.2.1 参数 k_b

参数 k_b 为 σ_3＝100kPa 时的体积模量 K_t，但无量纲。与参数 k 一样，由于各种原因，k_b 值可以相差数倍到几十倍。k_b 值对位移和应力的影响见图 6～图 8。

A 点垂直位移 S_{VA} 的变化率与 k_b 值的变化率相反，B 点水平位移 S_{HB} 的变化率与 k_b 值的变化率相同，当 k_b 变化较大时，位移也相应地变化较大。

当增加 50%时，A 点垂直位移 S_{VA} 和 B 点水平位移 S_{HB} 分别减小 10.64%和增大 13.73%；k_b 值减小 50%时，分别增大 32.89%和减小 26.30%。

3.2.2 参数 m

体积模量 K_t 与围压力 σ_3 成指数关系，m 为指数，其影响见图 6～图 8。由图可见，m 值增减对位移和应力的影响不大；当 m 增加 25%时，A 点垂直位移 S_{VA} 和 B 点水平位移 S_{HB} 分别增大 3.42%和减小 2.11%；m 值减小 25%时，分别减小 2.86%和增大 1.85%。

3.3 各参数综合对比

将上述 7 个参数对 A 点垂直位移 S_{VA}、B 点水平位移 S_{HB} 和 C 点竖向应力 σ_V 的敏感性分析结果绘制成图，详见图 6～图 8。

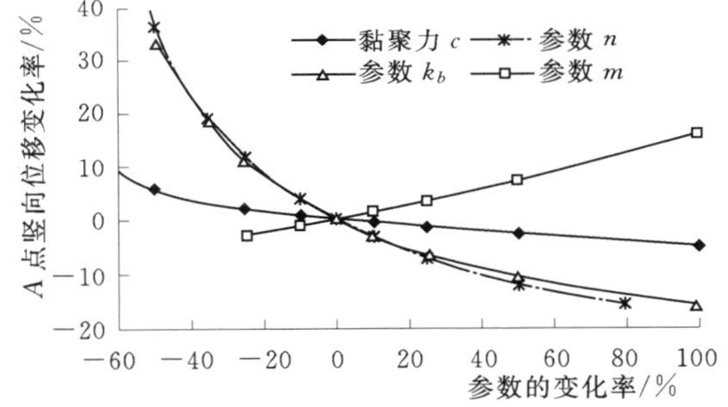

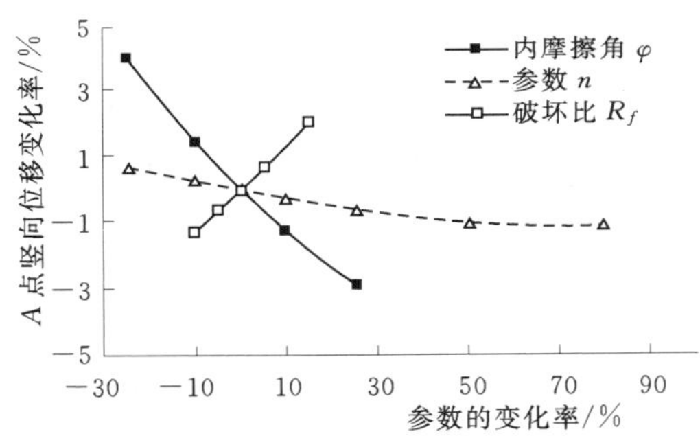

图 6 参数变化率对 S_{VA} 的影响

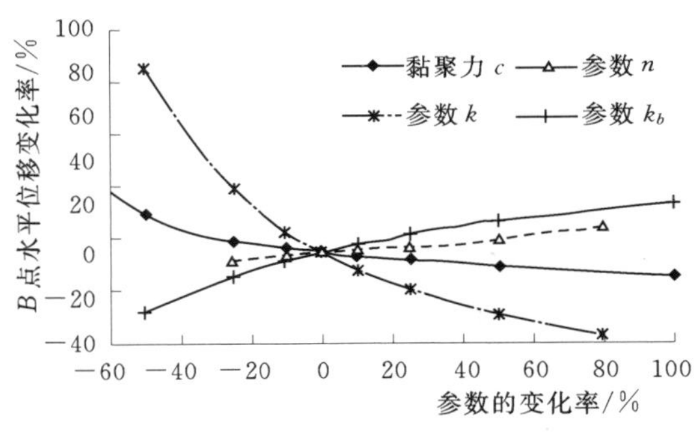

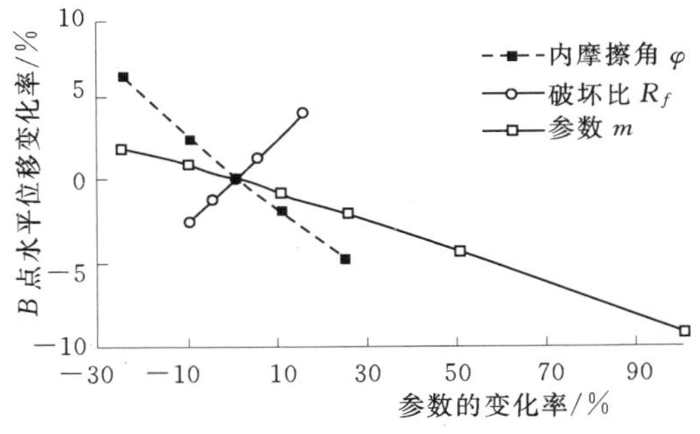

图 7 参数变化率对 S_{HB} 的影响

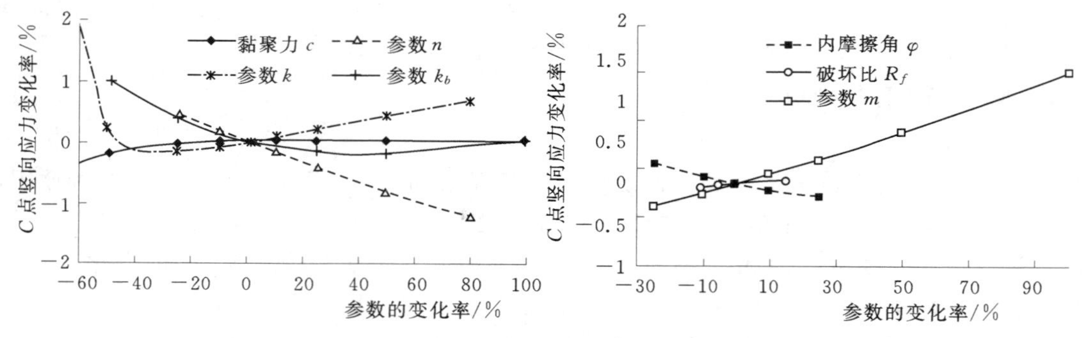

图 8　参数变化率对 σ_V 的影响

(1) A 点垂直位移 S_{VA} 随参数 m、R_f 增减而正向增减，随参数 c、φ、k、n、k_b 变化成反向增减。

(2) B 点水平位移 S_{HB} 随参数 k_b、R_f 和 n 增减而正向增减，随参数 c、φ、k、m 变化成反向增减。

(3) C 点竖向应力 σ_V 随参数 φ、n、R_f 和 k_b 增减呈反向增减，随参数 c、k、m 成正向增减，但是 σ_V 的变化幅度都非常小，这是因为应力的大小主要由填筑土容重大小决定，而与各参数变化关系不大。

(4) 相比而言，参数 k 和 k_b 对位移的影响最大，其变化率可达 30%～40%，甚至 80%以上，而且 k 和 k_b 值本身的变化范围就很大，因此对结果的影响很重要。

参数 c、φ 和 R_f 对位移的影响次之，多数变化率在 10%以内，虽然在本次的敏感性分析中，它们对位移和应力的影响不大，但是参数 c 和 φ 是土体的两个重要的抗剪强度指标，直接关系到土工结构的强度问题，而对于参数 R_f，当它增大时，本身就意味着变形增大和相应的应力水平提高。

因此，以上 5 个参数对计算结果的敏感性都很高，在参数的确定过程中需要给予重视，一方面要求提高试验设备及量测仪器的精度；另一方面要求严格按原本物理力学意义进行试验，同时试验人员还要对计算结果作出合理的分析判断后，根据各种情况提出建议值，使计算结果与工程实际观测尽可能地吻合等。此外，参数 n 和 m 对位移的影响最小，因此，在整理分析时的局限性要小一些。

4　结论

$E-K$ 模型中的 7 个参数，除 k 值以外，由于试验条件和试验方法等原因，相差 10%～40%是经常发生的事情。本文已证明，参数的上下浮动可引起位移 5%～50%的变化率。至于 k 值，其物理意义是围压力 $\sigma_3=100$kPa 的初始模量 E_i[4]。在模型推导过程中明确定义 E_i 为 $\varepsilon_a \to 0$ 时的模量，即要求在试验过程中尽可能地量测到微小应变，土体处于弹性状态时的模量。大多数土体应变 ε_a 从 10^{-6} 增至 10^{-2} 时，模量 E 将减小几十倍。多数试验由于量测设备精度不够，数据整理方法问题，实际上只取到 $\varepsilon_a=10^{-3}$ 数量级的 E，与其物理力学意义有一定差距。仅就现行整理分析方法而言，也因经验方法的差异，所得结果经常相差几倍甚至几十倍，引起计算位移相差一倍之多。多数情况下表现为计算结果远大于实测值，主要是 k 值取值偏小。这就是模型计算结果与实测结果相差较大的主要原因之一。

本构模型中的这些问题，类似于土压力、土坡稳定分析、地基承载力等土力学的普遍问题，参数的影响或带来的误差可能比某些理论和方法要大得多。要克服上述不利因素，须提高仪器的量测精度，改变整理分析方法，对理论、试验和计算的全面深刻理解以及经验的积累。

参考文献：

[1] 钱家欢，殷宗泽．土工原理与计算［M］．北京：中国水利水电出版社，1996：54－56.

[2] 郑颖人，等．岩土塑性力学原理［M］．北京：中国建筑工业出版社，2002：180－183.

[3] SIGMA/W for finite element stress and deformation analysis（User's guide）［M］．Geostudio，2004.

[4] 李广信．关于 Duncan 双曲线模型参数确定的若干错误做法［J］．岩土工程学报．1998，5（1）：116.

库岸堆积体斜坡降雨过程稳定性研究

吴海真[1,2]

1. 河海大学水利水电工程学院；2. 江西省水利科学研究院

摘　要： 在对降雨入渗及其引发边坡失稳机制的研究基础上，基于饱和-非饱和渗流数学模型，运用延伸的摩尔-库伦强度理论，对某库岸堆积体斜坡进行了降雨过程的饱和-非饱和渗流场计算和稳定分析。结果表明降雨是导致库岸堆积体滑坡的主要外动力因素之一，并得出了在采用实际可能降雨特征时，库岸堆积体的最小安全系数发生在雨后一定时间内的结论。这与实际工程情况相吻合，可更合理的解释因降雨引发的库岸堆积体滑坡问题。

关键词： 库岸堆积体；降雨入渗；饱和-非饱和渗流；稳定分析；实际降雨特征

1　概述

库岸堆积体是一类由滑坡堆积、崩塌堆积、残积层、冰溃堆积、坡积物等组成的松散体，即主要受重力作用堆积而形成的第四系覆盖层斜坡，在我国尤其是三峡库区和西南地区分布广泛。其物质组成为典型的土石混合体，是介于岩体和土之间的一类特殊地质体，使得人们较难把握其稳定性及其变形发展趋势，故在国内水电建设工程中发生滑坡并酿成重大灾害的情况时有发生。研究表明，降雨是影响库岸堆积体变形发展的主要外动力因素[1-2]。因此，研究因降雨入渗引发库岸堆积体滑坡的机制和规律具有重要意义[3]。在对降雨入渗及其引发边坡失稳机制的研究基础上，对影响边坡降雨入渗过程的条件及入渗参数进行分析研究，并基于饱和-非饱和渗流数学模型和实测土水特征曲线，在延伸摩尔-库仑强度理论基础上对某堆积体进行了降雨过程的饱和-非饱和渗流计算和稳定分析，得到了一些具有实用价值的结论。

2　降雨在边坡中的入渗过程分析

降雨入渗过程实质上是入渗水分在非饱和区运动的过程，即水在下渗过程中驱替空气的过程[4]。研究表明，雨水渗入边坡土体的强度主要取决于降雨的方式和强度以及土体的渗水性能。如果土体渗水性能较强，大于外界供水强度，则入渗强度主要决定于外界供水强度，在入渗过程中边坡表面含水率随入渗而逐渐提高，直至达到一稳定值；如果降雨强度较大，超过了岩土体的渗水能力，入渗强度就决定于岩土体的入渗性能，这样就会形成

本文发表于2008年。

地表径流积水。这两种情况可能发生在入渗过程的不同阶段。土体的入渗速率随时间而变化。一般来说，开始入渗阶段，边坡表层的水势梯度较陡，所以入渗速率较大，但随着雨水渗入土体中，土体的基质吸力下降，湿润层下移使基质吸力梯度减小，入渗速率逐渐减小最后接近于一常量而达到稳定入渗阶段。实际模拟计算中，降雨入渗边界虽然只是一个流量边界，但这个流量并不是不变的，在计算过程中需要根据含水率的变化不断的调整入渗流量，从而实现对降雨入渗问题的数值模拟。

据以上分析，可以确定降雨强度 $R(t)$ 与实际入渗率 $i_r(t)$ 的关系。设入渗面的外法线方向为 $n=(n_1, n_2, n_3)$，则降雨在入渗面法向上的入渗分量为 $q_n(t)=R(t)\cdot n_3$，由入渗面法向入渗能力 $q_r(t)=\sum_{i=1}^{3}q_{ri}(t)\cdot n_i$[$q_{ri}(t)$ 为整体坐标系中 x，y，z 3 个方向的入渗能力]，可确定入渗强度 $i_r(t)$ 与 $q_r(t)$ 及 $q_n(t)$ 的关系：

$$i_r(t)=\min\{q_r(t),q_n(t)\} \tag{1}$$

入渗过程中，累积入渗量、入渗率和岩土体含水率随时间的变化与地表处水的施加方式有关。为了求出入渗过程中土壤含水率分布，以及入渗率等随时间变化的定量结果，可以在一定的初始含水率分布条件下，根据入渗边界条件，求解水流运动基本方程。

3 雨水入渗引发边坡失稳的物理机制

如前所述，滑坡的发生与降雨关系密切早为人知，降雨入渗及地下水状态对边坡稳定的影响也是显而易见的，其作用主要有[2]：

(1) 降低岩土体特别是有泥质充填物的断层、层间错动带以及以节理裂隙等软弱结构面为滑裂面的岩土体强度。

(2) 地下水的静水压力一方面降低了滑裂面上的有效法向应力，从而降低了滑裂面上的抗滑力，另一方面岩土体非饱和区暂态附加水荷载又增加了边坡岩土体的下滑力，非饱和区饱和度增加，基质吸力（即毛细压力）降低，从而使边坡的稳定条件恶化，这也是雨季边坡失稳的重要原因[4]。

为定量研究因降雨入渗而导致的岩土体抗剪强度降低问题，引入非饱和土的抗剪强度理论[4-5]，该理论为定量计算因水分入渗而引起的岩土体软化的强度提供了一种计算方法，即 Fredlund 于 1993 年提出的岩土体吸水软化抗剪强度公式：

$$\tau_f=c'+(\sigma_n-u_a)\tan\varphi'+(u_a-u_w)\tan\varphi_b \tag{2}$$

式中：c'，φ'分别为土的有效黏聚力和有效内摩擦角；(u_a-u_w) 为孔隙气压与孔隙水压力之差；$(u_a-u_w)\tan\varphi_b$ 为与基质吸力直接相关的抗剪强度，称为基质吸力的附加强度；φ_b 为因基质吸力上升而引起抗剪强度增加的曲线倾角。

从式 (2) 可以看出，非饱和土的抗剪强度除了与 c'、φ'及正应力有关外，还与基质吸力 (u_a-u_w) 有关。当土体饱和时，可以认为 $\varphi_b=\varphi'$，退化为传统的摩尔-库仑公式，故式 (2) 又称延伸的摩尔-库仑公式。降雨时，岩土体饱和度和孔隙水压力上升，基质吸力 (u_a-u_w) 减小，抗剪强度明显减小，从理论上解释了降雨条件下滑坡发生的机理。

4 降雨入渗条件下的饱和-非饱和渗流场研究

目前国内外进行岩土体渗流场以及渗流应力耦合计算分析时，常根据实际工程水文地质条件复杂程度而采用不同的计算模型，主要有[6]：①等效连续介质模型；②裂隙网络非连续介质模型；③岩土体裂隙双重介质模型；④离散介质-连续介质耦合模型。大量工程研究和应用表明，尽管依据裂隙实测资料的统计数字由计算机生成等效网络已有可能，但三维裂隙网络水力学分析仍很难进行，故用离散网络模型来研究裂隙岩体非饱和渗流并不合适。采用等效连续介质模型。

4.1 堆积体非饱和水力参数的确定

对于饱和岩土类多孔介质，目前已有成熟的试验方法来确定其饱和渗透张量。对于非饱和岩土类多孔介质，关键是确定非饱和相对渗透系数与基质吸力、体积含水量与基质吸力的对应关系曲线，这两个对应关系曲线一般通过试验获取[7]。基于对江西某大（2）型水库库岸堆积体斜坡（崩坡积体）基质吸力的现场观测和观测竖井中的碎（块）石黏性土土原状样（取样的深度为 1.0～5.0m）试验成果，得到该堆积体的试验参数如下：饱和含水量 $\overline{\omega}_s=19.24\%$；天然含水量 $\overline{\omega}_s=18.44\%$；天然状态饱和度 $\overline{S}=95.84\%$；饱和容重 $\gamma_{饱和}=22.5\text{kN/m}^3$；干容重为 $\gamma_d=17.5\text{kN/m}^3$；天然容重 $\gamma_{天然}=21.5\text{kN/m}^3$。堆积体的土水特征曲线根据测试结果依据 $V-G$ 模型拟合得到[7]，非饱和渗透系数依据土水特征曲线按下式确定[5]：

$$k(\theta)=\frac{k_s}{k_{sc}}A_d\sum_{j=i}^{m}\{(2j+1-2i)(u_a-u_w)\}_j^{-2},i=1,2,\cdots,m \tag{3}$$

式中：$k(\theta)_i$ 为用第 i 个间断体积含水量 θ 确定的渗透系数；A_d 为调整常数，一般取 1；k_s 为饱和渗透系数；k_{sc} 为计算饱和渗透系数，计算公式为：$k_{sc}=A_d\sum_{j=i}^{m}\{(2j+1-2i)(u_a-u_w)\}_j^{-2}$，$i=1,2,\cdots,m$；$u_a$，$u_w$ 分别为孔隙气压和孔隙水压力。

由此得到非饱和渗透系数与基质吸力的关系，见图 1（对数坐标）和图 2。

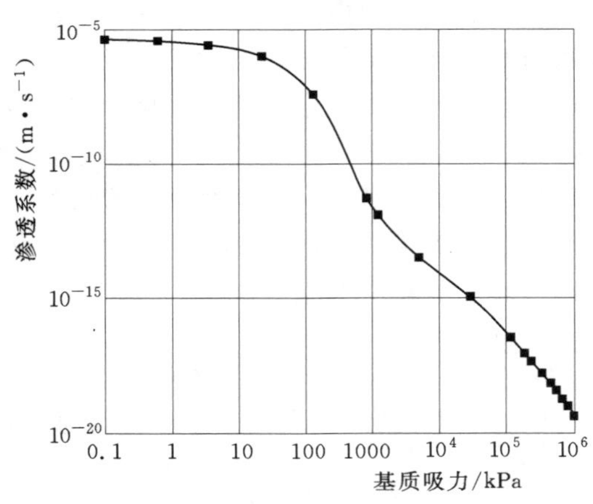

图 1 堆积体基质吸力—渗透系数关系图

图 2 堆积体基质吸力—体积含水量关系图

对于基岩裂隙介质非饱和等效水力参数，可采用均化方法求取[7]，关于应用裂隙网络方法来确定裂隙介质的非饱和水力参数的方法可参见文献［8-9］。采用均化方法确定部分岩层的非饱和等效水力参数。其中假设：①裂隙充分发育，裂隙介质存在表征单元体且体积不是太大；②流动随时间变化缓慢，也就是指岩块饱和度变化不大，岩块和裂隙间的水量交换瞬时完成。

取一体积为 V 的表征单元体，其中裂隙体积 V_1，岩块体积 V_2，$V=V_1+V_2$。假定垂直于水流方向的平面上裂隙内水头和岩块内水头相等，则按照流量等效和水头近似等效的原则，可依据裂隙和岩块各自的饱和及非饱和水力参数、各自所占的体积，通过体积加权平均来确定裂隙岩土体的等效非饱和水力参数，则裂隙介质的等效相对渗透系数 K_r、等效比容水度 C 及等效单元贮存量 S_s 分别为

$$K_r=\frac{k_{s1}k_{r1}V_1+k_s^2k_r^2V_2}{k_s^1V_1+k_s^2V_2},C=\frac{C_1V_1+C_2V_2}{V};S_s=\frac{S_s^1V_1+S_s^2V_2}{V} \tag{4}$$

其中 $$k_s^1=\sqrt[3]{k_{s1}^1k_{s2}^1k_{s3}^1},\ k_s^2=\sqrt[3]{k_{s1}^2k_{s2}^2k_{s3}^2}$$

式中：k_{s1}^1，k_{s2}^1，k_{s3}^1，k_{s1}^2，k_{s2}^2，k_{s3}^2分别为裂隙网络和岩块饱和渗透张量的 3 个主值；k_r^1，k_r^2 分别为裂隙和岩块的非饱和相对渗透系数；C_1，C_2 分别为裂隙网络和岩块的比容水度；S_s^1，S_s^2分别为裂隙网络和岩块的单元贮存量。

4.2 渗流场数学模型的建立及有限元求解

4.2.1 饱和非饱和渗流数学模型

基于地下水运动的连续性方程和 Darcy 定律，略去推导过程，在不考虑骨架及流体压缩作用下建立的非稳定饱和非饱和渗流基本微分方程为

$$\frac{\partial}{\partial x_i}\left[k_{ij}^sk_r(h_c)\frac{\partial h_c}{\partial x_j}+k_{i3}k_r(h_c)\right]-Q=[C(h_c)+\beta S_s]\frac{\partial h_c}{\partial t} \tag{5}$$

式中：h 为总水头；h_c 为压力水头；k_{ij}^s 为饱和渗透张量；k_{i3} 为饱和渗透张量中仅和第 3 坐标轴有关的渗透系数值；k_r 为相对透水率，在非饱和区 $0<K_r<1$，在饱和区 $k_r=1$；C 为比容水度，在正压区 $C=0$；β 为饱和-非饱和选择常数，在非饱和区等于 0，在饱和区等于 1；S_s 为弹性贮水率，在非饱和土体中 $S_s=0$，当忽略土体骨架及水的压缩性时对于饱和区也有 $S_s=0$；Q 为汇源项。

4.2.2 定解条件

（1）初始条件

$$h_c(x_i,0)=h_c(x_i,t_0),i=1,2,3 \tag{6}$$

（2）边界条件

$$h_c(x_i,t)|_{\Gamma_1}=h_{c1}(x_i,t) \tag{7}$$

$$-\left[k_{ij}^sk_r(h_c)\frac{\partial h_c}{\partial x_j}+k_{i3}^3k_r(h_c)\right]n_i\bigg|_{\Gamma_2}=q_n(t) \tag{8}$$

$$-\left[k_{ij}^sk_r(h_c)\frac{\partial h_c}{\partial x_j}+k_{i3}^3k_r(h_c)\right]n_i\bigg|_{\Gamma_3}\geqslant 0$$

$$\text{且 } h_c|_{\Gamma_3}=0 \tag{9}$$

式中：n_i 为边界面外法线方向余弦；t_0 为初始时刻；Γ_1 为已知结点水头边界；Γ_2 为流量边界；Γ_3 为饱和逸出面边界。

4.2.3 有限元求解

对于上述数学模型，微分方程和定解条件往往都是非线性的，通常采用数值法求解。应用 Galerkin 加权余量法求解上述数学模型。

将计算空间域 Ω 离散为有限个单元，对于每个单元，选取适当的形函数 $N_m(x_i)$，满足

$$h_c(x_i,t)=N_m(x_i)h_{cm}(t),i=1,2,3 \tag{10}$$

式中：$h_{cm}(t)$ 为结点压力水头值。

将 $h_c(x_i,t)$ 分别代入微分方程及边界条件，一般不能精确满足，分别会产生一定的误差，记残差值（余量）分别为 R 和 $\overline{R}$。按照加权余量法原理，选择权函数 $W(x_i)$ 及 $\overline{W}(x_i)$，取 $W(x_i)=N_n(x_i)$，在边界上，取 $\overline{W}(x_i)=-N_n(x_i)$，略去中间推导过程，对于离散后空间域 Ω 有：

$$\sum_{e=1}^{NE}\left[\iiint_{\Omega^e}\sum_{i=1}^{3}\sum_{j=1}^{3}k_r(h_c)k_{ij}^s\frac{\partial N_n}{\partial x_i}\frac{\partial(N_m h_m)}{\partial x_i}\mathrm{d}\Omega\right]+\sum_{e=1}^{NE}\left[\iiint_{\Omega^e}\sum_{i=1}^{3}k_r(h_c)k_{i3}^s\frac{\partial N_n}{\partial x_i}\mathrm{d}\Omega\right]$$
$$+\sum_{e=1}^{NE}\left[\iiint_{\Omega^e}N_n[C(h_c)+\beta S_s]\frac{\partial(N_m h_m)}{\partial t}\mathrm{d}\Omega\right]+\sum_{e=1}^{NE}\left[\iiint_{\Omega^e}N_n Q\mathrm{d}\Omega+\oint_{T2}q_n N_n\mathrm{d}S\right]=0 \tag{11}$$

令式（11）中

$$[A]=\sum_{e=1}^{NE}\left[\iiint_{\Omega^e}\sum_{i=1}^{3}\sum_{j=1}^{3}k_r(h_c)k_{ij}^s\frac{\partial N_n}{\partial x_i}\frac{\partial N_m}{\partial x_i}\mathrm{d}\Omega\right]$$

$$[B]=\sum_{e=1}^{NE}\left[\iiint_{\Omega^e}N_n N_m[C(h_c)+\beta S_s]\mathrm{d}\Omega\right]$$

$$\{P\}=-\sum_{e=1}^{NE}\left[\iiint_{\Omega^e}\sum_{i=1}^{3}k_r(h_c)k_{i3}^s\frac{\partial N_n}{\partial x_i}\mathrm{d}\Omega\right]-\sum_{e=1}^{NE}\left[\iiint_{\Omega^e}N_n Q\mathrm{d}\Omega+\oint_{T2}q_n N_n\mathrm{d}S\right]$$

则式（11）变为

$$[A]\{h_c\}+[B]\left\{\frac{\partial h_c}{\partial t}\right\}=\{P\} \tag{12}$$

采用向后有限时间差分格式对时间域的积分进行求解，其基本方程为

$$\left(\alpha[A]+\frac{[B]}{\Delta t^k}\right)\{h_c^k\}=\{P\}-(1-\alpha)[A]\{h_c^{k-1}\}+\frac{[B]\{h_c^{k-1}\}}{\Delta t^k},0\leqslant\alpha\leqslant1 \tag{13}$$

根据求解的稳定性，采用 t^k 时刻的压力水头值 $\{h_c^k\}$ 近似作为 Δt^k 时段的平均压力水头值，可得式（13）的隐式差分格式

$$\left([A]+\frac{[B]}{\Delta t^k}\right)\{h_c^k\}=\{P\}+\frac{[B]\{h_c^{k-1}\}}{\Delta t^k} \tag{14}$$

5 工程应用

5.1 基本情况

江西省某大（2）型水库库岸堆积体位于坝址左岸上游约 70～300m 地段，边坡底部横向宽度约 220.0m，纵向最大高度 120m，厚一般 12～15m，最厚近 20m，总体积约 30 万 m^3。经勘探查明，堆积体物质组成以碎（块）石为骨架，内充填黏性土，碎（块）石、砾含量达 40%～50%。堆积体下伏基岩顶部为石炭系岩层，一般为弱～新鲜岩体，岩层反倾；中下部为震旦系变余细砂岩夹板岩，其中顺坡向层间挤压带是边坡稳定的控制性结构面，见图 3。

为定量分析边坡在强降雨条件下的稳定性，采用前述理论和方法，对其进行了强降雨过程的渗流场计算和稳定分析，其中降雨强度和雨型有：雨型 1 为降雨强度 30mm/h，降

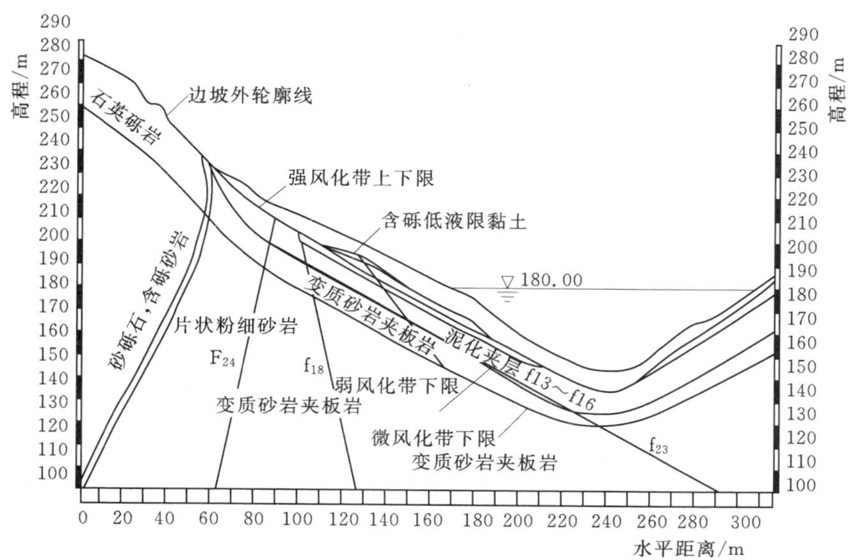

图 3　堆积体斜坡典型剖面图

雨持时 24h；雨型 2 的选取系参照文献［10］中的降雨区划而定，降雨持时 24h，其中前 12h 的降雨强度见图 4，后 12h 的降雨强度和雨型同前 12h，之后为无雨状态，相应库水位均取正常蓄水位 180.0m，初始稳定渗流场系根据钻孔稳定地下水位确定，见图 5。

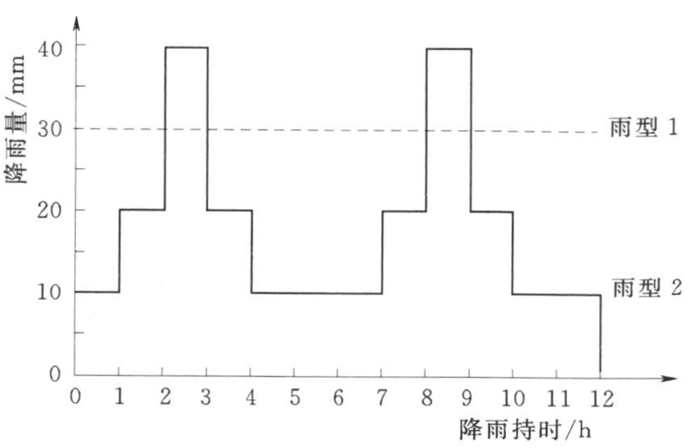

图 4　堆积体强降雨条件下的雨强和雨型分布图

基岩的非饱和水力参数参照 4.1 节所述方法，同时参照地质勘察中压水试验成果和相关地质描述确定，用于渗流计算分析的物理力学参数见表 1。抗滑稳定计算采用可考虑基质吸力、适用于任意滑裂面的 Morgenstern－Price 法。

表 1　　堆积体斜坡抗滑稳定计算物理力学参数选取表

参数指标	崩坡积体	断层破碎带	构造夹泥层	基岩（层面及节理）	基岩（强风化带）	基岩（弱风化带）
天然容重/(kN·m^{-3})	21.50	22.35	22.20	—	25.50	26.00
饱和容重/(kN·m^{-3})	22.50	23.05	22.39	—	26.20	26.50
黏聚力 c'/kPa	11.00	0	0	15.00	23.50	28.37
内摩擦角 φ'/φ_b/(°)	25.78/12.0	11.50/5.0	22.80/10.0	27.02/13.5	30.00	58.80
综合饱和渗透系数/(m·s^{-1})	4.45×10^{-6}	1.84×10^{-4}	3.50×10^{-4}	2.35×10^{-6}	1.25×10^{-6}	2.34×10^{-7}

5.2　计算结果分析

限于篇幅，仅列出雨型 2 情况下降雨初始时刻的稳定渗流场和强降雨 36h（雨后 12h）时刻的暂态渗流场，见图 5 和图 6。图 7 为堆积体斜坡安全系数随降雨过程的变化情况。

由图可看出：

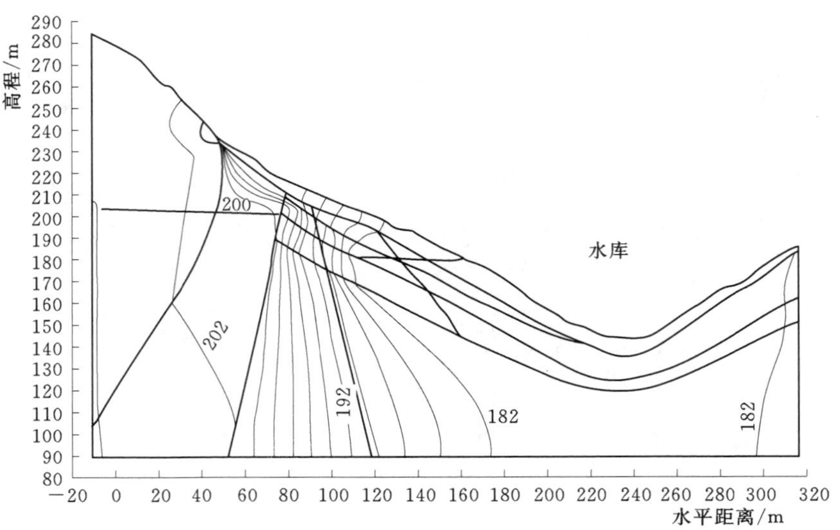

图 5　堆积体斜坡初始稳定渗流场

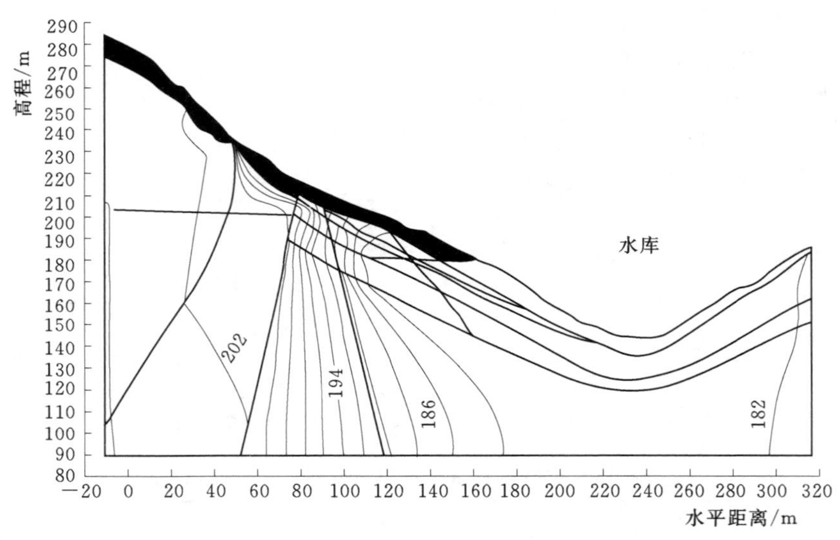

图 6　雨后 12h 时刻堆积体斜坡的暂态渗流场

（1）随着强降雨的进行，雨水不断入渗到堆积体中，非饱和带饱和度上升，暂态水荷载增加，基质吸力降低，同时地下水位略有上升，使滑裂面上的孔隙水压力增加，斜坡的安全系数随之逐渐减小。降雨停止后，因堆积体内的一部分孔隙水和基岩裂隙水缓慢排泄于库区和低洼地带，其安全系数有所回升。

（2）计算表明，在雨型 1 条件下，堆积体斜坡在强降雨 24h 时刻即降雨停止时刻安全系数达到最小值 1.005，较降雨前降低 6.34%；而雨型 2 条件下则是在强降雨 36h 时刻（即雨后 12h）安全系数达到最小值 0.989，较降雨前降低 7.83%，这一现象与多数自然

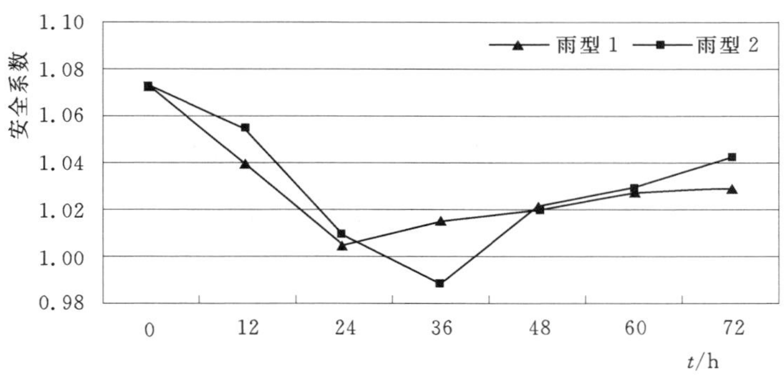

图 7　堆积体斜坡降雨过程稳定性分布图

边坡在雨后一段时间产生滑动失稳的情况相吻合。可见降雨特征（雨强、雨型）对堆积体的降雨过程稳定性有一定影响，故采用雨型 2 时，计算结果更符合工程实际。

6　结语

（1）库岸堆积体斜坡在我国许多水电站地区分布广泛，其稳定性问题关系到近坝库岸和大坝的安全。降雨是导致库岸堆积体变形发展和滑动的主要外动力因素之一，研究因降雨入渗引发库岸堆积体滑坡的机制和规律具有重要意义。

（2）基于饱和-非饱和渗流数学模型和实测土水特征曲线，运用延伸的摩尔-库伦强度理论，对某库岸堆积体进行了降雨过程的饱和-非饱和渗流场计算和稳定分析，得出其最小安全系数发生在雨后一定时间内的结论，这与多数自然边坡在雨后一段时间产生滑动失稳的实际情况相吻合，较为合理地阐述了因降雨引发的边坡失稳问题。同时研究表明，在模拟因降雨入渗引发边坡失稳过程时，应优先采用实际可能降雨强度和雨型，降雨特征（雨强、雨型）对边坡稳定的影响不容忽视。

参考文献：

[1]　田陵君，王兰生，刘世凯．长江三峡工程库岸稳定性［M］．北京：中国科学技术出版社，1992.
[2]　孙广忠．中国典型滑坡［M］．北京：科学出版社，1988.
[3]　吴宏伟，陈守义，庞宇威．雨水入渗对非饱和土坡稳定性影响的参数研究［J］．岩土力学，1999，20（1）：1-14.
[4]　包承纲．非饱和土的性状及膨胀土边坡稳定问题［J］．岩土工程学报，2004，26（1）：1-15.
[5]　Fredlunnd D G. 非饱和土土力学［M］．陈仲颐，译．北京：中国建筑工业出版社，1997.
[6]　戚国庆，黄润秋，速宝玉，等．岩质边坡降雨入渗过程的数值模拟［J］．岩石力学与工程学报，2003，22（4）：625-629.
[7]　Fr edlunnd D G，Xing A. Equation for the Soil - water Characteristic Curve［J］．CanGeotechJ，1994（1）：521-532.
[8]　胡云进，速宝玉，詹美礼．裂隙岩土体非饱和渗流研究综述［J］．河海大学学报，2000，28（1）：45-51.
[9]　王恩志，王洪涛，孙役．三维裂隙网络渗流数值模型研究［J］．工程力学，1997（增 2）：520-525.
[10]　江西省山洪灾害防治规划领导小组办公室．江西省山洪灾害防治规划报告［R］．2004，12.

联合运用改进的极限平衡法和动态规划法分析边坡稳定性

吴海真[1,2]，顾冲时[1]

1. 河海大学水利水电工程学院；2. 江西省水利科学研究院

摘　要：为了解决边坡关键滑动面全局优化处理问题，将动态规划理论引入基于有限元的改进极限平衡法。与工程上被广泛认可的一般极限平衡法比较表明，本文提出的方法可以获得更好的近似解甚至是整体最优解，关键滑动面更符合工程实际，有效地克服了传统边坡稳定分析方法的不足；与目前国内外同类方法相比具有明显的优越性，丰富了边坡稳定分析和确定关键滑动面的方法和理论。

关键词：边坡稳定；有限元；改进的极限平衡法；关键滑动面；动态规划

1　研究背景

目前，工程界对于边坡稳定分析评价常采用刚体极限平衡法，而土坡稳定分析计算中又以垂直条分法应用最为广泛，岩坡稳定分析计算中以 Sarma 法应用最为普遍。这些经典极限分析方法概念清晰，使用时间长，积累了丰富经验，但在理论上存在诸多不足，主要有：①没有考虑岩土体内部的应力应变关系，边坡失稳与变形无确定性联系，滑动面底部应力可能失真；②不能考虑岩土体与支挡结构的共同作用及其变形协调；③假定条件较多，如土条条分假设，条间力及滑动面假定等。

随着计算机技术的迅速发展，特别是有限单元法提出后，有力推动了边坡稳定分析方法的发展。有限元法已广泛应用于边坡稳定性分析，与极限平衡法相比，有限元法有如下优点：①可同时进行稳定分析与变形分析；②可考虑土的非线性应力应变关系；③可保证整个边坡同时满足力和力矩平衡条件；④可考虑先期应力状态和模拟开挖或填筑等施工过程。有限元法分析边坡稳定性一般可分成两类：有限元强度折减法和有限元与优化搜索法相结合的方法。早在 1975 年 Zienkiewicz 等[1]就通过等比折减土的黏聚力和内摩擦角正切值的方法分析边坡的稳定性，把使边坡达到临界稳定状态的材料强度折减系数值作为边坡的最小安全系数，并把该方法命名为“强度折减技术”。郑颖人等[2]采用有限元强度折减法进行土坡和岩坡的稳定分析。张鲁渝等[3]对有限元强度折减系数法计算土坡稳定安全系数的精度进行了研究。但是，有限元强度折减法在计算分析前对土的强度参数进行了折减，取得的应力与变形结果无法反映边坡实际的受力状态，因此该方法只能限于边坡最小

本文发表于 2007 年。

安全系数的评价，而在边坡失稳判据等方面仍值得深入研究。另一方面，确定临界滑动面位置及形状是边坡稳定分析的主要任务。一般的边坡稳定分析工具要求分析者在计算前输入滑动面的位置，这对工程技术人员的理论水平和工程经验提出了较高的要求。因此，从20世纪70年代后期开始，很多学者致力于临界滑动面自动搜索和优化的技术研究，提出了多种确定边坡圆弧或非圆弧临界滑动面的优化搜索方法。Zhu[4]提出了边坡临界滑动场的概念，陈昌富等[5]把自适应蚁群算法引入边坡稳定分析领域。但由于蚁群算法与临界滑动场本身无法直接应用于连续性问题，因此在一定意义上这两种方法难以找到边坡的全局最小安全系数。Zou等[6]提出了采用动态规划法或改进的动态规划法确定非圆弧临界滑动面及其对应最小安全系数的方法，但该动态规划法仍局限于常规的条分法计算。Goh[7]结合滑动楔体理论，运用遗传算法对多折线滑动面进行了搜索。Cheng[8]采用模拟退火法成功分析了多个较复杂边坡，但其对于如何利用模拟退火法搜索临界滑动面的方法介绍比较简单。

有限元法与临界滑动面搜索优化相结合的思想源于极限平衡法的临界滑动面搜索，以充分利用有限元法的优势对边坡的稳定与变形进行分析评价。本文采用动态规划理论与有限元相结合的方法确定边坡的临界滑动面，其中利用有限元法分析某状态下边坡的应力和孔隙水压力分布情况，按有效应力法计算给定滑动面的安全系数，同时引入动态规划理论，成功地解决了对边坡关键滑动面的优化处理问题，有效地克服了传统边坡稳定分析方法的不足。

2 改进的极限平衡法基本理论

传统的极限平衡法需要对边坡条块之间的作用力进行一定的假定，使之成为静定问题，由此计算的滑动面上应力并不精确。众所周知，有限元法能够同时满足力和力矩平衡，从而可以进行准确的应力分析。因此可利用有限元法分析的应力结果计算边坡的安全系数，也即基于有限元法得出边坡内部的应力场，然后基于传统极限平衡法安全系数的定义，采用优化方法搜索关键滑动面的位置并得到相应的最小安全系数。Farias等[9]提出了有限元法计算边坡安全系数的详细方法，本文基于Terzaghi有效应力原理对该方法进行了改进。

2.1 安全系数的定义

据毕肖普20世纪50年代重新定义的边坡稳定安全系数概念，每个土条的安全系数F_{si}等于该土条底部的抗剪强度与土体的剪应力之比[10]：

$$F_{si}=\tau_{fi}/\tau_{mi} \tag{1}$$

式（1）即为边坡沿滑动面的局部安全系数。对于整个滑动面，边坡稳定安全系数定义为滑动面底部的抗滑力之和与滑动力之和的比值：

$$F_s=\frac{\sum_{i=1}^{n}\tau_{fi}\Delta L_i}{\sum_{i=1}^{n}\tau_i\Delta L_i} \tag{2}$$

式（2）即为边坡沿滑动面的整体安全系数。对于圆弧滑动面，式（2）即成为沿滑动面的抗滑力矩与滑动力矩的比值。

2.2 边坡内任意点应力状态

首先采用弹塑性有限元方法计算边坡内的应力状态，得到单元高斯点应力（本文采用四节点等参单元）。为了得到土条底部中点的应力值，把单元高斯点应力投影成单元节点应力：

$$\{\sigma\}^e=[N]\{\sigma\}^g \tag{3}$$

式中：$\{\sigma\}^e$ 为单元节点应力；$[N]$ 为位移插值函数矩阵；$\{\sigma\}^g$ 为单元高斯点应力。

对所有节点进行绕节点平均，从而得到边坡内部各单元的节点应力值，进而可根据计算点的局部坐标关系，计算边坡内任意点的应力状态和孔隙水应力。

2.3 边坡滑动面应力状态

将滑动面所包含的土体进行条分，设任意土条底部长为 ΔL_i，则滑动面上一点的抗滑力 τ_f 和滑动力 τ_n 可由该点有限元分析的应力值 σ_x、σ_y 和 τ_{xy} 和孔隙水应力 p 计算，公式如下：

$$\begin{cases}\sigma_n=\dfrac{1}{2}(\sigma_x+\sigma_y)-\dfrac{1}{2}(\sigma_x-\sigma_y)\cos2\sigma-\tau_{xy}\sin2\alpha\\ \tau_n=\dfrac{1}{2}(\sigma_x-\sigma_y)\sin2\alpha-\tau_{xy}\cos2\alpha\end{cases} \tag{4}$$

$$\sigma_n'=\sigma_n-p \tag{5}$$

$$\tau_f=\sigma_n'\tan\varphi'+c' \tag{6}$$

式中：σ_x、σ_y 和 τ_{xy} 为土中一点正应力和剪应力，以拉为正；φ' 和 c' 分别为土体有效内摩擦角和有效黏聚力；σ_n' 和 σ_n 分别为沿滑动面上的有效正应力和总应力，以压为正；p 为孔隙水应力；α 为土条底部中点之切线与 x 轴正向夹角，逆时针方向转动为正。

由此可根据式（1）和式（2）计算边坡的局部安全系数和整体安全系数。

3 动态规划法确定边坡关键滑动面方法

3.1 动态规划法

动态规划法是 20 世纪 50 年代由 Bellman 和 Dantzig 发展起来的多阶段决策优化理论，其中的费用最小值问题表述为[11]：

$$J=\sum_{k=0}^{N}L(x(k),u(k),k) \tag{7}$$

其中 $x(0)$ 有固定值 C，系统方程为：

$$x(k+1)=g[x(k),k],k=0,1,\cdots,N-1 \tag{8}$$

应用动态规划方法研究上述问题就是希望把实际问题嵌入到一类似的问题中去。一般情况下，直接求解上述问题是困难的，但可以通过连接这一类问题中各组成部分的关系式即可求极值的嵌入递推方程进行求解。用函数 $I(x,k)$ 规定为最小费用，则嵌入递推方程为：

$$I(x,k)=\min u\{L(x,u,k)+I[g(x,u,k),k+1]\} \tag{9}$$

$$I(x,0)=\min_{u(0)}\{L[x,u(1),1]\} \tag{10}$$

基于嵌入递推方程，根据 Bellman 的阐述[11]，一个最优策略有这样的特性：不论初始状态和初始决策如何，相对于第一个决策所形成的状态来说，剩下的决策必定构成一个最优策略，此即动态规划法的基本原理。

3.2 数学模型的建立

本文基于条分法和有限单元法相结合的基本原理，引进动态规划理论确定边坡的关键滑动面，其数学建模过程如下：

引进辅助函数 G：

$$G=\sum_{i=1}^{n}(\tau_{fi}-F_s\tau_i)\Delta L_i=\sum_{i=1}^{n}(R_i-F_sS_i) \tag{11}$$

式中：R 为抗滑力；S 为滑动力。

可以设想，要使式（2）的安全系数 F_s 最小，等价于使式（11）中的 G 达到最小：

$$G_{\min}=\min(J)=\min\sum_{i=1}^{n}(R_i-F_sS_i) \tag{12}$$

为引进动态规划，用适当多的垂直线（共 $n+1$ 条，也即单元边界线）条分边坡，如图 1 所示。每一垂直线称为阶段（stage），初始阶段和最终阶段表示分析区域，滑动面与条分线的交点（单元节点）为状态点（state point），可以设想边坡临界滑动面必然是由所有各阶段的某些状态点连成的折线。取出任一两相邻阶段（即两条分线）i 和 $i+1$ 考虑，如图 2 所示，假定连接点 j 和 k 之连线 $\overline{jk}$是可能的临界滑动面，计算滑动面所穿过单元的 R_i 和 S_i，则

$$G_i(j,k)=R_i-F_sS_i \tag{13}$$

式中：$R_i=\tau_f\Delta L_i=\sum_{ij=1}^{ne}R_{ij}=\sum_{ij=1}^{ne}\tau_{f_{ij}}l_{ij}=\sum_{ij=1}^{ne}(c'_{ij}+\sigma'_{ij}\tan\varphi'_{ij})l_{ij}$；$S_i=\tau_i\Delta L_i=\sum_{ij=1}^{ne}S_{ij}=\sum_{ij=1}^{ne}\tau_{ij}l_{ij}$。

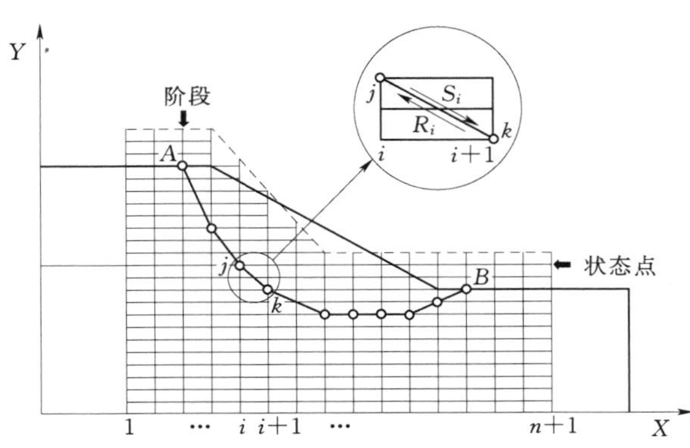

图 1　动态规划法的网格划分

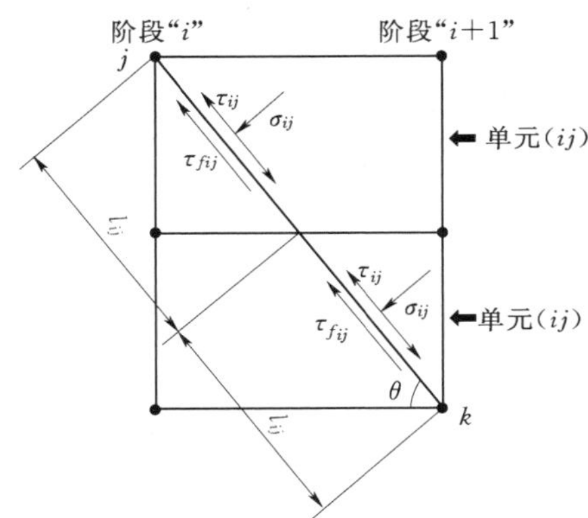

图 2　滑动面穿越相邻阶段示意

引入一个"优化函数"$H_i(j)$，表示从初始阶段到 i 阶段 j 状态即（i，j）之间 G 的最小值。根据 Bellman 的动态规划原理[11]，其嵌入递推方程为：

$$H_{i+1}(k)=H_i(j)+G_i(\overline{j,k}) \tag{14}$$

式中：$H_{i+1}(k)$ 为在 $i+1$ 阶段状态点 k 获得的优化函数；$H_i(j)$ 为在 i 阶段状态点 j 获得的优化函数；$G_i(\overline{j,k})$ 为滑动面穿越 i 阶段的状态点和 $i+1$ 阶段的状态点 k 所获得的函数值。

在初始阶段：$H_1(j)=0$，$j=1,2,\cdots,N_{p1}$，N_{pi} 为第 i 阶段的状态点数。

到最终阶段：$i=n+1, H_{n+1}(k)=H_n(j)+G_n(\overline{j,k})$

$$H_{n+1}(k)=G=\sum_{i=1}^{n}(R_i-F_sS_i),k=1,2,\cdots,N_{pn+1} \tag{15}$$

以上即为应用动态规划方法确定边坡关键滑动面的数学模型。

3.3 相关问题的处理

（1）关键滑动面与坡面交点的确定。据上述由动态规划法原理确定边坡关键滑动面的方法进行搜索，可以得到边坡的临界滑动面。同时因应力值已经由有限元方法计算得到，故安全系数 F_s 的计算是线性问题，不需要迭代。但在实际计算时，必须扩大计算范围，即在关键滑动面与坡面的可能交点部位的坡体以外也要划分足够多的状态点，此时只要令抗滑力 R 和滑动力 S 为零即可，而且初始及终了阶段的状态点都必须在坡外，这主要是为了确保得到完整的滑动面，见图 3。

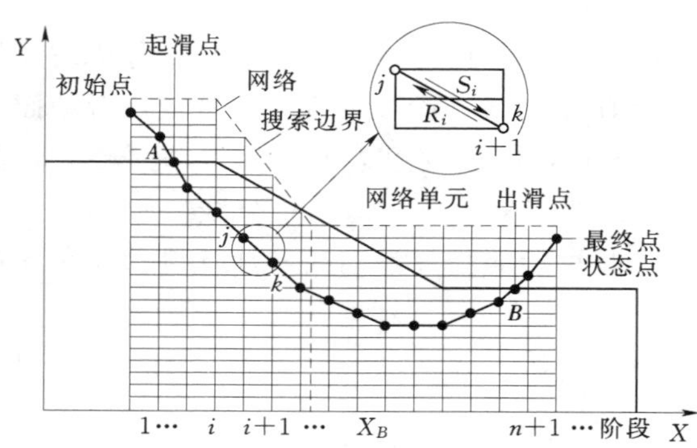

图 3 确定关键滑动面与坡面的交点示意

（2）搜索关键滑动面的起点设置。关键滑动面搜索起点的确定主要决定于初始阶段状态，如果初始阶段状态点总数为 N_{p_1}，则逐个以初始阶段状态点为起点进行分析计算，得到相应的安全系数和最小值，其相应初始阶段之状态点即为关键滑动面搜索起点。

（3）单元网格密度。单元网格密度决定垂直条分线之间的宽度和状态点的间距，从而直接影响计算的精度和搜索时间，这主要取决于边坡的地质条件、规模和工程本身的精度要求。通过大量算例证实，对于 4 节点矩形单元，当单元密度达到每 $10m^2$ 不少于 4 个节点时，计算精度较为理想，如果再增加节点，计算精度应还能提高，但此时耗费的机时也将成倍增长。

4 算例分析

根据以上理论和方法，编制了边坡稳定改进极限平衡法的计算程序，采用动态规划法确定边坡的关键滑动面位置和形状，可以考虑地下水和外荷载的影响，尤其适用于复杂地质条件下的边坡稳定性分析研究，程序计算速度快，收敛性好。

4.1 算例 1

图 4 所示为算例 1 的边坡剖面图。首先采用 Morgenstem－Price 法进行计算，对如图 4 所示的物理力学参数和剖面，取土条数 30，穷举法搜索关键滑动面，相应安全系数为 1.170；基于改进的极限平衡法基本原理，取弹 20MPa，泊松比 0.33，得到边坡的整体安

全系数为 1.130；若根据上述动态规划法的原理和步骤在改进的极限平衡法基础上再次进行关键滑动面的搜索计算，得到的边坡安全系数则为 1.05。3 种计算方法得到的关键滑动面位置如图 5 所示。

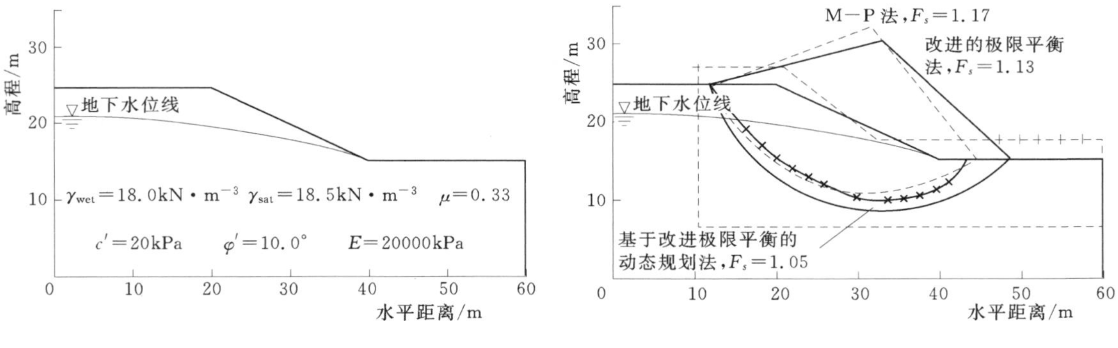

图 4　算例 1 的计算剖面　　图 5　算例 1 的计算成果

4.2　算例 2

计算一坡内具自由水面、坡外有水的情况，图 6 所示为算例 2 的边坡剖面及土层分布图，各土层的物理力学参数取值见表 1。

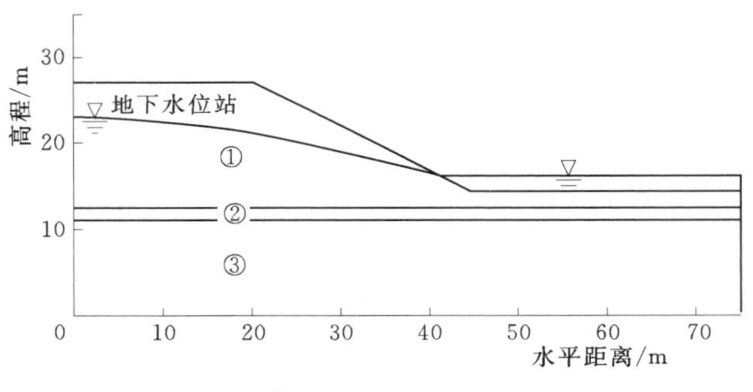

图 6　算例 2 的计算剖面

表 1　　算例 2 的物理力学参数取值

参数指标	天然容重/($kN\cdot m^{-3}$)	饱和容重/($kN\cdot m^{-3}$)	弹性模量/kPa	泊松比	黏聚力 c'/kPa	内摩擦角 φ'/(°)
①区（填土）	15.0	15.50	15000	0.33	20.0	30.0
②区（软弱层）	18.0	18.50	2000	0.45	0	10.0
③区（持力层）	19.0	20.00	100000	0.35	100.0	30.0

采用 Morgenstern－Price 法进行计算，边坡相应的安全系数为 1.14；采用基于改进的极限平衡法，动态规划法优化滑动面位置及形状，得到边坡的整体安全系数为 1.06。两种计算方法得到的相应关键滑动面位置如图 7 所示。可见采用改进的极限平衡动态规划法可获得更好的优化滑裂面及相应的近似解甚至整体最优解，滑裂面位置和形状也更符合实际。

需要说明的是，上述两个算例中 Morgenstem－Price 法与本文方法计算结果的差异，

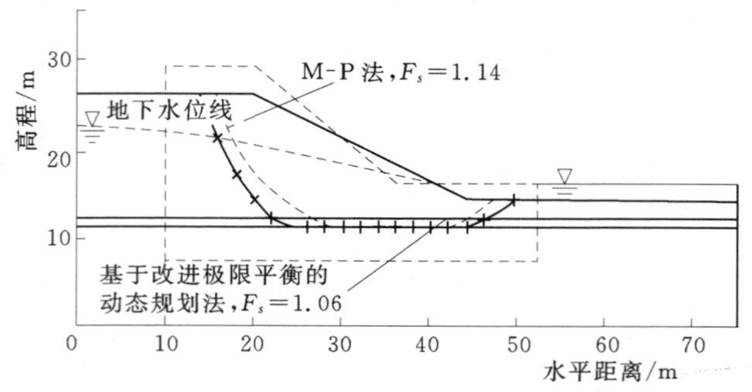

图 7　算例 2 的计算成果

并非 Morgenstern - Price 法本身所造成，而主要是由于优化处理后滑裂面形状和位置的不同所致。

5　结语

边坡稳定性分析一直是岩土工程中的重要研究课题，而极限平衡法又是边坡稳定分析理论中的重要内容，但在理论上存在诸多不足，其中边坡内部应力尤其是滑动面底部应力“失真”是突出问题之一；另一方面，如何合理的确定关键滑动面是提高边坡稳定分析准确性的关键。与传统的条分法相比较，本文提出的基于改进极限平衡的动态规划法的特点在于：

（1）把有限元与极限平衡结合起来，从而使该方法本身具有有限元法的一切优点，并在原有方法的基础上引入了有效应力原理，能够考虑坡内具自由水面、坡外有水的情况，可较为真实地反映边坡体内部的应力和变形。

（2）基于改进的极限平衡法，引入动态规划理论与之相结合，成功地解决了对边坡关键滑动面的优化处理问题。算例分析表明，与工程上被广泛认可的 Morgenstern - Price 法比较，本文提出的方法可以获得更好的优化滑裂面及相应的近似解甚至整体最优解。

（3）本文提出的方法和理论为复杂边坡稳定分析提供了新的思路，丰富了边坡稳定分析和确定关键滑动面的方法和理论。

参考文献：

[1] Zienkiewicz O C, Humpheson C, Lewis R W. Associated and non-associated visco - plasticity and plasticity in soil mechanics [J]. Geotechnique, 1975, 25 (4): 671 - 689.

[2] 郑颖人，赵尚毅．有限元强度折减法在土坡与岩坡中的应用 [J]．岩石力学与工程学报，2004，23（19）：3381 - 3388.

[3] 张鲁渝，郑颖人，赵尚毅，等．有限元强度折减系数计算土坡稳定安全系数的精度研究 [J]．水利学报，2003，(1)：21 - 27.

[4] Zhu D Y. A method for locating critical slip surfaces in slope stability analysis [J]. Canadian Geotechnical Journal, 2001, 38: 328 - 337.

[5] 陈昌富，龚晓南，王贻荪．自适应蚁群算法及其在边坡工程中的应用 [J]．浙江大学学报（工学版），2003，37（5）：566 - 569.

[6] Zou J Z, Willians D J, Xiong W L. Search for critical slip surfaces based on finite element method [J]. Canadian Geotechnical Journal, 1995, 32: 233-246.
[7] Goh A T C. Genetic algorithm search for critical slip surface in multiple-edge stability analysis [J]. Canadian Geotechnical Joumal, 1999, 36: 382-391.
[8] Cheng Y M. Location of critical failure surface and some further studies on slope stability analysis [J]. Computers and Geotechnics, 2003, 30: 255-267.
[9] Farias M M, Naylor D J. Safety analysis using finite elements [J]. Computers and Geotechnics, 1998, 22 (2): 165-181.
[10] 钱家欢，殷宗泽．土工原理与计算 [M]. 2版．北京：中国水利水电出版社，1995.
[11] 罗伯特·E拉森．动态规划法原理 [M]. 陈伟基，等，译．北京：清华大学出版社，1984.

区域水库群大坝安全评价体系的构建

傅琼华[1]，吴晓彬[1]，金　峰[2]

1. 江西省水利科学研究院；2. 清华大学水利水电工程系

摘　要：结合江西省水库群大坝安全评价工作，介绍了区域水库群大坝安全评价体系的构建过程，包括大坝安全评价技术标准的制定、单座水库大坝工程险度评价和风险评估、区域性水库群除险加固全面规划、水库群大坝安全风险排序、大坝病险原因分析及对策研究、除险加固方案的实施效果评价及区域性水库群除险加固动态管理信息系统的建立等内容。该体系在江西省水库群大坝安全评价中的应用效果良好，可为决策者提供科学依据。

关键词：群坝安全评价体系；风险评估；除险加固；区域性水库

水库大坝是关系国计民生的重要基础设施，病险水库是我国防洪工程体系的薄弱环节，近年来国家对病险水库的重视程度、投资力度、整治规模都是前所未有的。虽然国家投入了大量资金，但水库加固总数不到4000座，且重点加固的是大型水库和重要中型水库，大量小型病险水库尚无暇顾及，而我国近50年来发生溃决的水库有96.4%都是小型水库[1]。大坝安全鉴定是病险水库除险加固的前期工作，又是关键所在，评价工作做得充分与否直接关系到除险加固技术方案的科学性和资金投入的合理性。目前一些发达国家对大坝安全的风险管理十分重视，如美国科罗拉多州为了减少生命、财产损失，保障用水安全，在有限投入条件下确定合理蓄水高程，提出了全州大坝安全风险评估集成系统[2]。该系统的工作重点是风险评价、大坝破坏模式、灾害后果评估、风险管理等，2002年5月启动前期项目。美国华盛顿大学和美国陆军工程兵师团合作，对基于风险的大坝安全决策方法进行了全面总结，大坝安全的风险分析方法得到了很快的发展[3-4]。近年来，我国在大坝风险评估和管理方面也取得了一些进展[5]，但大都针对某一类型的大坝进行安全评价，鉴定大坝工程本身的安全性，如水库防洪安全、大坝结构稳定、抗震稳定、渗流稳定、泄水和输水建筑物安全、金属结构安全等是否满足现行规范或标准的要求。事实上，只要水库存在，必然给公众带来潜在威胁，如果水库大坝溃决，下游民众的生命财产必将受到极大威胁。因此，水库大坝的安全评价不仅仅是大坝自身的安全性状评价，还需考虑下游经济发展水平、防洪保护对象等因素。另一方面，对于一定区域内众多数量的病险水库，在有限的经济投入条件下，应全面权衡区域内各水库的病险程度，综合评价溃坝风险概率，使病险水库的除险加固按轻重缓急分类排队，更合理地安排使用加固资金，减少溃坝事故。

本文发表于2007年。

因此，建立一套完整的针对区域性水库群的安全评价体系是十分必要的，以便对整个区域内各类水库大坝的安全程度及风险进行综合性评价，客观地反映水库对区域发展的重要性及潜在的不安全性，及时采取除险加固措施，更好地保护国家和人民生命财产安全。本文以江西省水库群大坝为例，研究探索并逐步建立了区域水库群大坝安全评价体系。

1　区域水库群大坝安全评价体系的构成要素

水库大坝的安全问题涉及社会、经济、文化、自然等多个方面。仅从一个侧面考虑，难以理清相互之间的逻辑关系，也不可能透彻地分析水库大坝病险问题的症结所在，所以有必要采用系统分析的方法，从单座水库大坝安全评价及风险评估入手，围绕区域性水库群的安全状态及风险排序，探索水库大坝病险问题的深层原因，以寻求最终解决途径。构建区域水库群大坝安全评价体系的主要目的是评价区域内水库群大坝安全性状及水库群坝风险度排序，对评估风险高的水库优先考虑加固，针对病险症状采取有效的工程措施进行除险或采用非工程措施进行管理和控制，防患于未然。研究基本思路见图 1，研究技术路线见图 2。

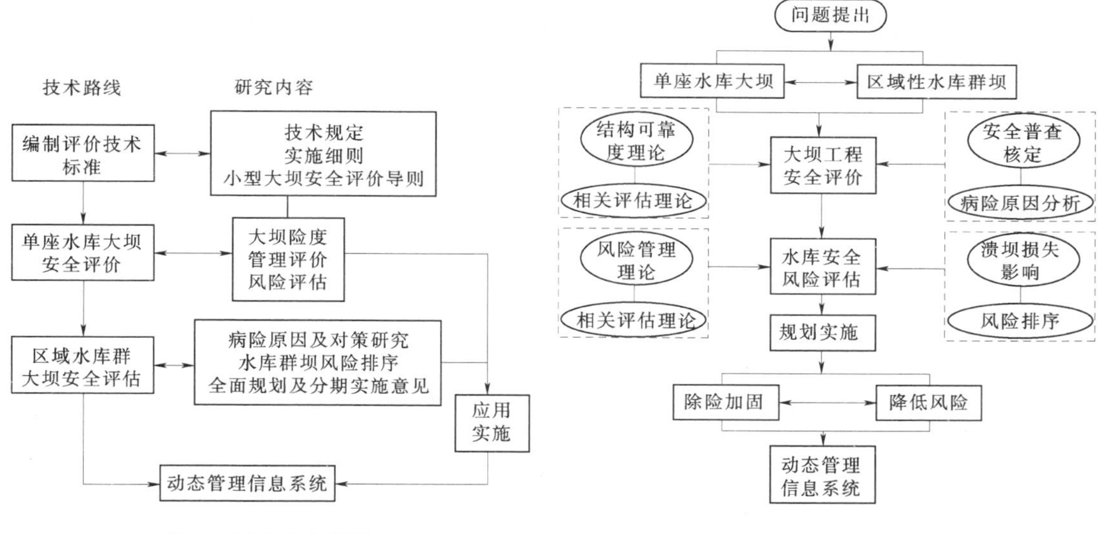

图 1　研究基本思路　　　　图 2　研究技术路线

区域水库群大坝安全评价体系的建立需从水库管理者、水行政主管部门和水库工程安全等多方面的需求出发，相比单座水库大坝安全评价起点更高，内容更全面，安全性状分析更加充分和深入，主要构成要素包括：

(1) 评价技术标准：要进行区域水库群大坝的安全评价，必须保证区域内水库的安全评价标准一致；因此，需要根据有关规程规范，结合区域水库大坝的特点和工程实际，编制并细化区域水库大坝安全评价技术标准与实施细则，指导和规范区域水库大坝安全评价工作。

(2) 单座水库大坝安全评价：根据编制的区域水库大坝安全评价技术标准及实施细则，采用先进的技术和理论，对区域内的单座大坝实施安全评价，这是进行区域内水库群大坝安全评价的重要基础，需要根据工程现状和大坝运行情况进行安全复核及评价，目的是评价现状大坝的安全富余度或工程可靠性是否可以接受。

（3）大坝风险评估：对大坝失事概率及后果进行估算，包括溃坝模式、溃坝路径、生命损失、经济损失及对社会与环境的影响，也就是研究溃坝所带来的人员伤亡和经济损失等的严重程度。评估建立在不确定性分析的基础上，损失后果与溃坝时间、预警系统、人口密度等密切相关。

（4）病险原因分析及对策研究：全面分析总结区域内病险水库的症结及影响因素，有针对性地提出经济实用的处理方法，指导区域内病险水库的除险加固。

（5）风险排序：在安全评价的基础上，进一步对水库病症进行核定，并根据大坝风险分析理论进行区域水库群坝风险评估排序。排序基本要素包括水库工程安全性状，水库下游防洪保护区的经济存量、人口密度和社会财富，防洪预警设施完备性及水行政主管部门对水库管理因素的排序等。

（6）除险加固规划：依据区域内水库大坝的安全普查，核定整个区域内的病险水库数量，考虑水库大坝的分级管理、资金来源和配套比例等方面因素，提出全面合理的病险水库群除险加固实施意见和保障措施。

（7）动态管理信息系统：为加强水库大坝安全信息量的管理，开发具有信息维护、信息查询与统计、地理信息系统、规程规范查询等功能的区域水库群除险加固信息动态管理软件，可以实现水库大坝安全鉴定、加固设计、加固施工、加固验收资料等的信息化管理，提高管理水平。

2 江西省水库群大坝安全评价体系的构建

江西省现有各类水库 9268 座，约占全国水库总量的 1/9，在全国各省中排名第二，且以小型水库为主。因此，评价体系的重点是制定小型水库安全评价标准、进行单座水库工程安全评价、水库群安全风险评估排序等。评价体系既能有效地分析单座水库大坝的病险程度及风险评估，又可系统地对区域性水库群大坝的风险进行排序，为决策者提供科学的决策依据，有病治病，无病防病。本项目主要进行了以下工作。

2.1 制定评价技术标准

根据有关规程规范，相继制定了江西省大坝安全评价技术规定及实施细则。主要包括：《江西省大、中型水库大坝安全评价报告编制若干技术规定（试行）》、《江西省小型水库大坝安全评价导则（试行）》、《江西省水库大坝安全鉴定管理办法》、《江西省水库大坝安全鉴定实施方案》、《江西省小（一）型水库大坝安全鉴定成果核查实施方案》、《江西省水库除险加固工程初步设计报告编制若干技术规定》等水库大坝安全评价技术规定和方法，指导和规范了江西省水库大坝的安全评价技术工作。各技术规定具有较强的操作性和实用性，从制度上、技术上对病险水库除险加固工作进行了规范管理。

经过 4 年的应用，上述评价技术标准得到了评价单位和各级水行政主管单位的好评，尤其是《江西省小型水库大坝安全评价导则》对小型水库安全评价内容、安全状况分类标准、评价报告编写提纲、小型水库地质勘察技术要求等作出了详细的规定，已成为江西省小型水库大坝安全评价的主要技术标准和依据，在应用过程中未出现明显失误和偏差，减少了病险水库除险加固的前期经费，保障了病险水库除险加固工作的顺利进行。

2.2 单座水库大坝安全评价方法

鉴于实际工程的复杂性，为满足工程安全评价多角度、多方面的需要，近年来在常规分析评价方法的基础上，采用了一些既能满足工程实际需要又切实可行的评价方法，并与常规计算方法的计算成果进行对比分析。主要有：①坝坡稳定可靠度分析（靖安县石马水库大坝）；②考虑降雨入渗及饱和-非饱和渗流作用下的边坡稳定分析（吉安市白云山水库库岸堆积体边坡）[6]；③边坡稳定分析关键滑裂面搜索与优化（吉安市白云山水库库岸堆积体边坡）；④三维有限元等效应力法拱坝坝体应力分析（井冈山市井冈冲高砌石拱坝）；⑤土石坝渗流多因素时变监控模型（铜鼓县大土段水库大坝）；⑥土石坝二维、三维有限元渗流计算分析（玉山县七一水库大坝）。多座水库大坝工程的实际应用表明，先进理论与方法的应用有效地解决了工程分析评价中的难点和疑点问题，已成为大坝安全评价的重要组成部分。

在单座水库大坝安全评价的基础上，根据大坝风险分析理论研究[7]，针对某个具体水库进行了溃坝模式和溃坝概率的研究和分析。对赣州市 5 座中型水库进行了典型实例研究，通过对大坝下游影响的调查分析了溃坝后下游生命损失、经济损失、对社会和环境的影响，进行大坝风险评估[8]。按专家确定的报警时间，各水库的风险指数从大到小的排序为灵潭水库、龙山水库、石壁坑水库、下栏水库、长龙水库。灵潭水库等 5 座示范水库的生命损失和经济损失均位于不可容许的高风险区，应尽快进行除险加固，同时应加强安全管理，降低水库风险指数。

2.3 大坝病险成因分析及对策研究

统计分析已鉴定为三类坝的 80 座水库，从防洪安全、渗流安全和结构安全等方面进行病险症状和成因分析，其中有 24 座（30%）防洪安全性不足，另有 43 座（54%）泄洪安全性不足，洪水不能安全下泄；73 座（91%）水库渗流安全性不足，评定为 C 级；另有 77 座（96%）水库大坝结构安全性不足，其中 47 座（59%）大坝坝坡稳定性不足，有 70 座（88%）水库输、泄水建筑物结构安全性不足。此外，水库普遍存在水（雨）情观测设施简陋，大坝安全监测设施不完善，管理、通信、交通设施简陋等问题。探讨总结了水库大坝安全影响因素，提出了针对性强、经济实用的处理对策。

2.4 群坝风险评估指数排序方法研究

经过对不同类别水库工程的现场调查，引入当前国际上先进的大坝风险分析概念和分析技术，通过对大坝风险度、溃坝后果、综合影响等因素的赋值量化研究，得到大坝风险评估综合分数，建立了群坝风险评估排序指数分析计算方法[9]，并对病险水库群大坝进行除险加固排序，其结构框图见图 3。

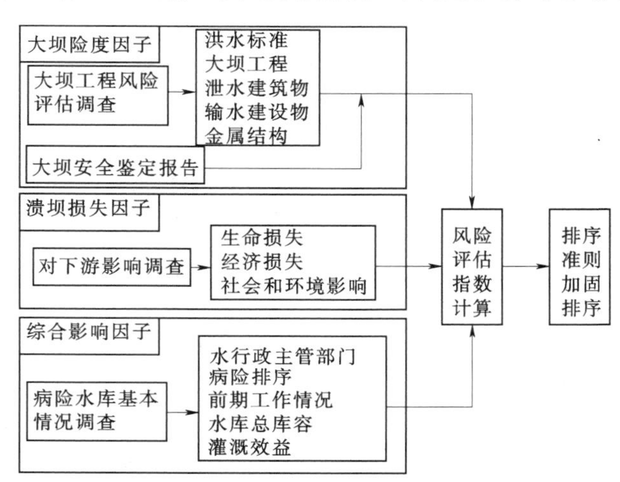

图 3　群坝风险评估指数结构框图

该评估方法主要包括三大影响因子：一是大坝险度因子，通过对大坝工程风险程度情况调查，依据安全鉴定报告对各建筑物安全性赋分，从而计算出大坝险度因子；二是溃坝损失因子，通过对大坝下游影响情况调查，分析溃坝后下游生命损失、经济损失、对社会和环境的影响，分别赋值计算，确定溃坝影响因子；三是综合影响因子，综合考虑水库的水行政主管部门对水库管理因素的排序、水库的总库容、灌溉效益、加固前期工作等因素，确定水库的综合影响因子。赋值量化的大坝险度因子、溃坝损失因子、综合影响因子的乘积即为水库大坝风险评估指数，按评估指数大小对群坝进行风险排序，风险指数越大表明水库危险性越大，应优先得到除险加固。

将该方法应用于242座不同类别的水库风险评估排序中，分析结果与实际情况基本相符。该方法计算简单、可操作性强，当水库数量多、分布区域广且时间和资金有限时非常实用，特别适合区域性水库安全主管部门权衡病险水库轻重缓急并有针对性地采取除险加固措施。

2.5 编制病险水库除险加固规划

对江西省小（2）型以上9268座各类水库安全状况进行全面普查分析与核定，截止到2002年年底有3488座水库（占水库总数的37.6%）存在不同程度的病险隐患。在分析病险症状及成因和加固效益评估的基础上，结合各地水资源状况和社会经济发展水平等因素，综合制定了2007年年底全面完成规划内病险水库除险加固目标应采取的组织、政策、技术、资金保障等措施，编制了《江西省病险水库除险加固规划报告》[10]，为水库大坝安全评价工作确定了明确的目标，有力地保障了除险加固工作的有序性和连续性。

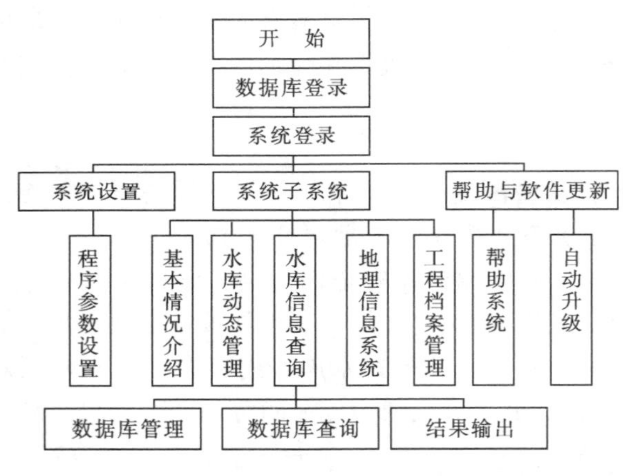

图4　水库除险加固动态管理系统基本结构
（C/S结构部分）

2.6 开发水库除险加固动态管理信息系统

建立了包括江西省水库大坝基本情况、安全评价、除险加固设计、除险加固施工、工程图纸和图片等6个方面的信息资料数据库，开发了一套完整的水库除险加固信息动态管理系统（图4），具有信息维护、信息查询与统计、规程规范查询等功能，实现了水库大坝资料的全面信息化管理，实现了人机对话和管理的程序化、自动化，为管理层和有关部门决策提供了科学、快捷的平台。

3 结语

从区域水库群大坝安全评价体系的构建过程来看，以下几项工作非常重要：①进行区域水库群大坝的安全普查，制定全面系统规划，保障安全评价工作的有序性和连续性；②将水库大坝安全评价技术标准的建立和大坝安全评价技术方法的研究与创新紧密结合，制定操作性、指导性强的技术标准，引进评价新方法，提高安全评价的准确性和针对性；③采用先进的大坝风险管理理念和技术，对大坝险度因子、溃坝损失因子、综合影响因子赋值量化评估大坝风险指数，提出实用、合理的群坝风险排序方法，增强水库大坝除险加固的科学性和合理性；④把数据库、地理信息系统、计算机多媒体等新技术引入到病险水库信息管理系统中，实现水库技术资料的全面信息化动态管理，为管理层决策提供科学、快捷的平台。江西省水库群大坝安全评价体系的实际应用取得了良好的效果，有力地促进了省级或区域性水库群病险水库除险加固工作的有序、顺利进行，提高了大坝安全管理水平，具有较强的现实意义和应用推广价值。

参考文献：

[1] 李雷，蔡跃波．我国水库大坝安全监测与管理的新动态［J］．大坝与安全，2005（6）：10-15.

[2] BYERS J B. Integration of risk assessment with a state dam safety program［R］. Denver：Department of Natural Resources，2002.

[3] HARRALD J R，RENDA－TANALI I，SHAW G L，et al. Review of risk based prioritization：decision making methodologies for dams［R］. Washington，D. C.：Gorge Washington University and US Army Corps of Engineers，2004.

[4] BOWLES D S，ANDERSON L R，GLOVER T F，et al. Portfolio risk assessment：a tool for dam safety risk management［C］// Proceedings of the 1998 USCOLD Annual Lecture. Buffalo，New York：USCOLD，1998.

[5] 李雷，王仁钟，盛金保，等．大坝风险评价与风险管理［M］．北京：中国水利水电出版社，2006.

[6] 苏立群，吴海真，祝德和．考虑降雨入渗及饱和渗流作用下的土石坝边坡稳定性研究［J］．江西水利科技，2006，32（2）：63-67.

[7] 王仁钟，李雷，盛金保．病险水库风险判别标准体系研究［J］．水利水电科技进展，2005，25（5）：5-8.

[8] 段智芳，傅琼华，李雷．病险水库除险加固排序在江西的示范应用研究［J］．江西水利科技，2007，33（1）：12-16.

[9] 傅琼华，段智芳．群坝风险评估指数排序方法的探讨［J］．中国水利水电利学研究院学报，2006，4（2）：107-110.

[10] 傅琼华，王纯．江西省病险水库除险加固规划［J］．水利发展研究，2004，4（8）：43-46.

井冈冲高砌石拱坝坝体应力计算分析

吴海真[1,2]，顾冲时[1]

1. 河海大学水利水电工程学院；2. 江西省水利科学研究院

摘　要： 多拱梁分载法概念清晰，使用时间长，积累了丰富的经验，但不能从整体上反映坝体应力分布情况，且基础变位采用简化方法；三维有限元等效应力法能基本消除应力集中现象，可进行各种复杂条件下拱坝的坝体应力分析，但计算成果的精度仍然受到其自身特点的影响。结合现行规范，分别采用多拱梁分载法和三维有限元等效应力法对井冈冲高砌石拱坝坝体应力进行了计算分析。结果表明，两种方法的计算成果总体趋势基本一致，坝体应力状态总体安全。说明了在高拱坝坝体应力分析中分别采用两种方法进行计算，并相互补充，相互验证的必要性。

关键词： 高砌石拱坝；多拱梁分载法；三维有限元；等效应力法；应力分析

目前，国际上拱坝坝体应力分析大致可分为两类：一类是以坝体的中心面作为简化计算模型，假定坝内应力沿上下游方向直线分布，在此基础上通过板壳单元或以结构力学假定为基础的拱梁分载法（又称试载法）求解中心面离散结点的线位移和转角位移，然后求解结点上的内力，最后按照线性分布假定求解坝内应力。但试载法也存在一些缺点，如基础变位一直沿用较为粗略的伏格特方法来确定；在计算中难以合理反映孔口对坝体应力的影响等。为克服这些缺点，许多研究人员进行了不懈的努力，如林绍忠、杨仲侯提出了对拱坝进行动力和非线性分析的分载位移法[1]。另一类则是广泛使用的三维有限元法。该方法不受结构力学假定的限制，在计算理论上也比试载法更为严密，不但可以比较合理地考虑拱坝的整体作用，还能够进行各种复杂条件下拱坝的坝体应力分析，解决了试载法中较难处理的各种问题。但是，有限元应力分析中存在着坝踵、坝趾的应力集中效应，这给应力评价及确定应力控制标准带来困难。傅作新教授提出了有限元等效应力法[2]，朱伯芳、董福品等人提出了基于有限元等效应力法的高拱坝应力控制标准[3]，周维垣等人提出了高拱坝的有限元分析方法和设计判据[4]。有限元等效应力法是基于有限元法的分析结果，将有限元所求得的应力合成为截面内力，然后求出对应的线性化应力。本文介绍多拱梁分载法和三维有限元等效应力法在井冈冲高砌石拱坝坝体应力分析中的应用情况。

1　工程概况

井冈冲水库坐落在井冈山主峰——五指峰东麓，位于赣江支流蜀水上游左溪上。坝址

本文发表于2006年。

控制流域面积 48.0km^2，水库正常蓄水位 727.0m（黄海高程，下同），设计洪水位（$P=2\%$）728.08m，校核洪水位（$P=0.2\%$）729.65m，死水位 670.00m，总库容 1990 万 m^3，是一座兼有发电、防洪、旅游、供水、养殖等综合效益的中型水库。枢纽工程主要建筑物有：大坝、放空洞、发电引水隧洞和电站厂房等。

大坝为细石混凝土砌石双曲拱坝，最大坝高 92.0m，坝顶高程 730.00m，坝顶宽 5.00m，坝底宽 24.00m，厚高比 0.26，最大倒悬度 0.252，坝轴线半径 160.00m，坝顶拱弧中心角 99.9°。大坝上游面设有混凝土防渗面板，底宽 3.00m，顶宽 0.80m。大坝坝基（肩）岩性以变余砂岩为主，呈弱风化—微新状，属中硬—坚硬岩。

2 基本计算参数

大坝防渗面板混凝土强度等级为 C20，坝体为 150 号二级配混凝土砌块石。坝址多年平均气温 15.5℃，历年最高气温 39℃，最低气温－7℃。主要计算参数如下：

混凝土容重 $\gamma_c=24.0\text{kN/m}^3$，弹性模量 $E_c=2.0\times10^{10}\text{Pa}$，泊松比 $\mu_c=0.167$，线性膨胀系数 $\alpha_c=1\times10^{-5}/\text{K}$。浆砌石体容重 $\gamma_{砌体}=23.00\text{kN/m}^3$，弹性模量 $E_{砌体}=6.0\times10^9\text{Pa}$，泊松比 $\mu_{砌}=0.22$，线性膨胀系数 $\alpha_{砌体}=8\times10^{-6}/\text{K}$。坝体导温系数，单位 3m^2/月。坝址基岩容重 $\gamma=26.40\text{kN/m}^3$，弹性模量 $E=9.0\times10^9\text{Pa}$，泊松比 $\mu=0.29$，线性膨胀系数 $\alpha=1\times10^{-5}/\text{K}$。拱梁分载法坝体综合弹性模量 $E_{砌体}=10.5\times10^9\text{Pa}$。

3 应力控制标准及荷载组合

3.1 应力控制标准

我国现行《浆砌石坝设计规范》（SL 25—91）规定[5]，拱坝应力分析以拱梁分载法作为衡量强度安全的主要标准，坝体应力控制标准如下：

拱冠梁（中央悬臂梁底）部位拉应力 1.5MPa，其他部位 1.4MPa（胶结料为 150 号二级配混凝土）。

采用拱梁分载法计算时，对于基本荷载组合，3 级拱坝容许压应力安全系数为 3.5；对于非地震情况的特殊荷载组合，容许压应力安全系数为 3.0。本次参照类似工程和有关规范，浆砌块石容许压应力在基本荷载组合时取为 5.1MPa，在特殊荷载组合时取为 6.0MPa。

由于《浆砌石坝设计规范》（SL 25—91）中未提供有限元计算成果的应力控制标准，为使坝体应力计算成果与现行规范相互配套，本次结合大坝的实际结构形态，现引入《混凝土拱坝设计规范》（SL 282—2003）中的应力控制标准如下[6]：采用有限元等效应力法计算时，3 级拱坝在基本荷载组合下容许拉应力不得大于 1.5MPa，对于非地震情况特殊荷载组合，容许拉应力不得大于 2.0MPa；容许压应力同拱梁分载法。

3.2 计算工况与荷载组合

因该坝为薄拱坝，依据规范可不计扬压力的影响，各计算工况和荷载组合如下：

工况Ⅰ：水库正常蓄水位＋相应下游水位（无水）＋设计正常温降＋泥沙压力（淤沙高程 663.0m，下同）＋坝体自重。

工况Ⅱ：水库死水位＋相应下游水位（无水）＋设计正常温升＋泥沙压力＋坝体

自重。

工况Ⅲ：水库校核洪水位＋相应下游水位（645.00m）＋设计正常温升＋泥沙压力＋坝体自重。

4 计算基本原理

拱梁分载法是当前用于拱坝应力分析的基本方法，该方法把拱坝看成由一系列水平拱圈和铅直梁所组成，荷载由拱和梁共同承担，各承担多少荷载由拱梁交点处变位一致条件决定。以拱梁上的外载和切割面上内力的合成力系为未知量，按拱梁交点变位协调条件求解拱梁荷载，荷载分配后，梁按静定结构计算应力，拱按纯拱法计算应力。

本次采用的全调整法是在五向调整法的基础上，以径向纤维直线假定取代传统的平截面假定来进行拱梁变形分析，进一步考虑绕径向弯曲分载与绕径向扭曲变位对梁变位计算的影响，能满足所有的拱梁变位协调条件、径向条形微元体的平衡条件以及坝体的边界条件，是传统试载法的提高与完善，其中温度荷载按照《浆砌石坝设计规范》（SL 25—91）所推荐的公式进行计算。

为消除有限元中因单元形态等因素引起的应力集中现象，本次三维有限元坝体应力计算中，采用了我国《混凝土拱坝设计规范》（SL 282—2003）提出的有限元等效应力法，用这一方法可基本消除应力集中现象。

此外，在本次应力计算中，坝体自重在拱梁分载法中按梁向考虑；而在三维有限元法中，仅考虑实际施工加载过程，不考虑施工中的坝体温度变化过程。

5 计算模型

5.1 拱梁分载法计算模型

针对井冈冲大坝的体形结构，将大坝划分为10拱21梁，拱梁布置见图1。规定应力以压为正，拉为负，径向、切向和竖向位移分别以指向下游、左岸和竖直向上为正。

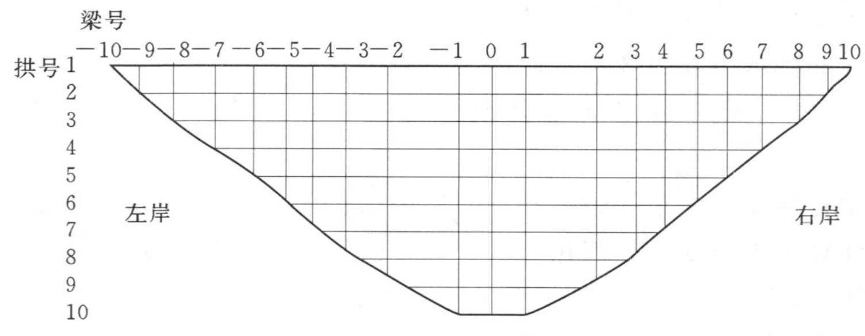

图1 大坝拱梁布置

5.2 三维有限元计算模型

根据大坝结构特点及其基础地质条件，有限元模型的计算范围，上游约1.5倍最大坝高，左右岸山体及下游取约1倍坝高，坝基以下约1.5倍坝高。单元划分：采用六面体八结点等参单元，图2为大坝的有限元网格空间布置图，共划分等参单元19469个，结点24162个。应力以拉为正，压为负；方向规定：X方向以指向左岸为正，Y方向以指向下游为正，Z 方向铅直向下为正。

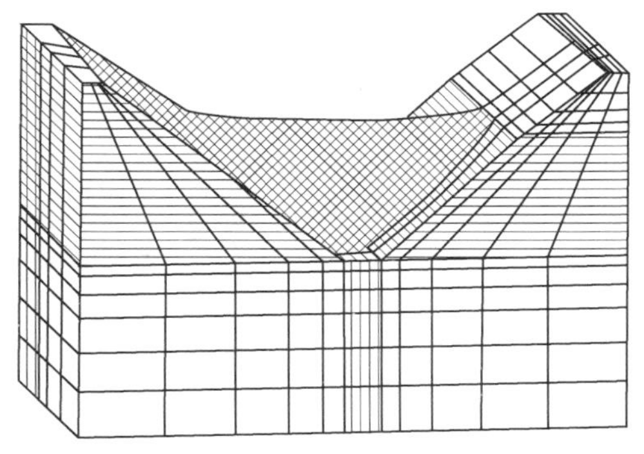

图 2 大坝三维有限元网格

6 计算成果分析

6.1 拱梁分载法计算成果

拱梁分载法计算成果表明，大坝主拉应力由工况Ⅱ控制，而主压应力则由工况Ⅲ控制。拉应力最大值为 1.05MPa，发生于 660.0m 高程下游坝面左拱端（工况Ⅱ），略超出容许应力值 1.00MPa，其次为 1.03MPa 和 1.02MPa，分别发生在工况Ⅲ条件下的 8 号拱左拱端（上游坝面 660.0m 高程）和 7 号拱右拱端（上游坝面 670.0m 高程）。拱冠梁底部的拉应力最大值为 0.96MPa（上游坝面）。各种工况下，上游坝面主压应力最大值为 3.29MPa（工况Ⅰ，拱冠梁约 710.0m 高程），下游坝面主压应力最大值为 4.07MPa（工况Ⅲ，670.0m 高程右拱端），详见表 1。限于篇幅，仅示出工况Ⅲ的计算成果图，见图 3～图 6。可见大坝的主压应力和主拉应力均能满足规范要求。

表 1　大坝上下游坝面最大主应力（拱梁分载法）　MPa

工况	上游坝面最大主拉应力	上游坝面最大主压应力	下游坝面最大主拉应力	下游坝面最大主压应力
Ⅰ	0.91（10 拱 0 梁）	3.29（3 拱 0 梁）	0.70（9 拱 0 梁）	3.87（6 拱−5 梁）
Ⅱ	0.47（9 拱 0 梁）	3.15（8 拱−3 梁）	1.05（8 拱−3 梁）	1.13（6 拱 0 梁）
Ⅲ	1.03（8 拱−3 梁）	2.67（1 拱 0 梁）	无拉应力	4.07（6 拱−5 梁）

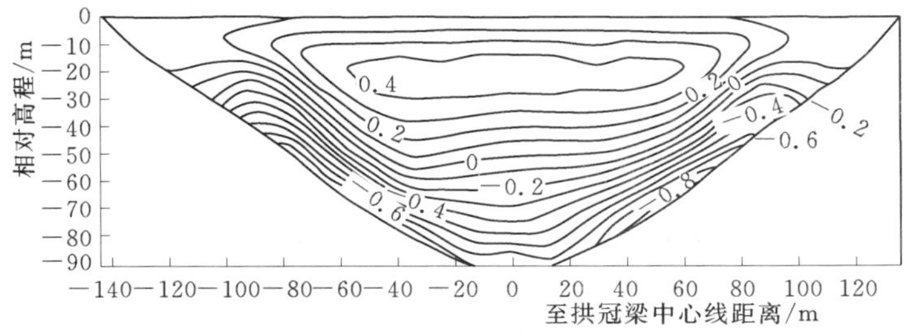

图 3 上游面第 1 主应力（工况Ⅲ，单位：MPa）

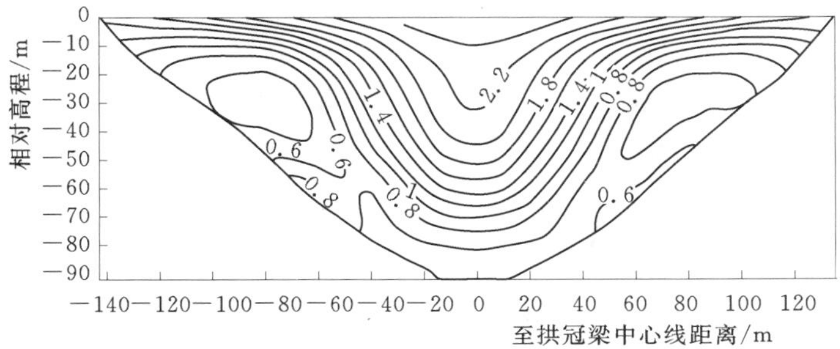

图 4 上游面第 2 主应力（工况Ⅲ，单位：MPa）

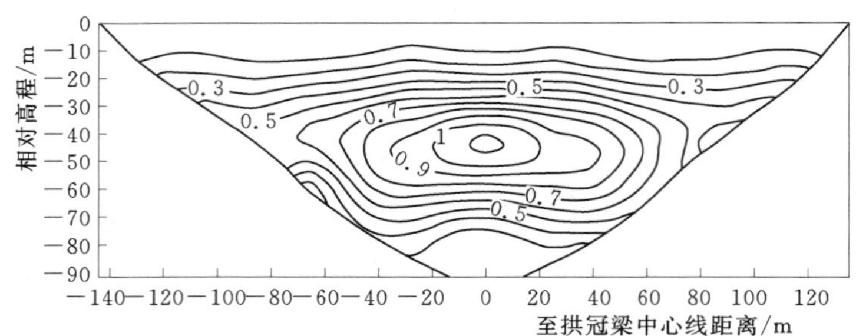

图 5 下游面第 1 主应力（工况Ⅲ，单位：MPa）

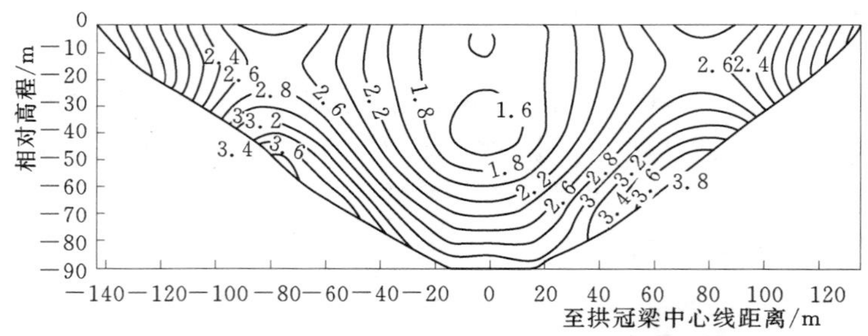

图 6 下游面第 2 主应力（工况Ⅲ，单位：MPa）

6.2 有限元等效应力法计算成果

本次采用三维有限元等效应力法的计算成果如下（限于篇幅，仅示出工况Ⅲ拱冠梁剖面的计算成果等值云图，见图 7、图 8）。

工况Ⅰ条件：该工况下大坝上游坝面周边均出现了拉应力，其余基本为压应力区；这其中两拱端拉应力约 0.20MPa，最大拉应力出现在拱冠梁底部附近，约 1.29MPa。该工况下压应力极值出现在上游坝面的约 2/3 坝高附近和下游坝面 1/3 坝高以下的周边，最大值分别为 1.60MPa 和 2.80MPa。

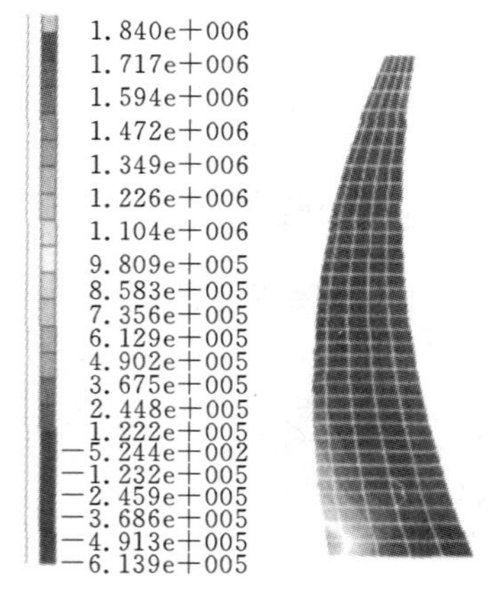

图 7 拱冠梁剖面大主应力等值云图（工况Ⅲ，单位：Pa）

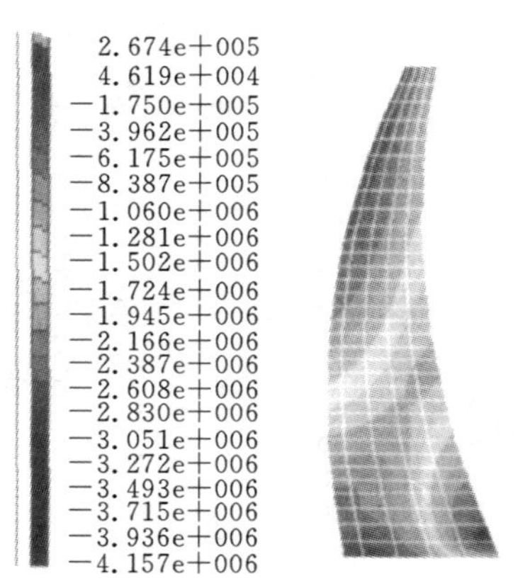

图 8 拱冠梁剖面小主应力等值云图（工况Ⅲ，单位：Pa）

工况Ⅱ条件：在该组荷载作用下，大坝将整体向上游变位。坝体最大主拉应力约为0.43MPa，发生在下游坝面周边的半坝高处；最大压应力值为4.21MPa，发生在上游坝面拱冠梁底部。

工况Ⅲ条件：大坝拉应力主要出现在3个特征区：上游面约1/3坝高以下、两拱端下游面附近以及拱冠梁下游面坝顶附近，其中拱冠梁底部最大值为1.84MPa。大坝压应力极值发生在拱冠梁上游面半坝高处、拱冠梁上游面顶部以及下游面约1/3坝高附近的周边，最大压应力分别为0.614MPa、3.05MPa和4.16MPa，均小于允许值6.0MPa。

可见大坝的主压应力和主拉应力均能满足规范要求。

7 结论

（1）多拱梁分载法概念清晰，使用时间长，积累了丰富经验，但不能从整体上反映坝体应力分布情况，基础变位采用简化方法等；三维有限元等效应力法能够进行各种复杂条件下拱坝的坝体应力分析，解决了试载法中较难处理的各种问题，但计算成果的精度仍然受到其自身特点的影响，如单元形态、网格密度、材料本构等。在高拱坝中分别采用两种方法进行计算分析，相互补充，相互验证，有一定的必要性。

（2）结合现行规范，分别采用多拱梁分载法和三维有限元等效应力法对井冈冲高砌石拱坝坝体应力进行了计算分析。结果表明，拱梁分载法与三维有限元等效应力法的计算成果总体趋势基本一致，坝体应力状态总体安全。

参考文献：

[1] 林绍忠，杨仲侯．分析拱坝应力的分载位移法［J］．水利学报，1987（1）．

[2] 傅作新．水工结构力学问题的分析与计算［M］．南京：河海大学出版社，1993.

[3] 朱伯芳，董福品．高拱坝应力控制标准研究［J］．水力发电，2001，（8）．

[4] 周维垣．高拱坝的有限元分析方法和设计判据研究［J］．水利学报，1997（8）．

[5] SL 25—91 浆砌石坝设计规范［S］．北京：水利电力出版社，1991.

[6] SL 282—2003 混凝土拱坝设计规范［S］．北京：中国水利水电出版社，2003.

有限元强度折减法在土石坝边坡稳定分析中的应用

吴海真[1,2]，顾冲时[1]

1. 河海大学水利水电工程学院；2. 江西省水利科学研究院

摘　要： 本文在指出传统边坡稳定分析方法的缺点和不足的同时，重点研究了有限元强度折减法的理论背景、突出优势、计算结果的影响因素、边坡失稳判据方法，探讨了该方法应用于土石坝边坡稳定分析中的可行性。采用摩尔-库仑屈服准则、非关联流动法则，在已知渗流场的前提下，基于有效应力原理逐步进行强度折减，直至满足力和位移不收敛的失稳判据，从而得到相应强度折减法的土石坝坝坡稳定安全系数。结果表明，无论是关键滑裂面位置还是最小安全系数，有限元强度折减法的计算结果与刚体极限平衡法的计算结果都相当接近，从而表明此法在土石坝边坡稳定分析中的可行性，拓宽了该法的应用范围。

关键词： 土石坝；边坡稳定；有限元；强度折减法

目前，对边坡稳定性分析评价多采用传统的刚体极限平衡法，其中土坡稳定分析计算又以条分法应用最为广泛。这些经典极限分析方法概念清晰、使用时间长、积累经验丰富，但在理论上存在诸多不足：①未考虑岩土体内部的应力应变关系，无法分析边坡破坏的发生和发展过程，边坡失稳与变形无确定性联系；②未考虑岩土体与支挡结构的共同作用及其变形协调；③假定的条件较多，如土条条分假设、条间力及滑裂面假定等。工程实践表明，边坡失稳总是与边坡的变形破坏存在一定的联系，随着计算机软硬件及非线性弹塑性有限元计算技术的发展，采用理论体系更为严密的有限元法分析边坡的稳定性已经成为可能。有限元法如果能保证足够的计算精度，就能有效地克服刚体极限平衡法的许多不足，其主要优点有：①能对复杂地形地貌和地质条件的边坡进行模拟计算；②能考虑岩土体的非线性本构关系以及变形对应力的影响；③能模拟边坡的渐进破坏及其滑裂面形状；④能模拟岩土体与支护的共同作用；⑤求解安全系数时，不需假定滑裂面的形状，自动形成破坏面，也无需进行条分。鉴此，本文就近年来发展的有限元强度折减法在土石坝边坡稳定分析中的应用作一探讨。

1　强度折减法的理论背景

20 世纪 70 年代，英国科学家 Zienkiewicz 提出采用增加外荷载或降低岩土材料强度

本文发表于 2006 年。

的方法计算岩土工程的安全系数，这是较早的极限分析有限单元法，但由于计算力学尚处起步阶段和缺乏相应的失稳判据等原因，长期以来并没有得到岩土界的广泛认可。近年来，采用强度折减法计算边坡的稳定性得到了重视，主要原因有：①计算力学的飞速发展；②有限元强度折减法能充分利用有限元法的突出优势，并有效克服传统极限平衡法的不足；③计算结果合理可信。

边坡稳定性分析有限元强度折减法的基本思想是：在理想弹塑性有限元计算中将边坡岩土体抗剪切强度参数逐渐降低直到达到破坏状态为止，程序可自动根据弹塑性计算结果得到破坏滑动面（塑性应变和位移突变地带），此时的折减系数 F 就是边坡的安全系数。于是有：

$$c_e'=\frac{c'}{F}\varphi_e'=\tan^{-1}\left\{\frac{\tan\varphi'}{F}\right\} \tag{1}$$

式中：c_e'为折减后的黏聚力；c'为有效黏聚力；φ_e'为折减后的内摩擦角；φ'为有效内摩擦角。

传统的边坡稳定极限平衡法采用摩尔-库仑屈服准则，安全系数定义为沿滑动面的抗剪强度与滑动面上实际剪应力的比值。可以证明，有限元强度折减法在本质上与传统方法是一致的。根据有限元强度折减法的基本思想，为使强度折减法边坡稳定计算顺利进行必须明确两点：①塑性增量本构由屈服函数、流动法则和硬化规律三个基本部分组成。计算塑性应变增量，首先需确定材料的屈服条件并选取材料所服从的流动法则（关联流动还是非关联流动），以确定塑性势函数，然后再确定材料的硬化规律。对于理想弹塑性材料其硬化参数为零（本文以理想弹塑性材料为研究对象）。②边坡失稳判据，即以何种标准作为边坡已经失稳或程序停止迭代的依据。

2 屈服准则和流动法则

本文计算采用的是理想弹塑性模型。对于岩土体材料，目前流行的有限元软件如ANSYS、MARC等均采用了广义米赛斯屈服准则，在国外称为德鲁克-普拉格准则(D－P准则)，表示为[1]：

$$f=\alpha I_1+J_2-k=0 \tag{2}$$

式中：I_1、J_2 分别为应力张量的第一不变量和应力偏张量的第二不变量（这是一个通用表达式，通过 α、k 的变换就可在有限元中实现不同的屈服准则）；α、k 为与岩土材料内摩擦角 φ 和黏聚力 c 有关的常数，当 α、k 满足下列表达式时

$$\alpha=\frac{2\sin\varphi}{\sqrt{3(3-\sin\varphi)}};k=\frac{6\cos\varphi}{\sqrt{3(3-\sin\varphi)}} \tag{3}$$

屈服面在 π 平面上为不等角度的六边形外接圆，即为摩尔-库仑屈服准则。

而当 α、k 满足下列表达式时

$$\alpha=\frac{\sqrt{3}\sin\varphi}{3\sqrt{3+\sin^2\varphi}};k=\frac{\sqrt{3}3\cos\varphi}{\sqrt{(3+\sin^2\varphi)}} \tag{4}$$

屈服面在 π 平面上为不等角度的六边形内切圆，在国内特指此圆为 D－P 准则，此时塑性区最大。

根据笔者的使用经验，摩尔-库仑屈服准则较为可靠，它的缺点在于三维应力空间中的屈服面存在尖顶和棱角的不连续点，导致数值计算不收敛；而 D－P 准则在偏平面上是一个圆，更适合数值计算。通常取摩尔-库仑屈服准则的外角点外接圆、内角点外接圆或其内切圆作为屈服准则，以利于数值计算。本文算例采用了适用于平面应变问题的摩尔-库仑屈服准则。

有限元计算中采用关联还是非关联流动法则，取决于 ψ 值（剪胀角）。当 $\psi=\varphi$ 时，为关联流动法则；当 $\psi\neq0$ 时，为非关联流动法则；当 $\psi=0$ 时，正好与郑颖人[2]提出的广义塑性力学理论相符。

文献［3］的研究结果表明，对同一边坡同一屈服准则，采用关联流动法则的计算结果比采用非关联流动法则的计算结果稍大，相对误差随内摩擦角增大而增大，一般为 3%～5%。文献［4］的研究结果表明，满足非关联流动法则的算例，其结果显示出较好的精度。

3　边坡失稳判据

采用有限元强度折减法分析边坡稳定性的一个关键问题，是如何根据有限元计算结果来判别边坡是否处于破坏状态。目前的失稳判据主要有两类：第一类是在有限元计算过程中采用力和位移的不收敛作为边坡失稳的标志[2-9]；第二类以广义塑性应变或等效塑性应变从坡脚到坡顶贯通作为边坡破坏的标志[10-13]。

大量计算实例表明，上述两种判据得到的安全系数相差不大。但从理论上分析，边坡发生塑性贯通并非意味着失稳，因为岩土体还有自身应力重分布的性能，只有具备发生塑性流动的边界条件，边坡才可能发生破坏，故第一类失稳判据应该更具合理性。

4　工程应用

石马水库大坝位于江西省靖安县境内，大坝坝顶高程 97.60m，坝顶宽 5.90m，最大坝高 23.0m。坝体填土多为黏土质砂，少量为含砂低液性黏土，呈硬塑—可塑状；坝基覆盖层为冲洪积粉黏土质砂；坝基主要发育有全—强风化黑云母花岗闪长岩及微新鲜岩体。大坝典型剖面及主要土层组成见图 1，主要计算参数见表 1。

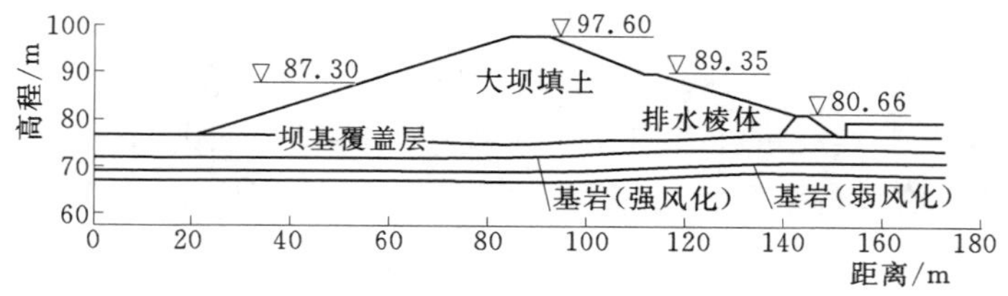

图 1　石马水库大坝典型剖面图

表 1　坝坡抗滑稳定计算物理力学参数选取表

部　位	天然重度 /(kPa·m⁻³)	饱和重度 /(kPa·m⁻³)	c' /kPa	φ /(°)	渗透系数 /(cm·s⁻¹)
坝体	17.45	18.80	15.00	22.50	2.75×10^{-4}
坝基覆盖层	18.94	19.53	10.50	23.10	7.62×10^{-4}

续表

部　　位	天然重度 /(kPa·m^{-3})	饱和重度 /(kPa·m^{-3})	c' /kPa	ψ /(°)	渗透系数 /(cm·s^{-1})
排水棱体	20.00	21.00	0.00	32.00	1.00×10^{-2}
全风化基岩	—	25.00	45.00	37.50	5.85×10^{-4}
强风化基岩	—	26.50	60.00	41.50	5.10×10^{-5}

主要计算内容和步骤为：①据表1所示各土层渗透系数和相应边界条件，采用有限元法（三角形六节点单元，共划分常应变单元1258个，节点1795个）进行渗流计算；②对于各工况条件下的渗流场，基于有效应力原理，采用Morgenstern－Price法[14]（土条数30，条间作用力为半正弦函数）进行计算，得到相应刚体极限平衡法的坝坡稳定安全系数；③基于摩尔-库仑屈服准则、非关联流动法则和第一类失稳判据，在基于有效应力原理基础上的自重应力场条件下逐步进行强度折减（单元网格同渗流计算），直至满足第一类失稳判据为止，从而得到相应强度折减法的坝坡稳定安全系数。限于篇幅，本次仅列出稳定渗流期的下游坝坡抗滑稳定计算结果。大坝在正常蓄水位（工况1，上游水位93.30m）和校核洪水位（工况2，上游水位95.35m）条件下，下游坝坡抗滑稳定的刚体极限平衡法（Morgenstern－Price法）计算结果和有限元强度折减法计算结果见图2～图4。

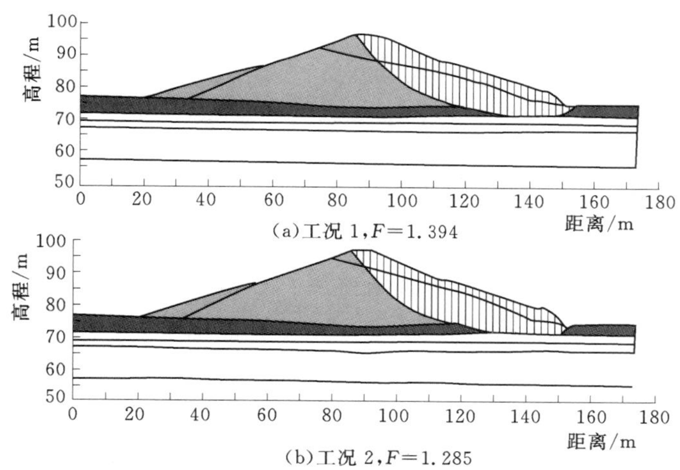

图2　下游坝坡抗滑稳定计算结果（刚体极限平衡法）

由以上计算结果可见，在土石坝的坝坡抗滑稳定计算中，无论是关键滑裂面位置还是最小安全系数，有限元强度折减法与刚体极限平衡法（Mogerstern－Price法）的计算结果相当接近。工况1条件下安全系数分别为1.380和1.394，相差仅1.01%；工况2条件下安全系数分别为1.256和1.285，相差2.26%。但就关键滑裂面而言，有限元强度折减法穿越坝基覆盖层更多一些，即呈下挫状，这与土石坝坝坡的真实滑动面性状更为接近。由此可见，有限元法若能保证足够的计算精度，其计算结果是合理可信的。它既能有效克服刚体极限平衡法的许多不足，又能充分利用有限元法的突出优势，

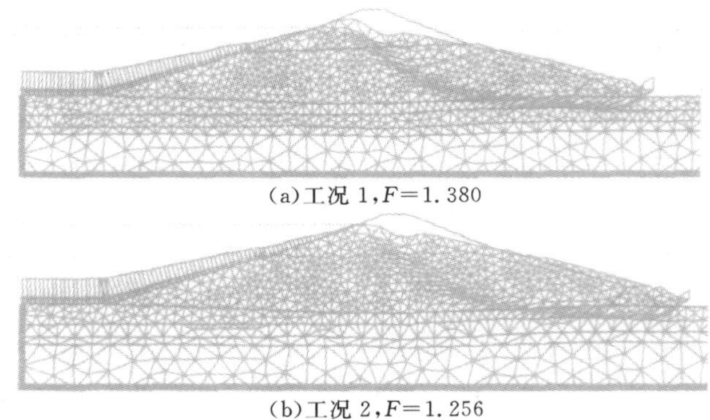

(a)工况 1,F=1.380

(b)工况 2,F=1.256

图 3　下游坝坡抗滑稳定计算结果，网格变形图

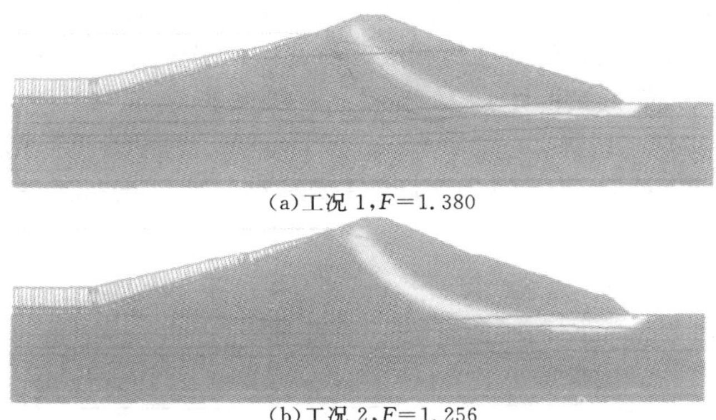

(a)工况 1,F=1.380

(b)工况 2,F=1.256

图 4　下游坝坡抗滑稳定计算结果，应变增量云图

如考虑了岩土体的非线性本构关系以及变形对应力的影响，能模拟边坡的渐进破坏及施工开挖过程、模拟岩土体与支护的共同作用、自动形成破坏面等，这是传统极限平衡法无法做到的。

采用有限元强度折减法计算边坡稳定时步骤较多，从而使计算精度的影响因素也较为复杂。文献［3］的 4 组计算方案共计 106 个算例研究结果表明，强度折减法所得稳定安全系数比简化 Bishop 法平均高出约 5.7%，且误差离散度极小。这有力地说明了有限元强度折减法用于分析边坡稳定问题的可行性，但也须合理选用屈服条件并严格控制有限元法的计算精度，如边界范围的大小和网格的密度等。

5　结语

（1）有限元法能求出边坡的安全系数，对具有复杂地形地貌和地质条件的边坡可自动求出任意形状的临界滑裂面，并能模拟出边坡渐进破坏及施工开挖过程。

（2）本文将有限元强度折减法应用于土石坝边坡稳定计算分析中，拓宽了应用范围，得到了预期的结果。但如何在有限元强度折减法中考虑其他因素如地震荷载、岩质边坡中存在的大量节理等仍需进行深入的研究。

参考文献：

[1] 郑颖人，龚晓南．岩土塑性力学基础［M］．北京：中国建筑工业出版社，1989.

[2] 郑颖人，孔亮．塑性力学中的分量理论——广义塑性力学［J］．岩土工程学报，2000，22（3）：269－274.

[3] 郑颖人，赵尚毅．有限元强度折减法在土坡与岩坡中的应用［J］．岩石力学与工程学报，2004，23（19）：3381－3388.

[4] 张鲁渝，郑颖人，赵尚毅，等．有限元强度折减系数法计算土坡稳定安全系数的精度研究［J］．水利学报，2003（1）：21－27.

[5] 赵尚毅，郑颖人，邓卫东．用有限元强度折减法进行节理岩质边坡稳定性分析［J］．岩石力学与工程学报，2003，22（2）：254－260.

[6] 郑颖人，赵尚毅，邓卫东．岩质边坡破坏机制有限元数值模拟的可行性［J］．岩石力学与工程学报，2003，22（12）：1943－1952.

[7] Griffiths D V，Lane P A. Slope Stability Analysis by Finite Elements［J］．Geotechnique，1999，49（3）：387－403.

[8] 张鲁渝，刘东升，郑颖人．平面应变条件下的土坡稳定的有限元分析［J］．岩土工程学报，2002，24（4）：487－490.

[9] 徐干成，郑颖人．岩土工程中屈服准则应用的研究［J］．岩土工程学报．1990，11（2）：95－101.

[10] 连镇营，韩国城，孔宪京．强度折减有限元法研究开挖边坡的稳定性［J］．岩土工程学报，2001，23（4）：406－411.

[11] 栾茂田，武亚军，年廷凯．强度折减有限元法中边坡失稳的塑性区判据及其应用［J］．防灾减灾工程学报．2003，23（3）：1－8.

[12] 郑宏，李春光，李焯芬，等．求解安全系数的有限元法［J］．岩土工程学报，2002，24（5）：626－628.

[13] 周翠英，刘祚秋，董立国，等．边坡变形破坏过程的大变形有限元分析［J］．岩土力学，2003，24（4）：644－652.

[14] 陈祖煜．土质边坡稳定分析——原理、方法、程序［M］．北京：中国水利水电出版社，2003.

已建水库土坝下涵管的安全评价

段智芳[1]，林新川[2]，张万辉[3]

1. 江西省水利科学研究院；2. 江西省建设银行新余市分行；

3. 余江县锦北灌区管理局

摘　要： 江西省部门水库建设年代早，土坝下涵管存在诸多安全隐患，通过对涵管的检查、计算、原因分析，对涵管的安全及影响做出评价，重点反映目前水库大坝安全鉴定中坝下涵管安全评价中应注意的一些问题。

关键词： 土坝；坝下涵管；渗漏检查；计算分析；安全评价

1　引言

目前，江西省已建水库9000余座，约占全国总量的1/9，在全国各省中水库总量排名第二。但这些水库大坝大多建设于20世纪的一些特殊时期，限于当时的历史条件、技术水平等诸多因素，很多工程建设标准偏低，工程质量较差，运行投入不足。老化失修严重，导致病险水库大量存在，不仅影响水库综合效益的发挥，而且对下游生命财产安全构成较大风险。特别是江西省绝大多数土石坝的坝下埋管，安全隐患特别严重。坝下埋管施工质量差，很多小型水库的坝下埋管是陶管，也有竹筋混凝土管和素混凝土管；有的埋管周围填土碾压不密实，和坝体的结合部无截渗环等防渗措施；还有因各种原因，改变了其设计运行状态，用做压力管道。据不完全统计，由于坝下埋管渗流破坏或埋管破坏后的库水直接冲刷引发的安全事件多达150例。

为全面了解一座水库大坝的安全状况，对大坝工程存在的问题及时分析研究，进而找出解决问题的办法，已形成水库大坝安全鉴定制度，并作为一种大坝安全管理的手段。因此，大坝安全鉴定工作是正确评价水库大坝安全状况，合理全面进行病险水库除险加固的一个重要基础工作。本文根据作者参与大坝安全鉴定工作、大坝现场检查及对鉴定成果审核所做工作，通过对坝下涵管的检查、计算、原因分析，对涵管的安全及影响做出评价的同时，重点反映目前水库大坝安全鉴定中坝下涵管安全评价中应注意的一些问题。

2　坝下涵管安全问题及危害

坝下涵管的安全问题主要是管身结构问题和渗漏问题，管身结构的安全问题也必将导致管身出现渗漏现象。涵管渗漏的破坏性较大，受管身的材料性能、施工质量的影响，涵

本文发表于2006年。

管的渗漏问题主要是沿管壁外的渗漏和穿过管壁的渗漏。此外，进口和闸门漏水也是常见的问题。

（1）涵管存在沿管壁与坝体接触面的渗漏，当渗漏发展成漏水通道时，造成集中渗漏，在建筑物下游坡面造成渗水漏洞险情，危及建筑物及土坝安全。

（2）涵管管壁裂缝或存在断裂、漏水，存在穿透管壁的渗漏，带失坝体土颗粒，严重时在管轴线坝顶附近出现塌坑，涵管出口可见黄泥水。

都昌县长垅水库南塘副坝坝下涵管，建于1959年，断面为0.4m×0.4m，由于当时材料紧缺，涵管侧墙、底板为混凝土，顶板钢筋预制盖板。1970年水库扩建，南塘副坝加高6m，而坝下涵管未做处理，1977年、1980年在涵管轴线上游坝坡距启闭台约4m处两次出现塌陷现象，塌陷坑深4m，直径约2.2m，管口见浑水流出。1980年将原涵管拆除，重建钢筋混凝土坝下涵管，内径为0.8m，但施工条件差，人工振捣不实。1999年汛期，在涵管轴线上游坝坡距启闭台约5m处出现直径为2m的塌坑，坑顶水流旋流而下，涵管出口则流出浑水。后经检查发现涵管管壁有贯穿性裂缝，漏水严重。由此可以分析认为，由于大坝加高，增加了原涵管管身的土压力荷载，改变了涵管的设计运行工况，导致原涵管盖板断裂；重建的钢筋混凝土坝下涵管裂缝的产生与涵管施工质量控制不严有关。由于管身存在断裂或贯穿性的裂缝，而涵管外坝体浸润线的存在，使涵管外存在一定的外水压力，涵管自身为无压灌溉管；因此，在外水压力的作用下，涵管裂缝周围的坝体土产生渗透破坏，带失坝体土颗粒，渗透破坏发展到一定程度时，在对应的坝面形成了塌坑。

（3）进口和闸门的渗漏。在已建涵管中，绝大多数均存在进口和闸门渗漏问题，漏水的主要原因有：闸门变形，止水老化失效，对于进口采用阶梯卧管取水的涵管，进口卧管的基础发生沉陷或卧管的质量不好，出现断裂漏水。

因此，坝下涵管前两种漏水易产生坝体土的接触渗透破坏，一旦形成通道，就会造成集中渗漏，水流带走大量坝体土颗粒，会使管轴线部位的坝坡出现塌坑，大坝与涵管间存在渗透破坏现象，危及大坝安全。进口和闸门的渗漏虽不直接影响坝体的安全，但会影响水库蓄水量，影响水库灌溉效益，若漏水量过大，则可能会影响涵管的水流流态，危及涵管安全，进而影响大坝安全。

3 坝下涵管的现场检查及分析

在水库大坝安全评价过程中，为了对涵管自身的安全性状及对大坝安全的影响做出正确而全面的评价，首先必须了解坝下涵管的现状运行情况及做相应的现场检查。

3.1 现场检查

从涵管运行过程中渗漏水的表象，进一步分析渗漏的原因及危害。对有条件进管检查的，还应做管身混凝土的强度等级、碳化深度等现场检测，绘制涵管现场检查展开图，标明断裂、裂缝、蜂窝麻面、露筋、渗漏等问题所在的位置，如图1所示。

根据检查漏水情况，对涵管危害性做出评价。

3.1.1 漏清水

涵管内漏清水有两种情况：一是闸门漏水；二是穿过管壁或沿管壁的渗漏。无压涵管漏水多半发生在管顶或管两侧，主要是由于砌体坐浆不严或管口接头不好，这种渗漏多半

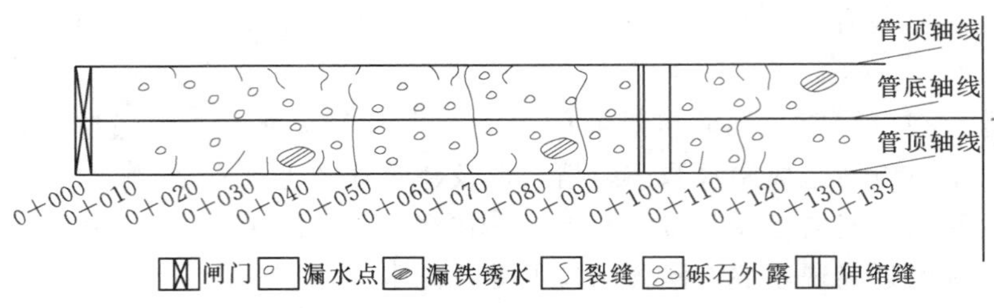

图 1 幸福水库东涵管裂缝展开图

表现为阴湿和漏清水，短时间内不会危及坝体安全，但长期渗漏有可能发展成漏浑水。对于有压涵管，不论是穿过管壁的渗漏还是沿管外壁的渗漏，对坝体的危害均较大。

3.1.2 漏浑水

涵管一旦漏浑水，说明已形成漏水通道，出现坝体土颗粒被水流带出的现象，大坝与涵管间存在渗透破坏现象。如发现涵管内漏浑水，有条件的地方，应进管检查漏水情况和漏水点。这种漏水主要是因为涵管管壁存在漏水孔洞或裂缝等薄弱环节，在渗透水流作用下形成漏水通道。随着漏水严重，坝坡上将出现塌坑，如果塌坑位于库水以下，一般会在相应水面上形成漩涡。

3.1.3 漏铁锈水

有的涵管在管壁有黄色、红色或黑色的松软物质，外表很似黏土，这是土料中含有的矿物质被水溶解后随水流渗漏到管壁上，遇到空气后氧化，最后沉淀、凝结成溶乳状物质，一般称为“铁锈水”。尤其在江西省，由于绝大多数大坝的筑坝土料是红黏土，红土中的矿物成分高，在坝下游及滤体外也经常会出现这种类似的氧化物。铁锈水的出现表明，漏水对坝体填土有一定的破坏作用。

3.1.4 管壁有游离石灰析出

有的涵管内壁凝结一层白色物质，这是由于水泥标号低，砂浆质量差，在渗水作用下，游离石灰被析出。这种渗漏对砂浆的强度有一定的影响。

通过对涵管的不同运行工况下渗漏量的变化，分析涵管渗漏的原因及危害。涵管的运行状态一般与涵管的渗漏量是密切相关的，通过分析涵管不同运行工况下渗漏量的变化，可以查明涵管渗漏的原因。莲花县河江水库坝下灌溉输水涵管，2002 年 4 月 22 日以来，发现大坝排水棱体坡脚渗漏点渗漏量明显增大，即在此设置了一量水堰进行渗漏量观测。据观测资料可知，大坝坝脚渗漏与库水位相关不密切，但与坝下涵管的运行状态密切相关，坝下涵管过水时渗漏量明显增大，每当涵管出口泄水闸关闭，涵管承压，则该处渗漏量又明显增大。由此可见该坡脚渗漏点渗漏主要是由于坝下涵管的渗漏引起的。经过对涵管建设过程的调查、了解发现，该涵管建于 1968 年，涵管浇筑均是人工操作，局部部位有蜂窝麻面。原在坝中设有启闭塔，由于施工不当等原因使该启闭塔断裂并倾斜，无法安装启闭设备，不能使用，后将进口段涵管和启闭塔封堵，重新在坝前修建圆形钢筋混凝土竖井，竖井后用内径 1.1m 钢筋混凝土管与原涵相接。进管检查还发现，涵管管壁裂缝众多，虽经 1998 年补强加固，但仍有几条裂缝，漏水严重，尤其是 1989 年新建启闭塔与老涵管接头处，漏水呈喷射状，可见该坡脚渗漏点渗漏主要是由于坝下涵管裂缝的渗漏引起的。

因此，只有充分了解分析涵管运行中出现的问题，对涵管进行认真细致的现场检查，才能对涵管的安全性状做出正确判断，进而分析其对大坝安全的影响。

4 坝下涵管的结构复核

由于涵管位于坝下，其结构安全与否直接影响到大坝的结构安全，其结构不安全引发的管身断裂、裂缝又影响大坝、管身间的渗透稳定，所以对坝下涵管按现行规范进行结构安全复核，是坝下涵管安全评价的重要内容。

坝下涵管的结构安全复核包括管身钢筋混凝土结构强度的复核及抗裂验算两大部分，在复核过程中应注意以下几点：

（1）复核计算中，特征水位应选取本次安全鉴定阶段的计算成果。

（2）对管身混凝土强度等级、管壁厚度的取值应结合现场检测、现场检查的成果考虑老化、剥蚀等多方面因素的影响慎重选取。

（3）复核计算中，土压力大小应结合本次地勘提供的土体参数进行计算，不能盲目地套用原设计参数。

（4）对涵管的原设计图纸与涵管现状应仔细核对。由于管理单位管理的疏忽，原始设计图纸严重缺失，或水库经多次扩建加固设计，中间成果较多而很多并未实施，这就要求鉴定承担单位的设计人员认真、谨慎。在已进行的安全鉴定中，已多次出现过涵管结构计算图与涵管现状严重不符的现象，有的涵管明明是圆管，却被设计单位按原设计图纸的方管进行复核；有的管径与现状严重不符。

5 结语

通过对坝下涵管的现场检查、结构复核计算，最后应对涵管的渗漏原因、严重程度、运行安全性做出评价结论。由于坝下涵管隐患多，建成后维修困难，出现安全问题后影响大，后果严重，近几年的大坝安全鉴定显示，坝下涵管的问题已严重影响到大坝安全，受到大家的普遍关注。建议目前的除险加固工作中，对于有条件的地方，尽量不要采取坝下涵管的取水方案，而以输水隧洞或虹吸管等型式替代。对于已经存在渗透破坏的坝下涵管，封堵要彻底，不要给工程留下隐患。

江西省水库大坝安全评价体系的建立及应用研究

吴晓彬[1]，杨丕龙[2]，傅琼华[1]

1. 江西省水利科学研究院；2. 江西省水利厅

摘　要：通过对江西省水库现状进行广泛的调研和分析，核准了全省病险水库的数量，编制了《江西省病险水库除险加固规划》；依据相关规范，编制了《江西省水库大坝安全评价技术规定》及《江西省大坝安全鉴定工作细则》；先后在200余座水库大坝安全鉴定中应用，并对评价方法进行了提高和创新；建立了江西省水库除险加固动态管理信息系统。形成了基于江西省水库工程特点的、较完整的、科学的江西省水库大坝安全评价体系，对水库大坝安全评价工作有较强的指导作用。

关键词：水库大坝；安全评价；除险加固

1　引言

截至2002年年底，江西省共建成各类水库9268座，数量居全国第二位。由于历史原因，许多水库建设标准偏低，管理体制不顺，维修养护经费短缺，老化失修严重。据统计分析，全省有病险水库3488座。病险水库的大量存在，既严重威胁着广大人民群众的生命财产和公共设施的安全，也制约了水库效益的充分发挥。基于此，江西省政府办公厅印发了《江西省病险水库除险加固实施意见》，提出至2007年年底，用五年时间全面完成现有病险水库的除险加固任务。病险水库的除险加固，是加强和完善水利基础设施建设，提高水资源保障和调控能力，保证水库安全运行和效益的充分发挥，实现以水资源可持续利用以保障社会经济可持续发展的重要举措。从历年病险水库除险加固工作的实践来看，水库大坝除险加固规划和水库大坝安全鉴定是病险水库除险加固的前期工作，又是关键所在。水库大坝安全评价工作做得充分与否直接影响到除险加固技术方案选取的科学性和资金投入、分配的合理性及有效性。经过几年的研究探索，建立了基于江西省水库工程特点的、较完整的、科学的江西省大坝安全评价体系，即在全面普查江西省水库的安全状况，总结近几年江西省大坝安全鉴定工作的基础上，针对江西省病险水库数量多、分布广、加固任务重、资金来源不足等具体情况，研究江西省大坝安全评价模式；通过对全省水库大坝的安全普查，制定系统规划，将水库大坝安全评价技术标准的建立和大坝安全评价技术方法的研究、创新紧密结合，为管理层提供科学、快捷的决策平台，指导、促进了江西省病险水库除险加固工作的顺利进行，提高了江西省水库工程的管理水平。

本文发表于2005年。

2 编制病险水库除险加固规划以确立目标

在对全省 7690 座小（2）型水库、1329 座小（1）型水库、224 座中型水库和 25 座大型水库工程安全性调查与分析的基础上，确定病险水库评定原则，核定至 2002 年 12 月 31 日止，江西省现有 3488 座水库存在不同程度的病险隐患，占全省水库总数的 37.6%，其中大型水库 11 座，中型 153 座，小（1）型 775 座，小（2）型 2549 座，分别占各类水库总数的 44.0%、68.3%、58.3%和 33.1%。在分析水库大坝的主要病险症状、产生原因及加固效益的基础上，考虑分级管理、资金来源和配套比例等方面因素，编制了《江西省病险水库除险加固规划》。再结合各地水资源状况和社会经济发展水平等具体情况，综合制定了至 2007 年年底全面完成规划内病险水库除险加固目标应采取的组织、政策、技术和资金保障等措施，以全面指导江西省病险水库除险加固工作。明确了水库除险加固的行政领导负责制，从制度上确保了病险水库除险加固工作落实到人，避免了多龙治水，实则无龙治水的局面，保障了江西省病险水库除险加固工作的有序性。

3 对病险水库进行风险排序以提高针对性

引入和采用目前国际上先进的大坝风险分析概念和技术，通过对大坝风险度、溃坝后果、综合影响赋值量化的办法，最终得到综合风险评估分数，确定大坝危险程度；研究适合江西省病险水库除险加固排序的实用方法，并对病险水库进行除险加固排序，排序的依据是水库大坝风险的大小。大坝风险是指大坝对作用于生命、健康、财产和环境负面影响的可能性和严重性的度量，是不利事件可能性与危害后果的乘积。风险排序涉及同一水库主要风险因素或事件对溃坝概率影响的重要程度排序及不同水库间风险程度排序。

将病险水库大坝风险评估指数表示为安全程度风险赋分值与溃坝后果综合系数、综合影响系数值的乘积。水库大坝险度值的构成共分五大部分——洪水（14 分）、大坝工程（32 分）、泄水建筑物（18 分）、输水建筑物（25 分）、金属结构（11 分）。当出现正常缺项时，以该项的平均分赋分。分值由水库管理技术人员根据水库大坝实际情况进行赋值。溃坝后果综合系数包括生命损失、经济损失和社会及环境影响，根据南京水利科学研究院的研究成果，采用 Saaty 建议的 1～9 标度法（AHP 法）来确定各子层因素对母层因素的权重系数。经计算，生命损失的权重系数 $S_1=0.737$，经济损失的权重系数 $S_2=0.105$，社会及环境影响权重系数 $S_3=0.158$，则溃坝后果综合系数 L 为：$L=0.737F_1+0.105F_2+0.158F_3$（式中，$F_1$、$F_2$、$F_3$ 分别为生命损失、经济损失和社会及环境影响的严重程度系数）。再考虑各设区市水行政主管部门对辖内各水库的风险排序、水库总库容、灌溉效益、前期工作情况等，确定水库的综合影响系数；根据分析研究后确定的赋分原则及系数，由专家组成员各自分别对每座水库进行赋分，统计后以平均值计算，再计算病险水库大坝的风险评估指数。群坝风险排序是对不同水库的风险程度进行比较，确定风险顺序；风险指数越大，表明该水库存在的危险性越大，且溃坝损失严重，应优先得到除险加固。经过在江西省尚未除险加固的 114 座中型及 128 座重点小（1）病险水库除险加固风险排序中应用，证明该方法使当前病险水库除险加固决策更具科学性和层次性，使各级管理部门在资金投入和安排上有轻重缓急之分，增强了有限资金使用的有效性和针对性。

4 制定大坝安全评价技术规定以规范标准

根据有关规程规范，结合江西省水库大坝工程实际，强化和细化大坝安全评价技术工作及安全鉴定工作细则，制定了《江西省大、中型水库大坝安全评价报告编制若干技术规定（试行）》、《江西省小型水库大坝安全评价导则（试行）》、《江西省水库大坝安全鉴定管理办法》、《江西省水库大坝安全鉴定实施方案》、《江西省小（1）型水库大坝安全鉴定成果核查实施方案》等水库大坝安全评价技术规定和方法，指导和规范了江西省水库大坝安全评价的技术工作内容、方法与标准以及鉴定工作内容、鉴定程序、分级组织与管理、鉴定成果核查等，有效地提高了江西省水库大坝安全评价报告的质量。各技术规定具有较强的操作性和实用性，从制度上、技术上对病险水库除险加固工作进行了规范管理，使江西省水库大坝安全鉴定工作在国内赢得了好评。特别是《江西省小型水库大坝安全评价导则》，对小型水库安全评价内容、安全状况分类标准、评价报告编写提纲、小型水库地勘技术要求等作出了详细的规定，已成为江西省3000多座小型水库安全鉴定工作的主要技术标准和依据。

5 应用先进的技术方法开展水库大坝安全鉴定工作

近年来，江西省先后完成了11座大型、130余座中型及近250座重点小（1）型水库的大坝安全鉴定工作，坝型涵盖了土石坝、重力坝、堆石坝、拱坝和拦河闸坝等。针对水库工程具体问题的复杂性，在充分了解水库历年的运行情况和现场表观，经补充地质勘探，按水库大坝目前的工作条件、荷载组合及运行工况，分别对水库的防洪标准、大坝渗流安全、结构安全、金属结构及启闭设备安全、工程质量和运行管理等进行复核计算与评价，明确了水库大坝的病险状况，并查明了病险成因，确保了除险加固工程措施的针对性、明确性和有效性。

防洪标准复核是通过对水库现状的流域特征参数及水位—库容关系曲线复核后，采用单站实测雨量资料计算设计暴雨，或由《江西省暴雨洪水查算手册》查算设计暴雨，采用推理公式法和瞬时单位线法推求设计洪水，经过综合分析比较，选定合理的洪水成果以复核大坝的抗洪能力；并复核与评价洪水下泄的安全性。大坝渗流安全评价是通过对大坝观测资料的分析，建立统计模型，采用二维或三维有限元渗流计算方法，结合水库大坝运行表现情况，对土石坝的坝体、坝基及坝肩的渗流进行计算分析，并评价其渗流安全性；复核大坝防渗体与反滤排水体的设计标准和方法是否满足现行规范要求，施工质量是否满足设计要求等。坝坡稳定安全复核采用规范规定的简化毕肖普法等方法对大坝上、下游坡进行抗滑稳定计算，有条件时考虑降雨入渗及非饱和渗流作用下的坝坡稳定分析；大坝变形安全评价采用大坝的表面沉降、横向水平位移的观测资料进行整理分析，并对坝面裂缝进行计算分析，必要时采用有限元法对大坝施工填筑过程模拟仿真，进行应力应变计算分析；对近坝库岸的稳定性进行分析与评价。对重力坝分析其扬压力变化对大坝整体稳定的影响，对面板进行裂缝检测和面板挠度、应力计算分析，分析应力状态的安全性。对泄水、输（引）水建筑物结构安全按有关规范要求进行泄（过）流能力、水面线、消能、管道压力坡线等水力学复核计算分析和结构稳定、强度等的复核计算。所采用的均是国内先进的复核计算方法，特别是对宁都县团结水库、上饶县茗洋关水库等土石坝建立了有限元法计算模型进行应力应变计算分析；对兴国县长冈水库浆砌石重力坝、永新县龙源口水库浆砌石重力拱坝、铅山县铁炉水库混凝土重力坝等防渗面板的裂缝进行计算，并通过建立大坝二维或三维有限元应力应变模型，计算分析坝体内的应力分布，从应力角度分析判断产生裂缝的部位及危害性，并对其发展进行了展望，准确评价了大坝的病险症状。

6 建立江西省水库大坝除险加固动态管理信息系统以提高管理水平

为配合《江西省病险水库除险加固规划》的实施，提高江西省病险水库管理水平，针对信息技术发展的现状，通过资料收集、整编并建立江西省水库大坝基本情况、安全鉴定、除险加固设计、除险加固施工、工程图纸和照片等6个方面的信息资料数据库，开发了一套完整的水库除险加固信息动态管理软件，具有信息维护、信息查询与统计、地理信息系统、规程规范查询等功能，实现了水库大坝安全鉴定、加固设计、加固施工、加固验收资料等的全面信息化管理。把计算机高级语言编程技术、地理信息系统技术、计算机数据库技术、计算机多媒体技术引入到病险水库信息的计算机管理中，从根本上改变了信息管理的传统模式，实现了人机对话和管理程序化、自动化，为管理层和有关部门决策提供了科学、快捷的平台，大大方便了病险水库各种信息的综合管理。特别是将 DMBS 系统和地理信息技术用于病险水库信息管理在国内尚不多见。系统操作简便，内容丰富，功能强大，更新及时，使江西省水库信息管理水平踏上了一个新的台阶。

7 结语

大坝安全评价工作是当前病险水库除险加固实际中遇到的迫切需要解决的重大课题，针对江西省 9268 座水库大坝的安全状况，通过全面普查，编制系统的除险加固规划；积极探索水库群大坝风险排序实用方法；制定实用的水库大坝安全评价技术规定；总结大坝安全评价技术方法，并加以提高或创新；开发了病险水库除险加固动态管理信息系统，为管理层提供科学、快捷的决策平台，逐步形成了基于江西省水库工程特点的、较完整的、科学的“江西省水库大坝安全评价体系”。建立并完善了江西省水库大坝安全评价技术指标体系和工作体系，规范了江西省水库大坝的安全评价工作，推动了病险水库的除险加固工作，促进了水库大坝安全管理规范化、科学化。但水库大坝的安全是一个动态的、发展的概念，随着坝工技术的发展，人类对工程特性、安全含义等理解的变化，对水库大坝安全的社会、经济含义认识的不断提高，水库大坝安全的权重也在不断变化。特别是针对江西省病险水库数量多、分布广、加固任务重、资金来源不足等具体情况，研究建立一个可以满足未来 10～20 年发展需要的大坝安全总体评价体系，对全省水库大坝除险加固工作有着重要的指导作用，对全国病险水库大坝的除险加固工作也具有一定的推动作用。

水力自动翻板闸门泄量计算方法的探讨

周永门，蔡红波

江西省水利科学研究院

摘　要：针对安装有水力自动翻板闸门溢流坝（溢洪道）的泄量计算分析，提出一种较为简便的水力自动翻板闸门泄量计算方法。

关键词：水力自动翻板闸门；溢流坝（溢洪道）；泄量计算；水力计算方法

1　概述

水力自动翻板闸门是利用水力和闸门重量平衡的原理，增设阻尼反馈系统来达到闸门随上游水位升高而逐渐开启泄流；上游水位下降时，又逐渐回关蓄水。它具有原理独特、作用微妙、结构简单、制造方便；施工简便、造价合理；自动启闭、运行时稳定性良好和管理方便安全、省人、省事、省时、省力等特点，许多水库工程都采用水力自动翻板闸门进行泄洪。安装有水力自动翻板闸门的溢流坝（溢洪道）在进行泄量计算时，由于水流过闸门后又混合在一起，相互影响，计算比较复杂，一般是通过水工模型试验来确定库水位与泄量的关系。本文通过分析比较，探讨一种较简单可行的水力自动翻板闸门泄量计算方法。

2　泄量计算分析

水力自动翻板闸门过水时，水流的流态很复杂。当库水位超过闸门顶一定水头后，闸门开启过水。过闸水流分为两部分，第一部分是通过闸门顶部的水流；第二部分是从闸底部通过的水流。过闸后，这两部分水流又相互混合，很难将其截然分开，这样就无法直接套用实用堰、宽顶堰或薄壁堰上的闸孔出流公式进行泄量计算。为了计算简便，可将第一部分水流近似地看作堰流，此部分流量可按矩形薄壁堰自由出流公式进行计算；第二部分水流近似地看作为孔流，此部分流量可按宽顶堰闸孔自由出流进行计算（见图 1）。

矩形薄壁堰泄量计算公式为：

$$Q_1 = m_0 B \sqrt{2g} H_1^{3/2} \tag{1}$$

式中：B 为过流净宽，m；H_1 为堰上水头，m；m_0 为包括行近流速影响在内的薄壁堰流量系数，可按雷保克（Rehbock）公式计算：

本文发表于 2005 年。

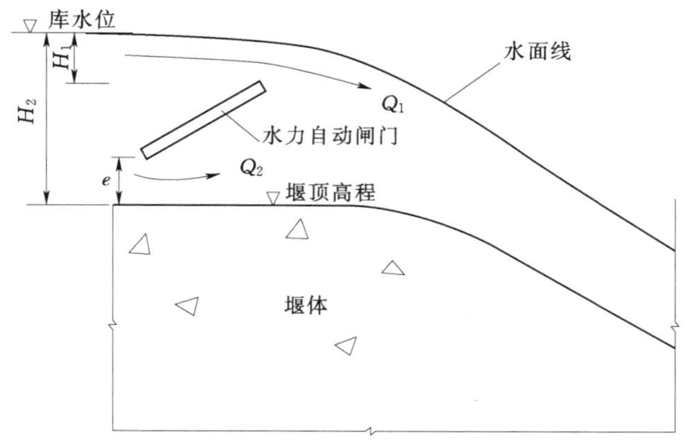

图 1　水力自动闸门泄量计算示意

$$m_0=0.4034+0.0534\frac{H_1}{a}+\frac{1}{1.610H_1-4.5}$$

式中：a 为上游堰高，m。

宽顶堰闸孔自由出流泄量计算公式为：

$$Q_2=\varphi\varepsilon' eB\sqrt{2g(H_0-\varepsilon' e)} \tag{2}$$

式中：B 为过流净宽，m；φ 为流速系数；ε'为平面闸门的垂向收缩系数；e 为闸门开启高度，m；H_0 为计入行近流速水头的堰上总水头，$H_0=H_2+\frac{V_0^2}{2g}$，m。

将相应库水位下的两部分泄量相加即可得到溢流坝（溢洪道）的总泄量 Q，即 $Q=Q_1+Q_2$，由此可算出库水位与泄量的关系。

3　泄流量计算实例

3.1　螺滩水库溢流坝泄流量计算

螺滩水库位于江西省吉水县富滩镇螺滩村，距吉水县城 25km，坐落于赣江水系赣江支流孤江下游。水库正常蓄水位为 72.5m，控制流域面积为 2160km^2，总库容为 4410 万 m^3，年发电量为 4300 万 kWh，是一座以灌溉、防洪、发电为主，兼有养殖、旅游等综合效益的中型水库。枢纽工程建筑物有浆砌石溢流式重力坝、溢洪道、发电引水系统、电站、右岸灌溉引水渠和灌区渠系工程建筑物等。其中溢流式重力坝堰顶高程为 69.5m，堰顶厚度为 10.5m，溢流净宽为 120m，堰顶设有 20 扇钢筋混凝土水力自动翻板闸门，尺寸为 6m×3m（宽×高）。当库水位为 72.73m 时（比正常蓄水位 72.5m 高 0.23m），闸门开始翻转，库水位为 73.09m 时（比正常蓄水位 72.5m 高 0.59m），闸门全开。闸门全开时，闸门顶高程为 71.807m，底高程为 70.913m。

分别用式（1）和式（2）计算溢流坝堰流的泄量 Q_1 和孔流的泄量 Q_2。式（1）中，过流净宽 $B=120$m，上游堰高 $a=71.807-69.5=2.307$m，假设一个库水位，则可由雷保克公式计算流量系数 m_0，再由式（1）计算出泄量 Q_1；式（2）中，过流净宽 $B=120$m，闸门开启高度 $e=70.913-69.5=1.413$m，流速系数 φ 查表取 0.95，底部为锐缘的平面闸门的垂向收缩系数 ε'可根据比值 e/H_0 由表查得，因溢洪道上游为水库，过水面积很大，行近流速 $V_0 \approx 0$，则 $H_0 \approx H_2$，假设一个库水位，则可由 e/H_0 查表得垂向收缩系数 ε'，再由式（2）计算出泄流量 Q_2。由此算出的库水位与溢流坝泄流量成果见表 1。

表 1　螺滩水库库水位—溢流坝泄流量关系计算

库水位/m	72.5	72.73	73.09	73.5	74.1	74.7	75.1	75.6	76.1	76.5
泄流量 $Q_1/(m^3 \cdot s^{-1})$	0	97	335	519	843	1231	1524	1929	2323	2655
泄流量 $Q_2/(m^3 \cdot s^{-1})$	0	0	740	790	859	923	964	1011	1056	1093
总泄流量 $Q/(m^3 \cdot s^{-1})$	0	97	1075	1309	1702	2154	2488	2940	3379	3748

3.2　楼梯磴水库溢洪道泄量计算

楼梯磴水库位于江西省莲花县西南部神泉乡瑶口村，距县城 14km，坐落在禾水上游文汇江大沙洲水上，坝址以上控制流域面积为 47.4km²，正常蓄水位为 200.4m，总库容为 1168 万 m³，年平均发电量为 62 万 kWh，是一座以灌溉为主，兼顾发电、养殖、防洪等综合效益的中型水库。枢纽工程主要建筑物包括：主坝、副坝、溢洪道、输水隧洞、输水涵管、放空涵管、电站和渠系建筑物等。溢洪道设置在主坝左端天然垭口处，全长 301m，由进口段、控制段、泄槽段、消能段组成，堰顶高程为 198.4m，溢流净宽为 30m，由水力自动翻板闸门控制，闸门高 2m，闸门顶高程为 200.4m。当水位超过正常水位（200.4m）0.1m 时，闸门开始翻转，溢洪道泄洪；当水位超过正常水位（200.4m）0.427m 时，闸门全开，全开时闸门与堰面夹角为 15°，此时闸门顶高程为 200.002m，闸门底高程为 199.484m。

同理分别用式（1）和式（2）计算溢洪道堰流的泄流量 Q_1 和孔流的泄流量 Q_2。式（1）中，过流净宽 $B=30$m，上游堰高 $a=1.602$m；式（2）中，过流净宽 $B=30$m，闸门开启高度 $e=1.084$m，流速系数取 0.95，由此算出库水位与溢洪道泄流量成果见表 2。

表 2　楼梯磴水库库水位—溢洪道泄流量关系计算

库水位/m	200.40	200.84	201.40	201.90	202.40	202.78	203.17	203.50	204.00
泄流量 $Q_1/(m^3 \cdot s^{-1})$	0	40	91	149	218	278	344	406	507
泄流量 $Q_2/(m^3 \cdot s^{-1})$	0	115	133	147	160	170	179	187	198
总泄流量 $Q/(m^3 \cdot s^{-1})$	0	155	224	296	378	448	523	592	705

3.3　计算成果分析

为分析本次泄流量计算成果的正确性，特将其与水力自动翻板闸门生产厂家通过水工模型试验所提供的库水位与溢流坝泄流量关系进行比较。生产厂家水工模型试验的库水位与泄流量关系见表 3 和表 4，计算成果与模型试验成果的差别见图 2、图 3 和表 5。

表 3　厂家水工模型试验提供的螺滩水库库水位—溢流坝泄流量关系

库水位/m	72.5	72.73	73.09	73.5	74.1	74.7	75.1	75.6	76.1	76.5
总泄流量 $Q/(m^3 \cdot s^{-1})$	0	95.2	959.8	1371	1793	2280	2592	3035	3514	3600

表 4　厂家水工模型试验提供的楼梯磴水库库水位—溢流坝泄流量关系

库水位/m	200.40	200.83	201.40	201.90	202.40	202.78	203.17	203.50	204.00
总泄流量 $Q/(m^3 \cdot s^{-1})$	0	145	218	293	376	440	506	590	710

表 5　　　　　　　　泄流量计算成果差别比较

水库	项目										
螺滩水库	库水位/m	72.50	72.73	73.09	73.50	74.10	74.70	75.10	75.60	76.10	76.50
	厂家提供泄流量/$(m^3 \cdot s^{-1})$	0	95.2	959.8	1371	1793	2280	2592	3035	3514	3600
	本次计算泄流量/$(m^3 \cdot s^{-1})$	0	97	1075	1309	1702	2154	2488	2940	3379	3748
	两者泄流量差占厂家提供泄流量的百分比	0	2%	12%	−5%	−5%	−6%	−4%	−3%	−4%	4%
楼梯磴水库	库水位/m	200.40	200.83	201.40	201.90	202.40	202.78	203.17	203.50	204.00	
	厂家提供泄流量/$(m^3 \cdot s^{-1})$	0	145	218	293	376	440	506	590	710	
	本次计算泄流量/$(m^3 \cdot s^{-1})$	0	155	224	296	378	448	523	592	705	
	两者泄量差占厂家提供泄流量的百分比	0	7%	3%	1%	1%	2%	3%	0	−1%	

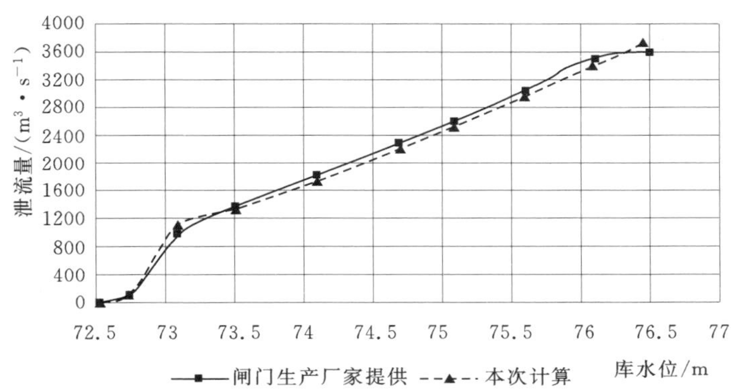

图 2　螺滩水库库水位与溢流坝泄流量曲线

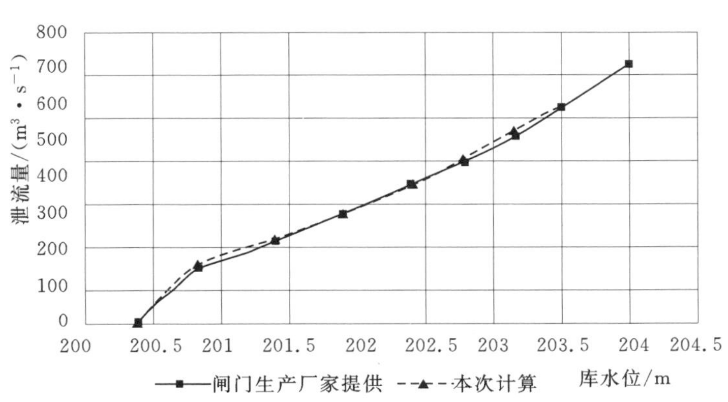

图 3　楼梯磴水库库水位与溢洪道泄流量曲线

从图 2、图 3 和表 5 可以看出，螺滩水库泄流量中除库水位 73.09m 时两者泄流量成果相差 12%外，其余泄流量误差均在 5%左右；楼梯磴水库中，两者泄流量误差均在 5%左右。笔者对江西省内其他一些类似工程进行计算分析比较，也得出同样的结论，由此可以看出，在计算水力自动翻板闸门泄流量时，将泄流量分成堰流和孔流两部分进行组合计算的结果可以满足精度要求。

4 结论

通过江西省螺滩水库和楼梯磴水库等水库的溢流坝（溢洪道）泄流量计算结果与水力自动闸门生产厂家水工模型试验提供的泄流量曲线进行比较，两者的泄流量曲线相差不大。因此，对采用水力自动翻板闸门的中小型水库溢洪道，在没有条件进行水工模型试验时，将通过翻板闸门的水流泄流量分成堰流和孔流两部分进行计算的方法能够满足一般精度要求，是一个可行的方法。但对于大型或重要工程，出于对工程安全性的考虑，建议应进行水工模型试验进行复核论证比较。

参考文献：

[1] 华东水利学院．水力学（第二版下册）[M]．北京：科学出版社，1987.

[2] 包中进，张锦娟．水力自动控制闸门的水力特性试验研究 [J]．浙江水利科技，2002，(5)．

七一水库心墙土坝变形应力研究

喻蔚然[1,2]，钱向东[1]，苏立群[2]
1. 河海大学土木工程学院；2. 江西省水利科学研究院

摘　要： 土坝应力—应变关系具有明显的非线性特征，本文采用了非线性有限元法对七一水库土坝进行变形和应力分析，全面了解坝体自施工以来的变形和应力变化规律。计算结果符合现状，并由此采取相应坝体加固措施，改善坝体的应力状态，可以解决坝体变形问题，达到除险加固的目的。

关键词： 心墙土坝；非线性有限元；变形；应力

1　概况

七一水库位于江西省玉山县信江支流金沙溪中游棠梨山西侧，1958 年 7 月 1 日开工，1960 年建成并投入运行。坝址以上控制流域面积 324.0km^2，总库容 2.29×10^{8}m^3。水库灌溉面积 0.933 万 hm^2，电站装机 9250kW，是一座以灌溉为主，兼有防洪、发电、养殖等综合效益的大（2）型水库。水库枢纽工程等级为二等，主要永久性建筑物为 2 级。正常蓄水位 94.0m（假设高程，下同），设计洪水位（P=1%）96.1m，校核洪水位（P=0.5%）96.61m。枢纽工程主要建筑物包括主坝、副坝、溢洪道、灌溉发电引水隧洞及电站等。

主坝为大体积心墙坝，坝顶高程 101.0m，最大坝高 56.68m，坝顶长 430.0m，坝顶宽 12.0m。上游坝坡坡比自下而上分别为 1：3.0、1：4.0、1：3.75、1：2.78，下游坝坡坡比自下而上分别为 1：3.0、1：2.75、1：2.75、1：2.75。上游坝脚设有两层黏土铺盖，总厚 6.3～9.0m，下游坝脚排水棱体顶高程 54.0m（见图 1）。

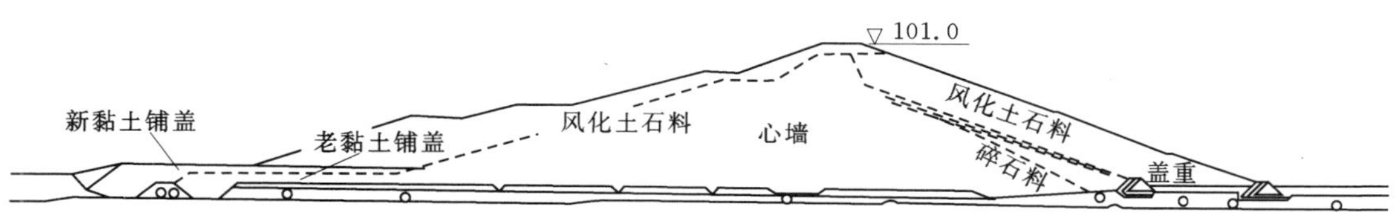

图 1　大坝典型断面分区图

主坝运行过程中在原施工导流沟段产生平行坝轴线的小裂缝，老河床段、坝右端靠近坝肩处产生垂直于坝轴线的裂缝，后仅做简单处理。目前在下游坝坡挡土墙顶面及坡面上

本文发表于 2005 年。

有数条裂缝产生并伸入坝体，缝宽最大达到1.2cm，并有逐年发展的趋势，故有必要进行变形分析。

2 计算方法与计算模型

2.1 本构模型及材料参数

土体本构模型主要有邓肯-张模型、剑桥模型和椭圆抛物双屈服面模型等[1]。邓肯-张模型是一种弹性非线性模型，它把土体的塑性变形当作弹性变形处理，且采用广义虎克定律不能反映土体剪缩性和应力引起的各向异性。剑桥模型是一种弹塑性模型，一般用于正常固结或弱固结黏土。本文计算中采用的是南京水科院沈珠江院士提出的双屈服面弹塑性模型[2]、[3]，该模型结合了剑桥模型和邓肯-张模型的优点，能较为全面地反映土料的应力变形特性，且计算参数的确定简单，应用方便。模型中的2个屈服面为：

$$\left.\begin{aligned} f_1-(p^2+4q^2)=0 \\ f_2-q^2/p=0 \end{aligned}\right\} \tag{1}$$

其中 $p=\frac{1}{3}(\sigma_1+\sigma_2+\sigma_3),q=\frac{1}{\sqrt{2}}[(\sigma_1-\sigma_2)^2+(\sigma_2-\sigma_3)^2+(\sigma_3-\sigma_1)^2]^{1/2}$。

模型的基本参数为切线杨氏模量 E_t 及切线体积比 μ_t，分别按下式计算

$$E_t=E_i(1-R_fS_l)^2 \tag{2}$$

$$\mu_t=1-2\nu_t \tag{3}$$

其中

$$E_i=KP_a\left(\frac{\sigma_3}{P_a}\right)^n \tag{4}$$

$$S_l=\frac{(\sigma_1-\sigma_3)(1-\sin\phi)}{2c\cos\phi+2\sigma_3\sin\phi} \tag{5}$$

$$\nu_t=\frac{G-F\lg(\sigma_3/P_a)}{[1-D(\sigma_1-\sigma_3)/(1-R_fS_l)/E_i]^2} \tag{6}$$

式中：E_i 为初始切线模量，是曲线 $(\sigma_2-\sigma_3)$—ε_a 的初始切线斜率；K、n 为 E_i 与周围压力 σ_3 关系式中的无因次基数、无因次指数；c、ϕ 为土料的黏聚力和内摩擦角，取自地勘土工试验成果；R_f 为破坏比，$R_f=\frac{(\sigma_1-\sigma_3)_f}{(\sigma_1-\sigma_3)u}$；$G$、$F$ 为初始泊松比 ν_t 与周围压力 σ_3 关系式中的两个无因次系数，$\nu_t=G-F\lg\left(\frac{\sigma_3}{P_a}\right)$；$D$ 为曲线 $-\frac{\bar{\varepsilon_r}}{\varepsilon_a}$——$\varepsilon_r$ 的斜率。

本文采用计算参数见表1。

表1　坝体土料计算参数表

坝料	K	n	R_f	C/kPa	ϕ_0	G	F	D	$K_{渗}$
风化土石料	500	0.57	0.72	10	21.7	0.29	0.098	7.26	1×10^{-3}
心墙土	400	0.55	0.76	21.5	17.0	0.26	0.072	3.00	3.6×10^{-5}
新黏土铺盖	350	0.52	0.78	25	22.0	0.24	0.065	2.60	1×10^{-6}

续表

坝料	K	n	R_f	C/kPa	ϕ_0	G	F	D	$K_渗$
老黏土铺盖	300	0.50	0.80	26	22.0	0.22	0.060	2.44	1.5×10^{-7}
碎石料	550	0.58	0.70	10	30.0	0.43	0.190	14.8	1×10^{-2}
棱体盖重	600	0.46	0.80	0	38.0	0.33	0.0	0.00	1×10^{-2}

2.2 计算方法[1]

土体的应力应变关系具有明显的非线性特性，有限单元法是开展土石坝非线性分析的最有效方法，既能模拟材料的非线性应力应变关系，又可以方便地考虑各种荷载及分期施工过程。由于土体弹塑性本构关系都是用增量形式来表示，所以计算方法也宜用增量法。增量法是将全荷载分为若干级微小增量，逐级用有限元法进行计算。对于每一级增量，计算时假定材料性质不变，作线性有限元计算，求得位移、应力和应变的增量，而各级荷载增量之间则认为材料性质可以发生变化。常用的增量法有基本增量法、中点增量法和迭代增量法。

用增量法进行弹塑性计算的步骤：

（1）假定材料为弹性材料，用弹性矩阵［D］形成［K］，求位移增量，进而计算弹性应力增量。

（2）用屈服准则判别应力的变化，以便计算时区别使用［D］还是［D_{ep}］。

（3）对于处于弹性区的单元，不作修正；对于处于塑性变形阶段的单元，修正应力增量，逐级进行计算。

2.3 计算网格

取大坝的0＋193断面作为研究对象，采用平面应变模型进行分析。将坝体断面分为3个子结构，上游坝壳和黏土铺盖作为第一子结构，下游坝壳和棱体盖重为第二子结构，中间心墙为第三子结构。由于第一子结构和第三子结构以及第二子结构和第三子结构之间土的性质差别不是很大，在接触面发生开裂或错动滑移的可能性不大，故未考虑在其间设置接触面单元。计算采用任意四边形等参数单元，对于边界上的三角形单元，可转换为四边形单元计算（见图2）。

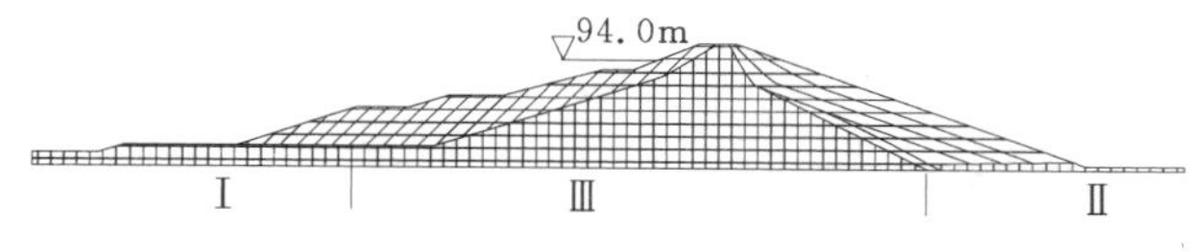

图2　坝体有限元剖分网格图

2.4 荷载及计算步骤

计算荷载主要为坝体自重、上游静水压力。孔隙水压力按《碾压式土石坝设计规范》（SL 274—2001）的要求以距离浸润线的铅直水头计。计算采用增量法，逐级施加荷载。计算过程共分为33级，其中主坝的施工过程分16级，共历时18个月。主坝竣工后一年为蓄水期，模拟分6次蓄水至正常蓄水位94.0m。而后再考虑10年的运行时间作为主坝沉降的稳定运行期，即认为10年后主坝沉降过程将趋于收敛。

3 计算成果及分析

计算成果包括施工期、蓄水期和运行期的水平、垂直位移、大小主应力和应力水平。计算成果见表2，运行期计算结果图如图3所示。

表2 有限元计算成果表

时　期	大主应力最大值/kPa	小主应力最大值/kPa	应力水平最大值	水平位移最大值/mm	沉降最大值/mm
施工期	989.3	460.8	0.95	81	434
蓄水期	991.4	461.1	0.95	86	433
运行期	991.2	461.1	0.94	86	421

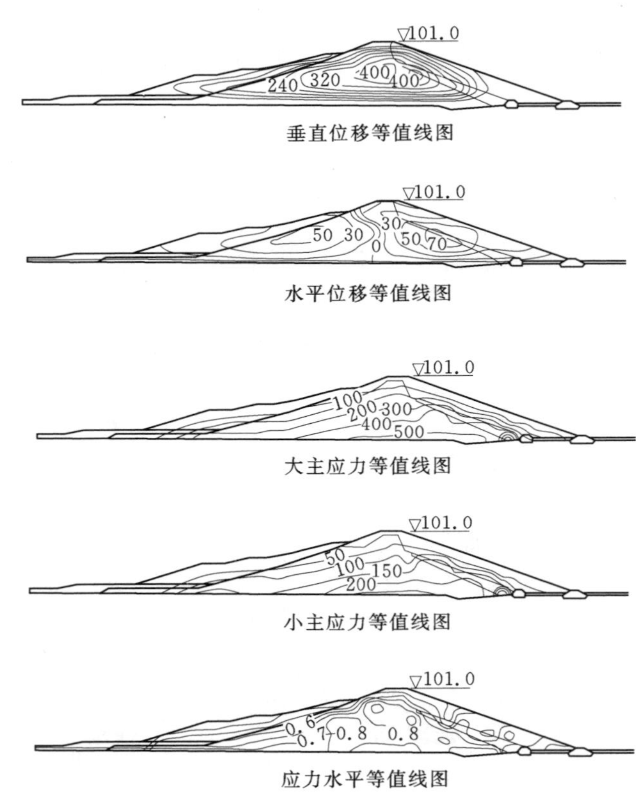

图3　稳定运行期有限元计算成果图

通过计算可以看出，七一水库大坝位移和应力分布存在下述规律。

3.1 位移分布

沉降是土坝最主要的变形指标之一，但主要的沉降过程是在施工期和蓄水期，根据其他类似工程经验，土坝沉降变形一般在竣工后的5～10年基本稳定。比较三个时期的沉降，前两个时期沉降较多，稳定期的沉降仍然有40cm，说明坝体沉降尚未达到收敛状态，但速率明显降低。三个时期的位移最大值也差别不大，累计最大沉降可达到1.3m。施工期和蓄水期仅30个月的时间，运行期有10年的时间，而前两期累计沉降超过0.8m，占总沉降量的70%，说明坝体的沉降是前快后慢，符合一般土石坝的变形规律。比较3个时期的沉降分布，可以发现沉降区变化不大，形状相似，说明坝体沉降比较均匀。分析整个沉降区，发现沉降值最大处位于半坝高处，随着向两边边界靠近，沉降值也逐步减小，形成一种类似水波由中心向外展开的形状。

水平位移分布大致处于对称的形式，但轴线偏上游侧，两侧各向上下游移动。由于受库水位的压力影响，上游坝体较下游坝体位移少，总体上位移值不大，估算累计最大水平位移值有25cm。

3.2 应力分布

由表2可以看出，主坝竣工时与10年运行期后的坝体应力最大值差不多，应力水平相当。3个时期的应力分布没有显著的变化，形状类似。不论在自重作用下或蓄水状态下，应力状态良好，坝体无拉应力区存在，大部分坝体应力在500kPa以下，但是在坝体下部碎石料区以及与心墙结合部位出现了应力集中现象，应力值从500kPa迅速升至991kPa，但并未发生破坏。其原因在于碎石料前面心墙刚度低，受库水压力影响向下游位移较多，而碎石料后面的盖重和棱体刚度较大，位移则较少，因此碎石料受两者挤压作用，故在下部碎石料区产生应力集中。所以有必要对该部分应力进行调整，可采用灌浆填充、挤压等方法改善应力分布。

整体上看，主坝应力水平较高，普遍在0.50以上，局部最大值达到0.95，说明主坝土体的强度发挥程度较高，但尚未达到破坏的程度。

由于坝体内部未埋设应变计，所以无法直观了解实际应力状况。但从坝体表面来看，在下游坝坡挡土墙处出现数条裂缝，缝宽最大达1cm，一直延伸至坝体内部。分析认为，坝体内部碎石料区应力水平较高且有应力集中，可能局部已产生应力破坏，致使坝体产生不均匀沉降，形成裂缝。

4 结语

本文采用“南水”模型对七一水库大坝进行非线性有限元应力应变计算和分析，了解了大坝不同时期的位移分布以及应力状态，分析了挡土墙裂缝成因。计算成果正确可靠，与现状基本吻合。从总体上看，坝体不存在严重变形问题，但局部应力较高，变形较大，因此有必要采取相应的方式进行加固处理，以确保大坝安全运行。

参考文献：

[1] 钱家欢，殷宗泽．土工原理与计算[M]．2版．北京：中国水利水电出版社，2003.
[2] 朱百里，沈珠江．计算土力学［M]．上海：上海科技出版社，1990.
[3] 孙振远．堆石料工程特性及面板堆石坝应力变形研究［R]．南京水利科学研究院，2003，6.

大坝安全鉴定及除险加固设计存在的问题和原因分析

吴晓彬[1]，李海辉[2]，章伟平[1]

1. 江西省水利科学研究院；2. 江西省水利厅

摘　要： 通过参与水库大坝安全鉴定及水库除险加固初步设计的技术审查工作，发现存在一些问题，并对产生问题的原因进行了分析。

关键词： 大坝；安全鉴定；设计；问题；原因

据统计，江西省有约3500座病险水库。近年来，江西省委、省政府加大了对病险水库的除险加固力度。作为江西省水库建设管理、大坝安全管理及安全鉴定技术审定和核查单位，参与了水库大坝安全鉴定和初步设计的技术审查工作。工作中发现当前水库大坝安全鉴定和初步设计中存在一些问题，直接影响了水库大坝安全鉴定和工程除险加固的质量。本文就当前水库大坝安全鉴定及除险加固工程初步设计中存在的问题和产生的原因谈一些粗浅的看法。

1　存在的问题

1.1　前期勘测工作不充分

1.1.1　坐标系统混乱

有的水库坐标系统不统一，未换算成黄海高程系统，给流域特征参数及库容曲线的量算、复核带来困难。有的工程溢洪道泄洪渠无坐标系统，有的涵管进水口无高程系统。

1.1.2　勘察工作量不足

勘察任务书不明确或不全面，未针对工程的病险情况及下一步除险加固的可能实施方案来布置勘察工作。某水库的除险加固比较方案未布置一个钻孔；有些工程在勘探期间不进行或很少进行原位压（注）水、标贯等试验；勘探断面及钻孔取样组数不满足规范要求，致使无法对岩土的物理力学指标进行有效、规范的统计分析，甚至导致成果失真。某水库除险加固新建隧洞，原钻孔反映围岩地质条件“很好”，结果开挖后，全长约200m的隧洞坍塌了150m，由不支护改为木支护，再改为钢支护，增加了工程费用，拖延了工期。

1.1.3　对工程原始资料的收集不重视、不认真，寻访力度不够

有些工程在安全鉴定及初步设计中无枢纽总平面图和主要建筑物平、剖面图，无建筑

本文发表于2005年。

物结构配筋图，或尺寸不详、标注不规范，有的连涵管的结构型式也出现偏差。某水库工程由于大坝坡度的不一致，造成安全鉴定与初步设计的结论迥然不同。

1.1.4 引用前人勘测设计成果没有复核和验证

某水库大坝初步设计报告称“本阶段共钻孔46个”，实际钻孔仅8个，其中还有2个在副坝上，对前人的勘探成果未进行补勘验证，施工中揭示的地层土质、砂砾卵石粒径与原资料完全不符，造成施工埋钻、卡钻，直接影响到施工工艺选型的正确性，降低了工效，延误了工期。有的工程水库水位—容积曲线呈等值增长（水桶形）或负增长（葫芦形），明显不合理，但报告称经过“验证”，仍用于调洪演算。

1.2 设计工作不细致

在工程除险加固初步设计中，不少工程项目没有进行严格的方案比较与论证，稍好一点的也只进行定性比较，致使方案论证不充分，加之审查把关不严，造成工程实施后无法施工而导致方案改变。某水库原设计为锯槽造混凝土墙防渗方案，因坝基卵砾石直径太大，超出了该工艺的适用范围而无法施工，后改为射水造墙工法。某水库坝下涵管原设计采用挂网喷浆，后因管径偏小，无法施工，改为钢衬套管加固，大幅度增加了工程投资，延长了工期。

设计人员未严格按有关规范进行设计，某些设计甚至连《工程建设标准强制性条文》也不执行。某中型水库，大坝为3级建筑物，其坝体土方填筑按强制性条文要求，压实度应为0.96，设计值却小于0.96；某水库压力钢管（内衬钢管）设计不进行抗外压稳定复核，导致失事；大部分坝下钢筋砼管只进行结构强度计算，未进行抗裂安全验算；压力管道的深式进水口未按规范要求设置通气孔；对受高速水流作用的坝下输水涵管、溢洪道陡槽、消力池的混凝土强度等级设计值小于C25（SL/T 191—1996《水工混凝土结构设计规范》规定：对有抗冲、耐磨要求的混凝土强度等级不低于C25）；金属结构的拦污栅扁钢平放；溢洪道的水力计算都没有依据工程布置条件、不进行流态判别，对流态相对复杂的也都没有进行物理模型试验，按堰流公式计算，造成泄量偏大，调洪水位偏低；对设计中的一些关键参数随意取值，甚至使用已废止的设计规范。

设计校审、签署不全，把关不严。一人包揽设计项目的校核、审查；一些明显的错误从设计开始，校核、审查、审定后依然存在，说明把关人员责任心不强，工作不细致。

2 产生问题的原因

（1）当前水库大坝的安全鉴定和除险加固设计工作任务繁重，不少工程主管部门和单位在进行任务委托时，不给予安全鉴定或设计承担单位合理的工作周期，加之安全鉴定和设计承担单位设计强度有限，同时，工程前期工作费用不足，经费偏少，某些设计单位压价竞争，合同价格较低，于是就尽量减少工作量，特别是地质勘测及试验量明显不足，影响了安全鉴定和设计质量。

（2）水利部颁发的《水库大坝安全鉴定办法》规定：大型水库和影响县城安全或坝高50m以上的中型水库的大坝安全评价，由具有水利水电勘测设计甲级资质的单位或者水利部公布的有关科研单位和大专院校承担。有的水库管理单位委托不具备相应资质的单位进行安全鉴定和除险加固设计工作。

（3）有的勘测设计单位的技术人员和主要技术负责人缺乏质量意识和责任心，未把“安全”放在最重要位置，只抓“进度”和“效益”，这是产生质量问题的根本所在。其实没有好的质量就没有长期的效益，也不能很好地锻炼技术人员，将会失去应有的市场，甚至可能导致质量事故和重大经济损失。

（4）水库安全鉴定与除险加固设计的审查由于任务重、时间紧，尽管有不少经验丰富的专家参与审查，仍有可能产生疏漏的地方。

（5）工程质量问题虽大多出现在施工图阶段，实际上初步设计阶段就已经存在一些问题。由于目前江西省水利行业的建管工作尚未能对设计质量实施有效的监控，地市的技术力量又难以对设计单位的施工图进行有效的审查，这也是造成工程质量问题的一个重要原因。

九江长江大堤实体护岸岸脚淘刷机理研究及防治对策

邵仁建，胡松涛，熊焕淮，王南海

江西省水利科学研究所

摘　要：本文对长江大堤常用的实体护岸及固脚技术进行了原型效果观测与分析，综述了前人研究的成果。针对实体护岸技术进行了概化模型试验，研究了岸脚淘刷机理，提出了护岸固脚设计的原则意见，通过比较、筛选，推荐了较佳的护岸固脚型式，为今后护岸工程提供了依据。

关键词：护岸；固脚；岸脚淘刷；试验研究

1　概述

江西省长江干流河段位于长江中下游结合部，江岸线全长 151.9km，江堤线全长 216km，其中长江干堤 123.89km。江岸总体走势为向南凹进的弧形，河道洲滩发育，受长江水流冲刷，河势在长期演变过程中，南冲北淤，主流不断南移（深泓距岸仅 70m），淘刷岸线和堤脚，因而造成崩岸频繁，对沿江堤岸极为不利。

长江中下游护岸工程按其平面形式，可分为平顺护岸、矶头群护岸和丁坝护岸三大类型。按护岸机理可分为实体抗冲护岸和减速不冲护岸。1998 年长江大水后，在长江江岸干堤加固整治工程中，九江长江大堤护岸工程除采用抛石护岸固脚外还采用了多种新技术、新材料。如采用了干砌石护岸、四面六边透水框架群护岸固脚、模袋混凝土护岸、软体沉排护岸、铰链排护岸、金属网兜装块石固脚等技术。

本文通过对不同实体护岸、不同固脚技术的原型效果观测与分析，在前人对岸脚淘刷研究的基础上，结合概化模型试验，对岸脚冲刷机理进行研究。

2　护岸固脚工程成果综述及原观资料分析

平顺护岸工程实施后，包括抛石和柴排，基本上没有改变近岸水流的流速场，因而纵横向输沙条件起初并没有改变。由于岸坡受到保护，横向变形得到控制，水流只可能从坡脚外未护河床上取得泥沙补给，于是沿护岸坡脚外河槽受到普遍刷深。当弯道曲率不大时，断面形态调整也不大；当弯道曲率较大时，崩岸剧烈，则护岸坡脚外冲刷较严重。同时因对岸边滩继续淤积，断面流速分布会有所调整，结果使得断面形态将变得比较窄深。

本文发表于 2003 年。

平顺护岸不存在由局部水流结构产生的局部冲刷坑。但护岸外沿岸脚部位会产生河槽下切，深泓内移现象，柴排工程有可能因此导致破坏。

护岸工程是在水流作用于近岸河床产生变形的情况下，通过自身的调整或加固而逐步达到稳定的。试验表明，平顺抛石护岸施工后，抛护在斜坡上的块石将不断调整，由不密实不均匀变得比较密实均匀。当抛石护脚以外河床继续冲刷时，块石失去稳定向深槽滑动，重新覆盖被冲刷的坡脚。当块石量不足时，就导致整个护岸下挫或护坡挫裂。块石在深槽冲深时下滑因坡度逐渐变缓和摩擦力的增大而停止，下滑中自然形成的坡度不缓于 1：2。同时，绝大部分块石下滑偏角不超过 10°，说明水流对块石的纵向推力与块石下滑分力相比是很小的。因此，块石在护岸滑挫后基本上仍在原断面附近而未流失，历年护岸工程数量可以累加。

护岸工程的兴建改变了冲积河流的边界条件，势必对河床演变产生一定的影响。护岸后崩岸受到控制，近岸河床向刷深发展，然而对岸边滩仍处于淤积状态并继续缓慢地淤高和增宽。当近岸河床逐渐达到最大冲刷深度，对岸边滩枯水位以下部分逐渐停止延伸并转为冲刷，护岸工程即趋于相对稳定，这说明护岸工程的稳定需要一个长期加固的过程。而边滩上部滩面仍继续缓慢地淤涨，直到使断面逐步达到冲淤平衡，这种变化结果使护岸段宽深比变小，长江中下游护岸河段河道断面形态都有不同程度朝窄深发展的现象。

通过大量实测资料分析可得，抛石护岸工程（设有备冲石）大部分因坡脚冲刷，引起底部泥沙大量走失，坡脚坍塌、块石滚动、致使边坡下沉，钢筋混凝土铰链排排脚无加固措施，在排脚处发生了较大冲刷，混凝土模袋由于采取了较好的固脚措施，具有一定的减速作用和适应地形变形能力，产生了较好的效果。四面六边透水框架群自身稳定性较好，减速促淤作用显著，固脚效果好。

几种护固脚措施断面图见图 1～图 4。

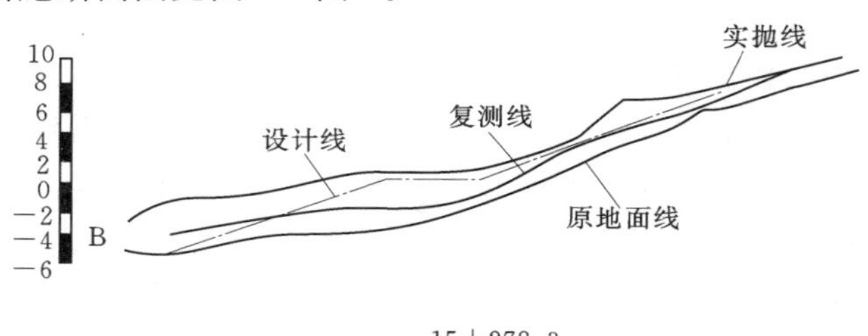

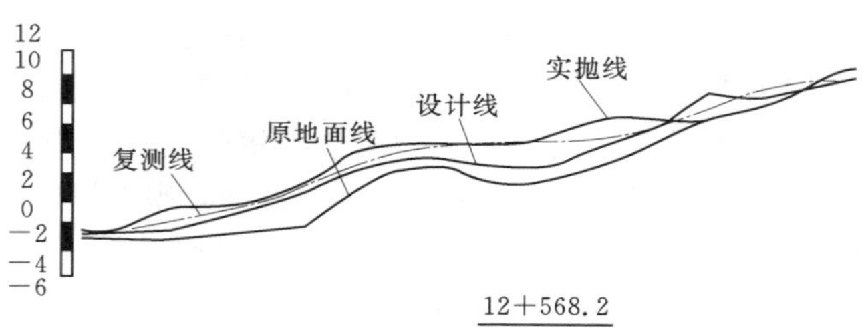

图 1　永安堤抛石断面图

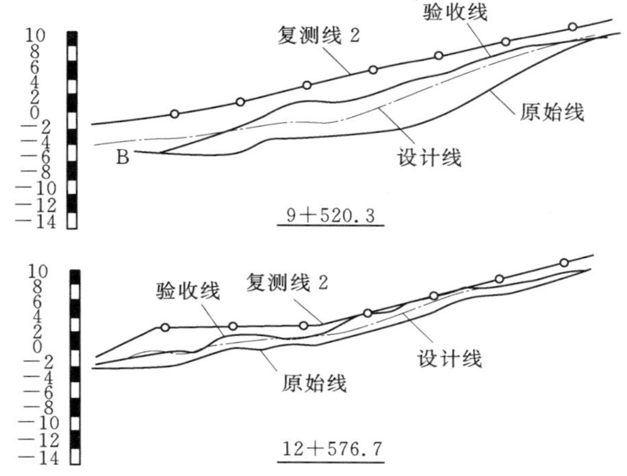

图 2　永安堤抛石护岸岸脚冲刷后用四面体框架群固脚断面图

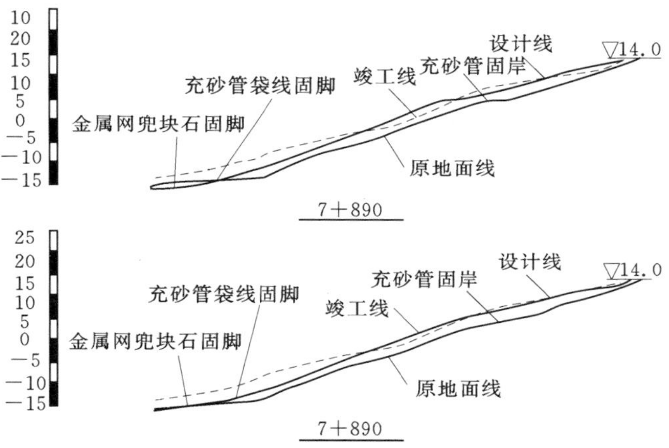

图 3　王以珍堤混凝土模袋断面图

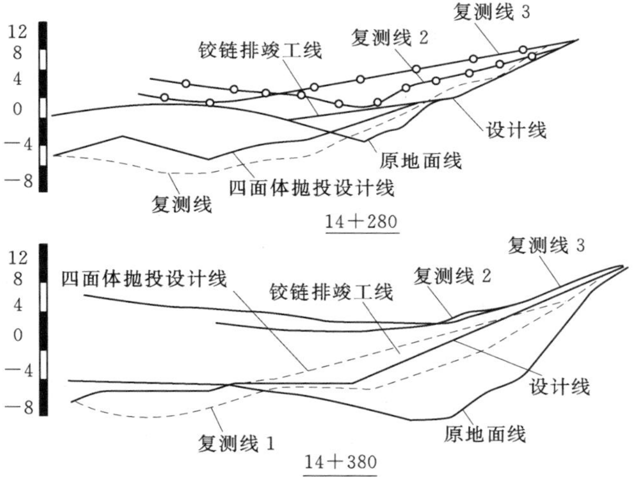

图 4　江新洲大堤铰链沉排断面图

针对平顺护岸岸脚冲刷，深泓逼近的问题。经过广泛的探索和较深入的研究，现主要采用固脚措施有防冲备填石、粗化河床（小颗粒石料固脚）、扩张金属网块石网兜、石笼、网罩、混凝土四脚锥、四面六边透水框架群减速促淤固脚等。

3 实体护岸技术岸脚淘刷机理研究

在江西省水利科研所与河海大学合作的模型试验中，选择3种常用的实体护岸型式，即混凝土模袋、钢筋混凝土铰链排及抛石，模拟长江九江段典型断面河段。根据长江水位、流速、床沙资料等进行实体护岸技术岸脚淘刷的动床概化模型试验。将其概化为刚性材料、柔性材料和散粒体3种护岸形式。研究内容为观测岸脚水流微观流态；泥沙起动流失过程；对岸脚部位的脉动流速压力进行观测；对岸脚冲刷机理进行分析。

刚性护岸（混凝土模袋或浆砌石）岸脚极易淘空，与其他形式护岸比较，在相同工况下岸脚淘刷的范围最大最深，不利于岸脚的保护，必须采取适当的保护基脚的措施才能有效。

柔性材料（钢筋混凝土铰链排）护岸，在洪水初期，对岸坡岸脚均有较强的保护作用，但随着洪水时间延长，岸脚处出现排块翻转和排体搭接处分离，水流从局部损坏处钻入排下形成局部冲刷并逐步扩展，加剧岸脚淘刷，也必须对基脚进行有效的保护。

抛石护岸在保证一定厚度，同时岸脚部具有足够的备冲石前提下，对岸脚、岸坡有较强的保护能力，有很好地适应地形的能力。但岸脚泥沙淘刷冲走，块石流失是一个长期的过程，必须不断地补充备冲石。

通过对护岸工程岸脚区微观流态的测试，较深入地揭示了实体护岸工程岸脚淘刷的机理。实体护岸工程岸脚的淘刷，是因为护岸工程实施后，由于边界条件的骤变，在护岸工程边缘其结合部的水流流态发生变化，水流流速脉动幅值加大，其脉动流速变幅最大点在距岸脚（0.10～0.15）H左右处，在顺水流方向和垂直水流方向的脉动幅值均较大，可达主流时均流速1.8～1.9倍左右。由于实体护岸外缘的不整齐，水流绕流将引起附近局部流速加大，时均值可增加10%，并伴随有螺旋流、回流等不稳定流态。流速时均值和脉动值的加大，加强了床沙的起动和流失，水流淘刷作用加强，形成冲坑，导致块石向外翻滚和护岸坡脚淘空。

较强的脉动流速使水流起动和输沙能力加强，增加岸脚部位的冲刷，还可能引起水体局部压力变化，引发一系列不同尺度的涡漩和回流。其产生、发展、扩大、汇合、破碎的水流流态，可能引发一系列类似猝发的水流现象，也将加剧岸脚部的床沙的起动和淘刷。因此，实体护岸岸脚部位可能出现较大的冲深和淘刷。值得注意的是这种冲刷在较小流速下也可发生，而且过程是动态的、长期的。

在以往的工程设计中，考虑了块石本身的斜坡稳定和抗冲能力，但计算冲刷深度时采用的是水流的平均流速，而冲刷的原因是增大的脉动流速的作用，使基底的泥沙起动和输移。所以往往按规范设计施工的工程仍然存在岸脚淘刷、工程失稳的问题。本次实体护岸工程岸脚冲刷机理研究中发现，形成岸脚河床淘刷的主要原因在于该部位脉动水流加强，顺水流方向和垂直水流方向可达均值的2倍，在《堤防工程设计规范》（GB 50286）中对平顺坝及平顺护岸冲刷深度计算（水流平行于岸坡）：

$$h_s=h_p+\left[\left(\frac{V_{cp}}{V_{允}}\right)^n-1\right]$$

式中：h_s 为局部冲刷深度，m，从水面算起；h_p 为冲刷处的水深，m，以近似设计水位最大深度代替；V_{cp} 为平均流速，m/s；$V_{允}$ 为河床面上允许不冲流速，m/s；n 为与防护岸坡在平面上的形状有关的系数，一般取 $n=0.25$。

上式中 h_s 采用的是断面平均流速，因此，在采用防冲备填石固脚时，应对计算冲刷深度的公式进行修正。

4 对策与建议

4.1 江岸整治工程是长期的动态工程

江岸的稳定不仅与局部江岸的防护工程有关，而且与水文条件，河势演变，人类活动等关系极为密切，而防护工程本身，只是防止河床侧向侵蚀。

长江上平顺式实体护岸工程，80%以上是抛石护岸。大量的工程实践和研究成果表明，护岸工程岸脚边都会出现不断冲深，脚坡不断变陡、深泓高程单向下切的现象，必须有足够的防冲备填石料，不断地加固岸脚，使护岸段坡脚形成一个块石防冲层，直至本岸岸脚地形不变化，而对岸边滩下部河床发生冲刷时，护岸工程方趋于稳定。这充分说明护岸工程的稳定需要一个长期加固的过程。护岸工程的外缘线是一个动态变化的过程线，在新护岸工程实施后，不是一劳永逸，可以置之不管了，还必须坚持每年汛后进行一次原型观测，进行必要的维修与防护，对护脚工程冲刷变异较大处，及时给予块石补充，使流失的块石得以补足，护岸工程不致在最终稳定前遭到破坏，而造成损失或灾害。

单纯地增强河床边界的抗冲能力，只是护岸工程初步稳定与加固的基础，要使河工长期处于相对稳定，其根本途径是不失时机地进行河道整治，消除节点与卡口的不利水流条件，改造不恰当的护岸方式，改善工程附近局部冲刷的水流条件等；控制河势，改善护岸河段维修工程与河道整治工程相结合，经过长期的、全河道上下游与左右岸统一规划和治理方能使岸线稳定除险。

4.2 护岸与固脚措施必须相结合

江岸保护工程由护岸与固脚两部分组成，护岸工程的稳定，取决于护岸材料的抗冲性及抗滑稳定边坡；固脚工程在于保护岸脚床沙不致淘空、流失，从而保护护岸工程的基础。清楚地认识到这两点，在设计与使用中，就会有的放矢，既节约了工程量，又保证工程的稳定性。经九江长江大堤江岸的整治工程实践，认为抛石护岸、铰链排护岸等（即散粒料护岸和柔性材料护岸）与四面六边透水框架群固脚相结合为最佳组合，既易实施，又较经济，护岸固脚效果显著，值得推荐。

4.3 几种效果较好的固脚技术

没有稳定的基脚，就没有稳定的护岸工程，无论哪种新技术护岸都必须考虑固脚问题。在上述的多项固脚措施中，通过实验研究与工程效果分析认为，下列几项技术比较实用、效果好、使用便捷。

（1）防冲备填石：这是一项较成熟的技术，只要备填石足够，不断地补充岸脚，护岸工程是可以相对稳定的，备填石的量取决于冲刷深度的估算。而本次研究发现，冲刷深度公式应加以修正，在进行防冲备填石设计时进行修正或运行时及时观测补充备填石料，都是可行的。

（2）小颗粒石料抛护岸脚河床：其护脚机理与块石不同，小颗粒石料铺护河床后，起到粗化河床的作用，保护了床沙，使床沙起动流速大大提高。因此，可以保护岸脚不致冲刷。

（3）钢丝网石笼类护脚：从工程实例来看，钢筋网石笼护脚工程，当护脚工程外沿被淘深后，整个钢丝网石笼可随之下沉护脚，石块不会流失，上部工程不致立即坍塌出险，经观测后可及时采取修复措施。

（4）减速促淤类：（含四面六边透水框架技术、挂柳、梢料等）透水框架固脚原理与上述措施不同，它改变岸脚的局部流态，大大降低岸脚的流速值，可降低流速40%～70%，有时框架群内近乎静水。因此，岸脚的冲刷变成了淤积，框架群外侧淤积泥在汛期可能会产生一些冲刷，但汛期后又会淤回，且框架群适应地形变形强，依然起着固脚作用。

4.4 建议堤岸工程管理部门对堤岸工程定期观测、长期维护

九江长江堤岸整治施工过程中，在抛石护岸工程外沿被冲刷的深泓处，又大都采用了四面六边透水框架群固脚，这是很及时有效的措施。从原观资料分析可见，岸线是稳定的。由于影响工程稳定的因素非常复杂，因此，建成之后如何管理好江堤江岸，这是更为艰巨的任务。要建立健全管理机构，建立管理规章制度，坚持原型观测分析，实时评估堤防的安全度等。对于江岸工程来说，至少每年汛后进行一次地形观测，与竣工工程进行对比分析，进行安全稳定分析，及时提出维修方案，及时维护，才能保证工程长期稳定运行。

湖口县双钟圩堤滑塌事故成因分析

张文捷[1]，熊冲玮[1]，吴晓彬[2]，周国栋[2]

1. 江西省水利科学研究所；2. 水利部稽查办

摘　要：本文简要介绍了湖口县双钟圩工程概况及滑塌基本情况，结合工程实际分析了本次圩堤滑塌的原因。此次事故的发生表明，对深厚的软弱地基，应采用适当的方法进行处理。

关键词：双钟圩；滑塌；成因分析

1　工程概况

江西省湖口县双钟圩位于湖口县城城区，呈半月状分布于鄱阳湖出口南滩上，圩堤自上石钟山至下石钟山，堤线总长度 1.351km（含上石钟山接头段），是鄱阳湖二期防洪工程 15 座重点圩堤之一，属新建 4 级堤防。工程主要由堤防工程、客运交通闸、汽运交通闸、排污泵站等组成，与长江干堤共同保护湖口县城，保护区面积 2.15km^2，保护人口 5.4 万。

2　滑塌基本情况

双钟圩建设工程于 1999 年 12 月 1 日正式开工，原计划 2000 年 5 月底完成主体工程。原设计堤身断面为土堤加防洪墙，外坡脚设抛石反压平台。2000 年 4 月 9 日清晨 6：40～7：05，施工中的双钟圩在桩号 0＋280～0＋560 堤段发生大规模突发性严重滑塌。据现场测量，最大水平位移约 60m，沉降约 10m，属深层滑动。据调查，自双钟圩开工以来，此前已先后发生了 3 次小规模滑动，位置分别在桩号 0＋150～0＋240、0＋780～0＋840、0＋325～0＋390 堤段附近。

3　滑塌成因分析

为弄清双钟圩滑塌的原因，我们收集、查阅了有关工程资料，进行了现场和室内试验，经计算、分析，认为主要是以下几个方面原因造成的。

3.1　工程地质条件差

（1）双钟圩位于鄱阳湖与长江交汇处，圩区主要地形地貌为江湖滩地和阶地。滩地地面高程一般为 10～14m，地形平坦，微向水面倾斜，倾角小于 5°，靠岸边滩地与阶地相

本文发表于 2003 年。

接，受岗阜残积作用，倾角稍大些。而双钟圩堤正好置于斜坡形滩地上，向湖心倾斜。

(2) 事故段堤基除局部分布有杂填土层外，主要为第四系全新统湖积冲积交替相沉积，总厚度17.4～22.7m，具二元结构。下部堤基土中软弱层（淤泥质黏土、淤泥层）分布广、厚度大，呈软可塑～流塑状，高压缩性，抗剪强度低，物理力学性状差。对桩号0+425断面采用陕西省水电勘测设计院改编的土石坝边坡稳定分析程序（STAB）进行了抗滑稳定计算（计算参数见表1）。

表1　主要计算参数

土层＼计算参数		黏聚力/kPa	内摩擦角/(°)	天然密度/(g·cm^{-3})	饱和密度/(g·cm^{-3})
人工填土		25	22	1.95	2.0
风化料		4	25	1.9	1.93
淤泥质黏土		11.5	9.5	1.81	1.81
淤泥		5.63	3	1.87	1.87
含淤泥质粉细砂		5.0	18	1.8	1.80
砾石	底部	0	31	1.74	1.94
	夹层	6	22	1.9	1.90
堆石		0	40	1.8	2.08

经计算，桩号0+425断面临水坡1月16日、4月9日两工况下抗滑稳定安全系数分别为1.01、0.64，均小于1.05（规范允许值），此堤整体处于不稳定状态。

3.2 堤基未作有效处理

依据SL 239《堤防工程施工质量评定与验收规程(试行)》，堤基清理后，应在第一次铺填时进行平整、压实，压实后的质量应符合设计要求，深厚的软弱堤基须另行处理。但据资料记载，滑塌堤段的上述工作均不符合规范要求。

3.3 施工加载速度过快

双钟圩“4·09”大滑塌段在事故发生前，堤身加载速度为5～6cm/d，远大于其他类似工程的加载速度（1～2cm/d）。河海大学土木工程学院相关专家在事故发生后对双钟圩进行了固结沉降及稳定分析计算，结果认为，堤外砂井加固后的填土控制速度应为1.9cm/d，可见，堤身加载速度过快是导致本次滑塌事故的又一重要原因。

安全系数计算公式为

$$F_s=\frac{\sum[C_{cu}L_i+\gamma B_i(H_{1i}+UH_{2i})\tan\varphi_{cu}\cos\theta_i]}{\sum\gamma B_i(H_{1i}+H_{2i})\sin\theta_i}$$

式中：C_{cu}为地基土的固结快剪指标，堤身填土的黏聚力，kPa；φ_{cu}为地基土的固结快剪指标，(°)；U为堤身荷载作用下地基中已形成的平均固结度，计算采用由太沙基一维固结理论在线性加载条件下的扩充解；B_i为第i条土条的宽度，m；γ为土体的容重，kN/m^3；L_i为第i条土条的底部弧长，m；H_{1i}为第i条土条在地基中的土条高度，m；H_{2i}为第i条土条在堤身的土条高度，m；θ_i为第i条土条底面对水平面的倾角，(°)。

具体分析计算结果见表2。

表 2　　桩号 0+752 断面加载过程试算结果

荷载填筑过程	时间/月	安全系数	加固方案
无荷载填筑	0	2.72	不作任何处理
一次性加载至 22.5m 高程	4	0.71	
加载至 14.5m 高程	1	0.924	堤外坡一定范围内打设砂井（ϕ7cm，砂井间距 2m）
	2	1.038	
	3①	1.121	
停工 2 个月	2	1.194	
加载至 16.5m 高程	1	1.036	
	2①	1.105	
停工 2 个月	2	1.134	
加载至 19.0m 高程	1	1.108	
停工 2 个月	2	1.163	
加载至 22.5m 高程	1	1.002	
	2	1.103	

注　加载起始高程为 12.5m。

①　表示最佳方案，按此方案填土速度为 1.9cm/d。

4　结语

通过钟圩滑塌事故原因分析，可以得到以下经验教训：

（1）新建圩堤，地勘工作务必细致、详尽；堤线布置时，应尽量避开不良地基。

（2）如遇软弱地基，则应充分分析论证所选方案的合理性，必要时应对地基进行加固处理，同时加强观测。

（3）软弱土层在外加允许应力不变的条件下，经过一段时间的排水固结后，它的不排水强度将有所增加。所以，为了弥补软土地基强度的不足，当施工期限许可时，分期施工往往是行之有效的方法。

井冈山市罗浮水库大坝安全论证

喻蔚然，邵仁建，龙国亮

江西省水利科学研究所

摘　要： 针对罗浮水库大坝病、险情的现状，按照有关规范的规定，做了分析和评价，并提出了建议。

关键词： 罗浮水库；大坝；安全鉴定

江西省井冈山市罗浮水库大坝于1970年11月开始动工，1973年才完成初步设计，技术人员缺少，施工队伍也是不固定的民工，因此罗浮水库是典型的“三边”工程。经过多年的运行，水库的主要建筑物都带病工作，不能在设计条件下安全运行，不能正常发挥水库的应有效益。为了彻底消除水库存在的隐患，恢复正常的运行状况，水库主管部门根据水利部颁发的《水库大坝安全鉴定办法》的要求，组织江西省大坝安全监测中心、吉安地区水电规划设计院、井冈山市水利局成立罗浮水库大坝安全鉴定课题组，对水库进行全面安全论证。

1　大坝运行安全管理及建筑质量评价

1.1　工程概况

罗浮水库位于赣江流域禾水水系牛吼江支流拿山河上，坝址以上控制流域面积$54km^2$，河长16.01km，多年平均流量$1.67m^3/s$，多年平均来水量5260万m^3，多年平均降雨量1778.4mm。

罗浮水库大坝正常蓄水位425.00m，校核洪水位426.20m，总库容1125万m^3。大坝平面布置如图1所示，大坝横剖面图如图2所示。坝址所在区域地震基本烈度为Ⅴ度以下。按《防洪标准》(GB 50201—94)，工程等别属于Ⅲ等，主要建筑物属三级。

枢纽工程由浆砌块石重力坝、引水隧洞、发电厂房和放空洞组成。溢洪道为开敞式溢流堰，以弧形闸门控制。

大坝最大坝高62m，坝顶高程432.00m，坝顶长111.5m，坝顶高10m，最大坝底宽51.2m，溢洪道孔口净宽24.0m，堰顶高程420.00m，大坝防渗面板为钢筋混凝土。

发电隧洞位于大坝左岸，进口段50m用混凝土衬砌，进口底板高程381.5m，出口底板高程361.44m，全长308.2m，洞径1.8m，进口设平板钢闸门。

发电厂房位于大坝下游左岸约300m弯道处的山坡上，装机容量2000kW（2台单机

本文发表于2001年。

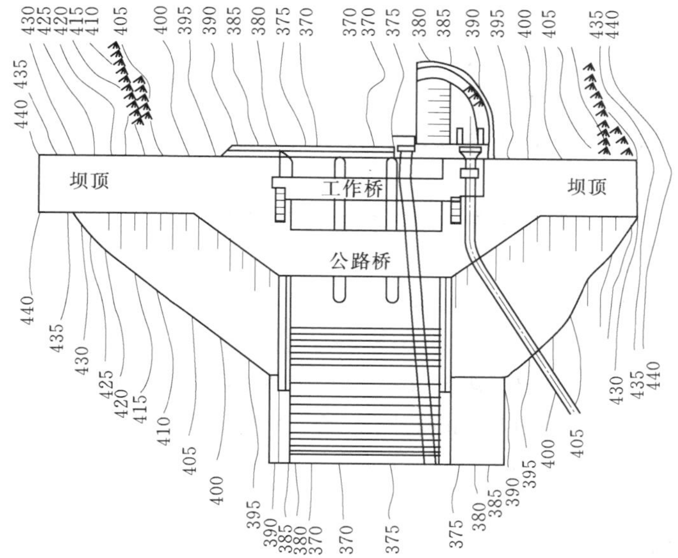

图1　大坝平面布置图

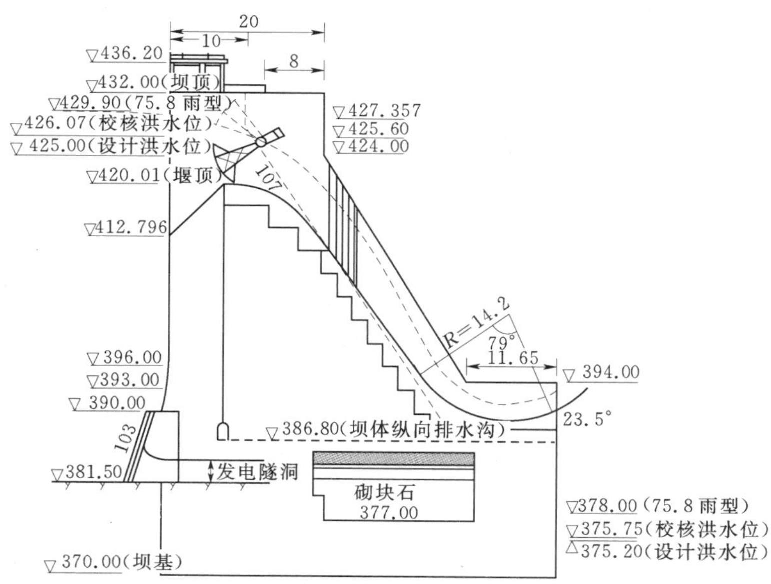

图2　大坝横剖面图

1000kW)，最大工作水头63.56m，最小工作水头28.56m，单机发电流量2.23m^3/s，厂房地面高程363.04m。

放空洞位于大坝左边底部。进、出口底板高程均为374.00m，长51.2m，断面为城门洞形，进口设平板钢闸门。

1.2　大坝运行管理及建筑物质量评价

1.2.1　防洪调度

罗浮水库的度汛方案，由市防汛办制定，并报地区防汛办审批后，下达给水管单位。

由于渗漏严重，进入 20 世纪 90 年代以后，库水位受到严格限制，主汛期控制水位在 420.50m 以下，后汛期控制在 422.50m 以下。

1.2.2 大坝建筑质量综合评价

根据坝区工程地质和水文地质基本资料，结合大坝施工、运行及现状质量检查，对罗浮水库作出如下评价：

（1）坝基：库区工程地质条件良好，坝基为花岗岩，完整坚硬，力学指标高，压水试验 $\omega<0.01$L/(min·m·m)，一般无渗漏问题。

（2）混凝土防渗面板：未设置齿墙与坝基连接，混凝土的抗渗性能不能满足要求，单位吸水率 $\omega=0.007\sim0.026$L/(min·m·m)，尤其在高程 401.10～406.40m，ω 高达 0.78L/(min·m·m)。坝面裂缝纵横交错，是大坝渗漏的主要原因。

（3）坝肩：坝体与坝肩未设置混凝土齿墙，结合不紧密，存在接触面渗漏。左坝肩 ω 为 0.0478～0.2464L/(min·m·m)，右坝肩 ω 为 0.039～0.051L/(min·m·m)，存在绕坝渗流。

（4）坝体：浆砌块石以花岗岩和硅质砂岩为主，质地坚硬，强度高。但坝体与坝基之间未设置混凝土垫层，坝体存在明显的砌缝和空洞，砂浆充填不饱满，大坝的整体性差。

（5）浇筑闸墩时因模板走样，影响了 3 扇闸门的安装，因而 3 扇闸门均不能全部开启，影响泄洪，危及大坝安全。

（6）上坝公路为简易的沙石路面，宽仅 2.5m，并通过一个窄小的隧道，不能会车。在紧急防汛期间，不能满足通畅的交通要求。

2 防洪标准复核

2.1 基本资料

根据《防洪标准》（GB 50201—94）以及《水利水电工程设计洪水计算规范》（SL 44—93），大坝按 50 年一遇洪水设计，500 年一遇洪水校核。

距水库坝址约 11km 的井冈山气象站具有 1959—1999 年共 41 年降水资料。因流域汇流历时短，所以应重点分析 24h 设计暴雨。罗浮水库最大 24h 设计暴雨值与井冈山站最大 24h 设计暴雨值基本相同，因而选用井冈山站的降水资料作为罗浮水库的设计洪水的计算资料。

2.2 设计洪水计算

本次洪水复核采用推理公式法、瞬时单位线法和经验公式 3 种方法推求设计洪水。计算结果见表 1。

表 1　　设计洪水推算结果

设计频率	暴雨途径/mm		经验公式
	推理公式	瞬时单位线	
0.2%	786	652	584
2%	482	433	369

从表 1 中可以看出，用推理公式法计算的成果最大。考虑到所用资料时间系列短，实

测资料综合分析所得成果有偏小的可能。为安全起见，采用推理公式计算成果。

2.3 调洪演算

2.3.1 调洪原则

汛期洪水起调水位采用正常蓄水位425.00m，对应下泄流量为492.2m^3/s，水库调洪原则：当入库流量小于492.2m^3/s时，闸门控制泄量，库水位维持在起调水位425.00m；当入库流量大于492.2m^3/s时，闸门全开，按泄流能力下泄；当库水位降到425.00m时，仍控制下泄流量，使库水位维持425.00m，以满足综合利用要求。

2.3.2 调洪计算方法及成果

调洪计算采用瞬态法，调洪计算结果见表2。

表2　　罗浮水库调洪计算结果

洪水标准/%	洪峰流量/($m^3 \cdot s^{-1}$)	最大流量/($m^3 \cdot s^{-1}$)	最高库水位/m	相应库容/10^6m^3
0.2	786	659	426.07	10.72
2	482	482	425.0	10.0

2.3.3 建筑物防洪能力复核

根据调洪结果，最高库水位为426.07m，现有的坝顶高程为432.00m，有5.93m的超高，因而大坝的坝顶高程满足防洪要求。

3 大坝结构安全分析

3.1 荷载、荷载组合及计算参数

荷载考虑以下5种：

（1）坝体自重和坝上永久设备自重。因缺乏资料，故用反算法计算得坝体容重为2.10t/m^3。

（2）扬压力。罗浮水库坝基未设防渗帷幕，也未设竖向排水，按《浆砌石坝设计规范》（SL 25—1991）计算方法，扬压力按无折减情况考虑，大坝坝体中设有竖向排水，则取折减系数0.2，同时考虑到竖向排水失效，扬压力上升的可能，对坝体扬压力还按三角形分布计算。

（3）上、下游静水压力。

（4）泥沙压力。

（5）动水压力。

荷载组合：包括基本荷载组合和特殊荷载组合。基本荷载组合按正常蓄水位和设计洪水位两种工况考虑，特殊荷载组合按校核洪水位考虑。

3.2 大坝抗滑稳定分析

3.2.1 抗滑稳定计算公式、计算断面及参数

按《浆砌石坝设计规范》（SL 25—91）推荐采用的抗剪公式计算。

计算断面选取：最大断面坝基高程370.00m，非溢流坝选取2个断面，坝基高程为

374.00m 和 390.00m，另对 400.00m 高程进行复核。

本次复核坝基面摩擦系数取 0.60。坝体空洞较多，砌石间胶结作用不强，则坝体摩擦系数取 0.55。

3.2.2 计算结果

抗滑稳定计算成果见表 3。

表 3　　抗滑稳定计算成果表

坝　段	高　程/m	扬压力折减系数	基本荷载组合		特殊荷载组合	规范允许值
			正常	设计		
非溢流坝段	374.00	无折减	1.06	1.05	1.00	基本荷载组合下 $K=1.05$；特殊荷载组合下 $K=1.00$
	390.00	无折减	1.37	1.37	1.27	
	400.00	0.2	2.03	2.03	1.85	
		无折减	1.67	1.67	1.49	
溢流坝段	370.00	无折减	0.99	1.12	1.08	
	389.00	0.2	1.34	1.83	1.77	
		无折减	0.95	1.16	1.10	
	400.00	0.2	1.49	1.49	1.40	
		无折减	1.05	1.05	0.97	

注　下划横线者为不能满足规范要求的数据。

3.2.3 成果分析及评价

非溢流坝段：不论是正常运行情况还是非正常运行情况，不论是坝基面还是坝体计算截面，计算所得到的稳定安全系数均大于规范允许值，满足规范要求，因此非溢流坝段是稳定的。

溢流坝段：坝基面 370.00m 高程在正常蓄水位下，不考虑扬压力折减系数时，所得到的稳定安全系数为 0.99，小于规范值 1.05，不能满足规范要求。不考虑扬压力折减系数时，在正常蓄水位 389.00m 高程计算稳定安全系数为 0.95，小于 1.05；在非常运行 400.00m 高程计算稳定安全系数为 0.97，小于 1.00，均不能满足规范要求。

3.3 大坝应力分析

3.3.1 计算方法和强度参数

按《浆砌石坝设计规范》（SL 25—91）规定，用材料力学方法进行分析。基岩的允许承载力取 $640kg/cm^2$。由于缺乏试验资料，根据规范的规定，浆砌块石抗压强度在基本荷载组合时取 3.8MPa，在特殊荷载组合时取 4.4MPa。

3.3.2 计算成果及分析

大坝应力计算成果见表 4。对成果分析可知，非溢流坝段：计算截面应力均为正应力，没有主拉应力。垂直正应力最大值为 0.603MPa，主应力最大为 0.899MPa，均小于砌体容许压应力，符合规范要求。溢流坝段：在坝踵的位置出现拉应力，其值达到 0.143MPa，不满足规范要求。坝趾为压应力，最大值为 0.998MPa，坝体计算截面的应力也均为压应力，不大于 1.0MPa，符合规范要求。其他均满足要求。

表 4　　罗浮水库大坝应力计算成果表　　MPa

荷载组合		坝段	高程	垂直正应力		剪应力	水平正应力	主应力	
				$\sigma_{y上}$	$\sigma_{y下}$	τ	σ_x	$\sigma_{1上}$	$\sigma_{1下}$
基本	正常	非溢流坝段	374.0	0.342	0.585	0.409	0.287	0.342	0.871
			390.0	0.396	0.341	0.289	0.167	0.396	0.508
			400.0	0.402	0.213	0.149	0.104	0.402	0.317
			400.0*	0.510*	0.244*	0.171*	0.119*	0.510*	0.364*
		溢流坝段	370.0	−0.08	0.998	0	0	−0.08	0.998
			389.0	0.361	0.176	0	0	0.361	0.176
			389.0*	0.331*	0.029*	0*	0*	0.331*	0.029*
			400.0	0.178	0.144	0.171	0.120	0.178	0.364
			400.0*	0.188*	0.094*	0.066	0.046*	0.88*	0.140*
	设计	非溢流坝段	374.0	0.342	0.573	0.401	0.281	0.342	0.854
			390.0	0.361	0.176	0	0	0.361	0.176
			400.0	0.178	0.144	0.171	0.120	0.178	0.364
			400.0*	0.510*	0.244*	0.171*	0.119*	0.510*	0.364*
		溢流坝段	370.0	−0.106	0.973	0	0	−0.106	0.973
			389.0	0.362	0.176	0	0	0.362	0.176
			389.0*	0.158*	0.145*	0*	0*	0.158*	0.145*
			400.0	0.178	0.244	0.171	0.120	0.178	0.364
			400.0*	0.188*	0.094*	0.066*	0.046*	0.88*	0.140*
特殊		非溢流坝段	374.0	0.296	0.603	0.422	0.296	0.296	0.899
			390.0	0.356	0.370	0.259	0.181	0.356	0.551
			400.0	0.369	0.235	0.165	0.115	0.369	0.351
			400.0*	0.482*	0.269*	0.188*	0.132*	0.482*	0.400*
		溢流坝段	370.0	−0.143	0.994	0	0	−0.143	0.994
			389.0	0.348	0.186	0	0	0.348	0.186
			389.0*	0.138*	0.153*	0*	0*	0.138*	0.153*
			400.0	0.165	0.253	0.177	0.124	0.165	0.377
			400.0*	0.052*	0.22*	0.154*	0.108*	0.052*	0.327*

注　坝基扬压力折减系数取 1.0；坝体扬压力折减系数取 0.2 和 1.0（用“*”表示）。

综上所述，非溢流坝段应力满足要求，而溢流坝段在坝踵处出现拉应力，不能满足要求。

4　渗流安全状况分析

4.1　渗流的主要问题

大坝主要渗漏问题表现在以下几个方面：

(1) 基础及接触面。坝体与基岩间无混凝土垫层，接触带处 ω 为 0.031～0.051L/(min·m·m)，有可能存在接触渗漏。坝基内没有设置排水管。

左岸坝肩处 ω 为 0.0478～0.2464L/(min·m·m)，右岸坝肩处 ω 为 0.039～0.26L/(min·m·m)，均超过规范要求。存在绕坝渗流薄弱带。

(2) 上游垂直防渗面板。防渗面板设有 7 条伸缩缝，有 6 条伸缩缝表面已裂开，缝宽 3～5mm，有两条裂缝偏离伸缩缝向坝基延伸；有 7 条不连贯的水平裂缝，均在水平施工缝处。伸缩缝用 30cm 宽的紫铜片嵌缝连接，紫铜片之间搭接未焊，形成漏水隐患。混凝土强度低，其 ω 为 0.007～0.026L/(min·m·m)，最大为 0.78L/(min·m·m)，达不到规范要求。

(3) 放空洞、发电隧洞、溢洪道。放空洞内有明显的渗漏孔洞 19 个，沿洞长方向有 19 个射流点，射流直径 1～2cm，还有许多小孔洞、蜂窝、麻面，钢筋出露，锈蚀严重。发电隧洞有 3 条贯穿环向裂缝，在渐变段末端还有一个 50cm×40cm×8cm 的冲坑，钢筋外露。溢洪道堰面大部分露筋，锈蚀严重，挑流鼻坎出口壁有漏水射出。

4.2 观测资料分析

4.2.1 渗流量观测

渗流量观测设施为三角量水堰，其安装及测量基本符合规范要求。有连续 4 年观测资料，渗流量与库水位相关系比较密切。1992 年 4 月测得的最大渗流量是 $4.733\times10^{-2}m^3/s$。由于量水堰所测的是廊道 386.8m 高程以上部分坝体渗漏量，对于廊道底部以下的渗漏无法测量，整个大坝的实际渗漏量比实测的要大。

4.2.2 绕坝渗流

根据压水试验结果，采用断面法计算两岸绕坝渗流总渗漏量 $4.49\times10^{-3}m^3/s$，由于影响裂隙渗流的因素很多，因而实际渗流量要比计算值大。

4.2.3 大坝渗流

在无压渗流条件下，对渗流起控制作用的是裂缝的几何尺寸及缝面糙率情况。在有压渗流条件下，只有面板裂缝宽度小于毫米数量级时，面板才能较好地发挥防渗作用，而当裂缝宽度接近厘米数量级时，面板的防渗功能将大为降低，比无压渗流条件下裂缝宽度的影响要不利得多。

4.3 渗流现状对大坝安全的影响

如不对防渗面板进行彻底处理，裂缝渗水将不断带走块石间的砂浆，加剧裂缝的发展，最终可能导致面板失效，影响坝的稳定性。

5 引水隧洞结构安全评价

5.1 基本情况

隧洞洞身长 180m，采用混凝土喷浆处理（洞径为 2.2m），其余 125.7m 为钢筋混凝土衬砌（洞径 1.8m）。后接 3 条支洞，支洞 1 长 16.8m，支洞 2 长 13.65m，洞径均为 1.0m，后接长 8m 直径 0.8m 的压力钢管通向厂房，支洞 3 为放空洞，洞径 1.6m。喷混凝土衬砌段下部 1/3 圆喷混凝土厚度偏大（约 5～10cm），严重老化，普遍成块剥落；进口闸门处左侧边墙及左侧洞顶有数股呈射流状漏水，水量较大。有 3 条闭合环向裂缝；渐变段末端底板有一冲坑（50cm×40cm×8cm），环向钢筋已出露。

5.2 工程地质

洞线进口在左坝段内埋管，穿过坝体后开始进入山体基岩，不存在进口边坡稳定问题。洞身段在0+060～0+070，此段裂隙发育，岩体完整性较差，属Ⅳ～Ⅴ类围岩；0+070～0+230属微风化～新鲜砂岩，裂隙不发育，洞上覆岩体厚度最大约150m，属于Ⅳ～Ⅴ类围岩。

花岗岩天然容重2.67～2.70g/cm^3，饱和抗压强度90～110MPa，弹性模量14～20GPa，砂岩天然容重2.69～2.74g/cm^3，饱和抗压强度70～100MPa，弹性模量15～25GPa。岩石弹性抗力系数取400kg/cm^3。

5.3 隧洞钢筋混凝土衬砌复核

（1）水锤压力计算。导叶启闭时间为3s，计算两种运行情况：一是丢弃全负荷时的水锤压力；二是增加全负荷时水锤压力。

（2）基本荷载组合。山岩压力、衬砌自重，设计水位时的内水压力（包括水锤压力）。

（3）复核结果。引水隧洞在计算荷载组合下，衬砌各断面均能满足配筋面积、抗裂安全系数、裂缝宽度等有关规范要求。

5.4 压力钢管管壁强度复核

计算结果满足管壁最小厚度要求，管壁环向应力满足设计要求。

6 大坝安全评价结论与建议

6.1 大坝结构安全

当考虑扬压力无折减时，正常蓄水位下溢流坝段在370.0m（坝基）、389.0m（坝体）高程以及校核洪水位下400.0m（坝体）高程稳定不能满足要求，其余情况均稳定。

在任何水位下，非溢流坝段没有产生拉应力，各计算截面边缘应力也小于砌体允许承载力，但溢流坝段坝踵出现了拉应力，不符合规范要求。其余部分所受压应力均小于砌体允许承载力，满足规范要求。

大坝水平位移总的趋势是向下游偏移，位移量属于正常变化范围。

6.2 渗流安全状况

罗浮水库大坝渗漏主要是由于上游垂直防渗面板存在裂缝造成的。过缝渗水承压，高速水流将胶结材料不断带走，降低浆砌块石间的黏聚力和摩擦系数，削弱坝体整体性，将进一步导致防渗面板裂缝的加大，坝体扬压力越来越大，将对大坝的结构安全和稳定带来非常不利的影响。

6.3 隧洞结构安全评价

经复核计算，衬砌各断面均能满足配筋面积、抗裂安全系数、裂缝宽度等有关规范要求，压力钢管管壁能满足最小厚度要求，管壁环向应力满足设计要求。

隧洞进口闸门后有数股水流射出，渐变段底板有一冲坑已露筋，3条贯穿环向裂缝及孔洞、蜂窝、麻面等隐患必须加固处理。

6.4 建议

（1）建议尽快对上游防渗面板进行加固处理，同时完善排水设施。

（2）廊道、发电隧洞、放空洞内混凝土孔洞、裂缝、蜂窝、麻面多，溢流堰面冲刷严重，应进行修补加固。

（3）对溢流堰的3扇弧形闸门进行校正或重新安装，启闭机进行更新改造。

（4）建立大坝的安全监测系统和水情、雨情测报系统。

参考文献：

[1] 中华人民共和国水利部．浆砌石坝设计规范（SL 25—91）［S］．北京：水利电力出版社，1991：2-12，30-58.

[2] 华东水利学院．水工设计手册（土石坝）［M］．北京：水利电力出版社，1984：180-204.

[3] 祁庆和．水工建筑物（第二版）［M］．北京：水利电力出版社，1986：9-45.

二、工程风险评估

SHUIGONGANQUANYUFANGZAIJIANZAI

基于事件树分析法的大坝可能破坏模式分析

董建良[1]，吴欢强[2]，傅琼华[1]

1. 江西省水利科学研究院；2. 广东珠荣工程设计有限公司

摘　要：水库大坝一旦失事，势必对下游地区造成重大生命财产威胁。运用事件树分析法原理，以油罗口水库大坝洪水荷载作为初始事件，对大坝可能的破坏模式及其溃坝概率进行了分析和计算。结果表明，该水库大坝最大可能破坏模式为坝体管涌破坏，且溃坝风险远大于安全可接受值，上述结论与水库大坝存在的主要病险问题具有较强的吻合性。研究结论在一定程度上为水库管理单位预防溃坝事件的发生提供了理论依据。

关键词：事件树分析法；破坏模式；溃坝概率；油罗口水库大坝

江西省是一个水利大省，水库众多，数量居全国各省第二位。这些水库是全省防洪体系与水利基础设施的重要组成部分，在防洪、灌溉、供水、发电和改善生态环境等方面发挥着巨大的作用。但由于先天建设不足、后期管理不善等问题，曾发生过多次溃坝事故，给人民生命财产造成重大损失。

油罗口水库下游 10km 为大余县城及南康市，涉及 0.25 万 hm^2 农田及京九铁路、赣韶高速、323 国道等重要设施，地理位置十分重要。2004 年的油罗口水库大坝安全鉴定认为，该水库大坝为“三类坝”，即属于病险水库大坝。为保证水库大坝的安全运行，对油罗口水库进行破坏分析研究，事先判断可能的破坏方式，是十分必要的。

1　水库大坝溃坝概率分析方法

水库大坝溃决概率计算一般采用历史资料统计法和事件树法。至于其他方法，如可靠度理论，由于其复杂性以及所需计算参数很难获得，实际应用起来困难，只有对非常重要的大坝进行详细的定量分析时才采用。而历史资料统计法对于不同的水库，没有太大的可参照性和可借用性，因此，本文决定采用事件树法对油罗口水库大坝的溃坝概率进行分析。

1.1　事件树法基本原理

事件树分析方法（Event Tree Analysis，ETA）是一种按时间顺序进行分析的方法[1]，即应用逻辑演绎法和图表法，对给定的初因事件，分析可能导致的各种事件序列的发生概率，从而评价事件的风险。在事件树分析方法中，事件树可表示为二元树，即每个

本文发表于 2013 年。

节点代表事件成功或失败的可能性。

对于某一座水库大坝来说，事件树分析法是指从某一荷载状态出发，采用追踪方法对构成大坝的各要素进行逻辑分析，分析在该荷载状态下大坝的可能溃决失事路径，从而构造相应的事件树以评价大坝总体的溃决概率。“所有可能的荷载”，在非汛期，包括正常高水位以下可能出现的各种水位荷载，在汛期，包括汛限水位或可能遭遇的设计洪水、校核洪水甚至更大洪水。在具体计算分析中应根据水库的实际运行情况，确定几个特征水位，使其能够代表“所有可能的荷载”。这些水位出现的概率是不同的。针对某一水位荷载，分析溃决事件发展的过程，形成溃决路径，并对每个过程发生的可能性采用专家经验法等赋予某一概率值，得出该水位荷载作用下这一溃决路径发生的概率。用以上方法可依次分析出该水位荷载作用下其他溃决路径及发生概率。

1.2 事件树分析法步骤[2]

(1) 计算每种荷载状态下每条溃决路径的溃坝概率，即各个环节发生的条件概率的乘积。设在某水位下大坝某一溃决模式中各环节的条件概率分别为 $P(i, j, k)$，$i=1, 2, \cdots, n$；$j=1, 2, \cdots, m$；$k=1, 2, \cdots, s$。其中，i 为库水位荷载，j 为破坏模式，k 为各环节。第 i 种荷载、第 j 种溃坝模式下大坝溃决的概率 $P(i, j)$ 为

$$P(i,j) = \prod_{k=1}^{s} P(i,j,k) \tag{1}$$

(2) 同一荷载状态下的各种破坏模式一般并不互斥，因此同一荷载状态溃决的条件概率应采用 de Morgan 定律计算。设第 i 个荷载状态下有 n 个溃决模式：A_1，A_2，…，A_n，其概率分别为 $P(i, 1)$，$P(i, 2)$，…，$P(i, n)$，则 n 个溃决模式发生的概率 $P(A_1+A_2+A_n)$ 为

$$\max(P_1,P_2,\cdots,P_n) \leqslant P(A_1+A_2+\cdots+A_n) \leqslant 1-\prod_{i=1}^{n}(1-P_i) \tag{2}$$

de Morgan 定律就是上式中事件并集概率的上限。$P(A_1+A_2+\cdots+A_n)$ 即为第 i 种荷载下的大坝破坏概率 $P(i)$。

(3) 对所有特征水位逐一重复上述步骤，即可以得到所有可能荷载下的所有可能溃决路径和溃决概率。由于一般情况都认为不同荷载状态下的各条件概率是互斥的，因此大坝溃决概率等于各种荷载状态下溃决概率之和，即：

$$P=P(1)+P(2)+\cdots+P(n) \tag{3}$$

2 工程实例分析

2.1 油罗口水库工程概况[3]

油罗口水库位于江西省赣州市大余县城以西 10.0km 的章江上游，集水面积 557km²，总库容 1.19 亿 m³，是一座以防洪为主，兼有供水、发电、灌溉等综合效益的大 (2) 型水库。水库正常蓄水位 220.00m，设计水位 222.29m (500 年一遇)，校核洪水位 223.70m (5000 年一遇)。工程于 1969 年动工兴建，1971 年基本建成，后虽经多次加固处理，但由于种种原因，水库一直带病运行，远未能发挥设计效益，并对下游造成巨大的威胁。

2.2 可能破坏模式分析

影响油罗口水库大坝安全的因素较多，根据现场掌握的油罗口水库大坝存在的病险问题，结合武汉大学编制完成的《江西省赣州市大余县油罗口水库大坝安全评价报告》，参照前人总结的我国大坝主要破坏模式[4]，对造成水库大坝破坏的初始事件进行逐一筛选，考虑将洪水作为可能导致水库大坝破坏的初始事件做进一步的破坏分析研究，并按照破坏原因和机理，归纳出了如下几种破坏路径状况。

（1）状况1。由坝体渗漏引起溃决，其破坏路径为：坝体集中渗漏→继续大渗漏→渗漏发展成管涌→干预失败→坝体溃决；坝体集中渗漏→继续小渗漏→渗漏发展成管涌→干预失败→坝体溃决；坝体集中渗漏→继续大渗漏→渗漏发展成管涌→干预失败→塌陷/坝顶沉陷溃决；坝体集中渗漏→继续小渗漏→渗漏发展成管涌→干预失败→边坡失稳漫顶溃决。

（2）状况2。由坝基渗漏引起溃决，其破坏路径为：坝基集中渗漏→继续大渗漏→坝基冲刷发展→干预失败→坝体溃决；坝基集中渗漏→继续小渗漏→坝基冲刷发展→干预失败→坝体溃决；坝基集中渗漏→继续大渗漏→坝基冲刷发展→干预失败→塌陷/坝顶沉陷溃决；坝基集中渗漏→继续小渗漏→坝基冲刷发展→干预失败→边坡失稳漫顶溃决。

（3）状况3。由坝基渗漏引起溃决，其破坏路径为：坝体边坡逐渐破坏→形成管涌→管涌发展→干预失败→坝体溃决；坝体边坡逐渐破坏→形成管涌→管涌发展→干预失败→边坡失稳漫顶溃决。

（4）状况4。由发电引水隧洞渗漏引起溃决，其破坏路径为：发电引水隧洞集中渗漏→继续大渗漏→渗漏发展→干预失败→坝体溃决；发电引水隧洞集中渗漏→继续小渗漏→渗漏发展→干预失败→坝体溃决。

（5）状况5。由溢洪道渗漏引起溃决，其破坏路径为：溢洪道翼墙集中渗漏→继续大渗漏→渗漏发展→干预失败→坝体溃决；溢洪道翼墙集中渗漏→继续小渗漏→渗漏发展→干预失败→坝体溃决。

2.3 各可能破坏模式年溃决概率分析

根据油罗口水库大坝可能破坏模式分析结果，结合洪水重现期，本文采用手工划分法把洪水重现期划分为6种荷载状态。根据破坏路径构造各分支事件树，利用事件树分析法，计算同一荷载状态下各种破坏的溃决概率，然后计算该荷载状态下的溃决概率，最后得到洪水条件下大坝的溃决概率，见表1。

表1　　油罗口水库大坝溃决概率

荷载状态（以洪水重现期表示）	频率	破坏模式	
		破坏概率	年溃坝概率
5000～10000年	f_1	p_1	f_1p_1
1000～5000年	f_2	p_2	f_2p_2
500～1000年	f_3	p_3	f_3p_3
100～500年	f_4	p_4	f_4p_4
50～100年	f_5	p_5	f_5p_5
1～50年	f_6	p_6	f_6p_6

注　p_1，p_2，p_3，p_4，p_5，p_6 为各洪水荷载状态下各可能破坏模式溃坝概率之和；f_1p_1，f_2p_2，f_3p_3，f_4p_4，f_5p_5，f_6p_6 为各洪水荷载状态下各可能破坏模式年溃坝概率之和。

每种洪水荷载状态发生概率 f_1，f_2，…，f_6 为该洪水频率范围内洪水出现概率，见表 2。

表 2　各洪水重现期区间频率计算结果

洪水重现期/年	洪水频率	库水位/m	洪水频率区间	平均库水位/m	f
1	1	218.50	1/1～1/50	219.69	0.9801
50	0.02	220.88	1/50～1/100	221.10	0.0100
100	0.01	221.31	1/100～1/500	221.80	0.0080
500	0.002	222.29	1/500～1/1000	222.50	0.0010
1000	0.001	222.71	1/1000～1/5000	223.21	0.0008
5000	0.0002	223.70	1/5000～1/10000	223.91	0.0001
10000	0.0001	224.12			
					$\sum f=1.0$

这里以 10000～5000 年一遇洪水荷载状态为例，构造了各种可能破坏模式的事件树，如图 1～图 3 所示（部分事件树略）。

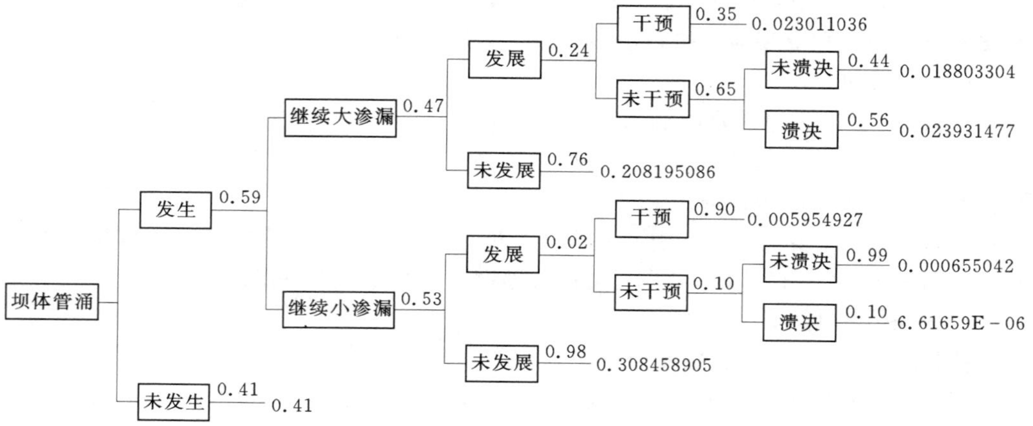

(a)

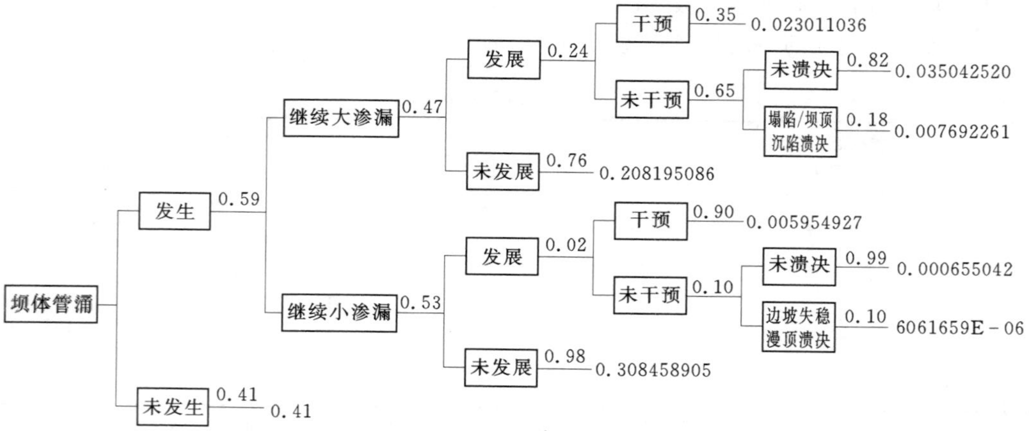

(b)

图 1　坝体管涌破坏事件树

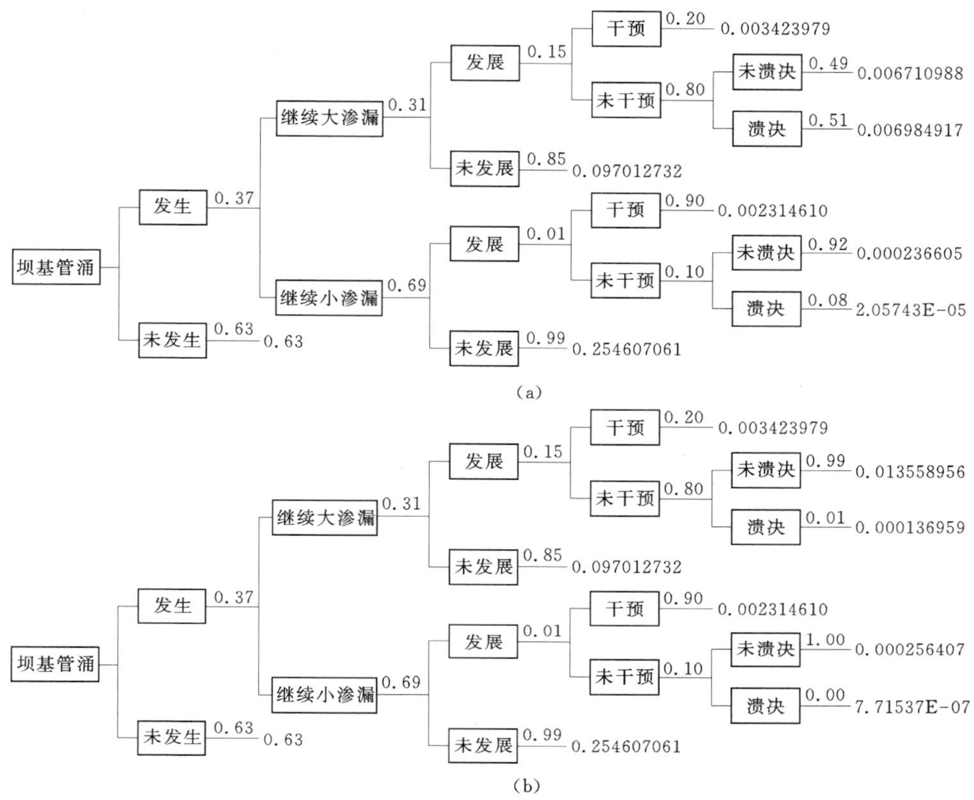

图 2　坝基管涌破坏事件树

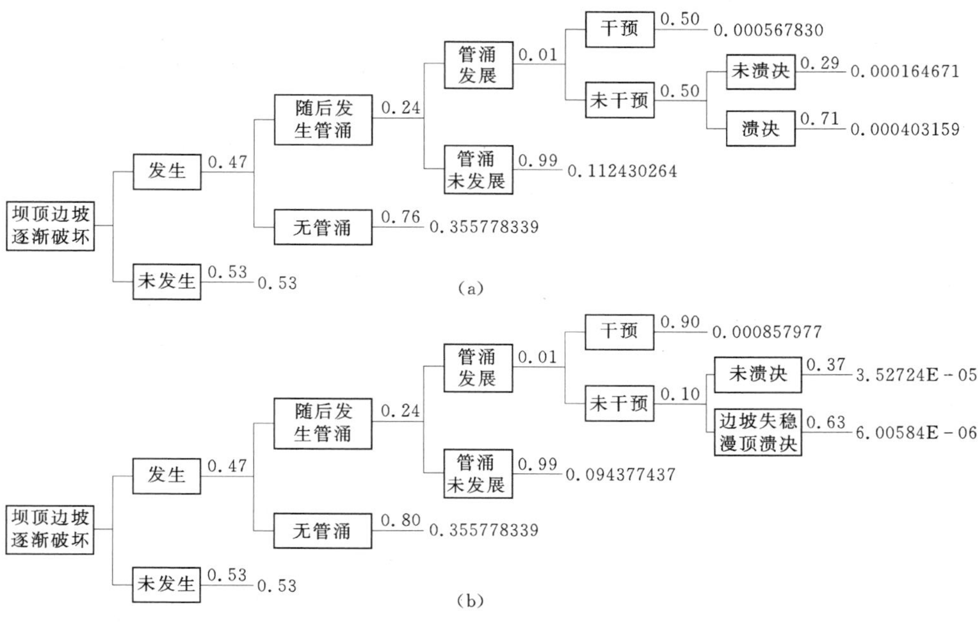

图 3　坝顶边坡逐渐破坏事件树

图1～图3中，“发生”系指开始形成集中渗漏并开始冲刷；“继续”系指继续冲刷；“发展”系指冲刷通道扩大，形成管涌；“干预”系指人工干预成功，“未干预”系指未采取人工干预或人工干预失败；“溃决”系指在未采取人工干预或人工干预失败情况下溃决。

事件树各环节条件概率 $P(i, j, k)$ 的计算，首先应找出导致这一环节事件发生的全部因素，后依据李雷等提供的我国大坝破坏事件发生的定性描述和概率对应表[2]，把各种因素对事件的影响大小转换成概率，利用各种因素之间的关系及发生的概率，计算出各环节的条件概率。

通过对事件树的整理分析，应用 de Morgan 定律，10000～5000年一遇洪水荷载状态下各种破坏模式的年溃坝概率见表3。

表3　10000～5000年一遇洪水下各种破坏模式年溃决概率

破坏模式	破坏概率	年溃坝概率
坝体管涌	2.393×10^{-2}	2.393×10^{-6}
坝基管涌	6.985×10^{-3}	5.675×10^{-7}
边坡破坏后管涌	4.032×10^{-4}	4.032×10^{-8}
发电引水隧洞管涌	6.637×10^{-7}	6.637×10^{-11}
溢洪道管涌	3.191×10^{-7}	3.191×10^{-11}
塌陷/超高不足	7.828×10^{-3}	7.828×10^{-7}
边坡失稳后漫顶	6.745×10^{-5}	6.745×10^{-9}

以上计算仅针对10000～5000年一遇洪水事件，其余洪水荷载状态下计算方法完全相同，只是事件树中各子事件赋值概率不一样，由于篇幅有限，这里不再累赘。将各洪水荷载状态下相同破坏模式的年溃决概率相加，可得到各种破坏模式总的年溃决概率。经整理分析，油罗口水库大坝各种破坏模式的年溃决概率见表4。

从表4可知：①该水库大坝洪水引起的年溃坝概率为 2.36×10^{-3}，远大于大坝安全所能接受的值 10^{-4} 或 2×10^{-4}[5]，具有除险加固的必要性。② 该水库大坝最有可能发生的破坏模式为坝体管涌产生的破坏，其次为坝基管涌产生的破坏，与《江西省赣州市大余县油罗口水库大坝安全评价报告》评价鉴定结论大坝坝身、坝基和两岸坝肩漏水较为严重具有较强的吻合性。除险加固前坝体渗漏隐患部位应特别引起水库管理单位的高度重视，严格落实好日常的安全检查及维护工作。

表4　油罗口水库大坝各种溃坝模式的年溃决概率

破坏模式	年溃坝概率	所占百分比/%	破坏模式	年溃坝概率	所占百分比/%
边坡失稳后漫顶	7.864×10^{-6}	0.333	塌陷/超高不足	4.929×10^{-4}	20.857
坝体管涌	1.515×10^{-3}	64.090	溢洪道失事	1.506×10^{-8}	0.001
坝基管涌	3.099×10^{-4}	13.114	发电引水隧洞失事	2.673×10^{-8}	0.001
边坡破坏后管涌	3.793×10^{-5}	1.605	合计	2.360×10^{-3}	100

3　结语

随着经济社会的发展，水库大坝安全与公共安全更加息息相关，水库安全已成为公众关注的焦点。通过本文的分析，在一定程度上为油罗口水库大坝管理单位有重点、有目的地组织制定水库大坝日常管理、维护和防洪减灾工作提供了技术依据，从而有助于提高水库大坝的安全管理水平及应对突发事件和溃坝灾害防控能力，降低溃坝事件的发生，保证下游生命和财产安全，维持社会经济可持续发展。

参考文献：

[1] 曾声奎，赵廷弟，张建国，等．系统可靠性设计分析教程［M］．北京：北京航空航天大学出版社，2001.

[2] 李雷，王仁钟，盛金保，等．大坝风险评级及风险管理［M］．北京：中国水利水电出版社，2006.

[3] 周和清，等．江西省大余县油罗口水库除险加固工程初步设计［R］．武汉：长江勘测规划设计研究院，2005.

[4] 彭雪辉．风险分析在我国大坝安全上的应用［D］．南京：南京水利科学研究院，2003.

[5] 马福恒．病险水库大坝风险分析与预警方法［D］．南京：河海大学，2006.

River2D 模型在溃坝洪水演进数值模拟中的应用

胡国平[1]，姜世俊[2]

1. 江西省水利科学研究院；2. 南昌大学

摘　要： River2D 模型能模拟溃坝洪水在水库下游地区的演进过程。为模拟水库溃坝洪水演进，以油罗口水库为例，利用 River2D 模型模拟大坝溃决后洪水在下游的演进，预测洪水淹没范围、水深及流速等洪水风险信息。

关键词： River2D；水动力学模型；溃坝洪水；洪水风险图；油罗口水库

1　引言

大坝安全问题与人类生命财产安全息息相关。我国水库溃坝问题日渐突出，大坝一旦溃决，将给下游人们带来严重的灾难。因此，进行溃坝洪水模拟分析，预测溃坝洪水的演进过程，可以为事故早期预警等提供参考信息，对保护人民生命财产安全等具有十分重要的意义。

采用实用有效的科学模型模拟溃坝洪水对下游的影响，以评估溃坝影响、最大程度地保障公众的生命财产安全具有现实意义。目前，数学模型是模拟分析并预测溃坝洪水演进过程的主要研究手段，用于进行溃坝分析和下游洪水演进的模型计算软件有 DAMBRK、FLDWAV、MIKE、River2D 等模型。国内对 River2D 模型在溃坝洪水模拟中的应用还处在初步探索阶段，其具体应用实例较少[1,2]。

本文就 River2D 模型的理论基础与应用过程作简要介绍，并模拟油罗口水库在 5000 年一遇校核洪水情况下副坝溃决后洪水在水库下游的演进过程。

2　River2D 模型原理

River2D 模型是二维平均深度有限元模型[3]，主要由水动力学模型和栖息地模型组成，下面仅对水动力学模型理论进行详细介绍。

2.1　基本假定

River2D 水动力学模型是基于二维瞬时流的圣维南方程，其基本假定如下[3]：

（1）沿垂向的水压力分布为静水压力。

（2）水深方向上的水平流速分布为常数。

（3）忽略科氏力和风应力。

本文发表于 2013 年。

2.2 控制方程

质量守恒方程：

$$\frac{\partial H}{\partial t}+\frac{\partial q_x}{\partial x}+\frac{\partial q_y}{\partial y}=0 \tag{1}$$

X 方向动量守恒方程：

$$\frac{\partial q_x}{\partial t}+\frac{\partial}{\partial x}(Uq_x)+\frac{\partial}{\partial y}(Vq_x)+\frac{g}{2}\frac{\partial}{\partial x}H^2=gH(S_{ox}-S_{fx})+\frac{1}{\rho}\left(\frac{\partial}{\partial x}(H\tau_{xx})\right)+\frac{1}{\rho}\left(\frac{\partial}{\partial y}(H\tau_{xy})\right) \tag{2}$$

Y 方向动量守恒方程：

$$\frac{\partial q_y}{\partial t}+\frac{\partial}{\partial x}(Uq_y)+\frac{\partial}{\partial y}(Vq_y)+\frac{g}{2}\frac{\partial}{\partial y}H^2=gH(S_{oy}-S_{fy})+\frac{1}{\rho}\left(\frac{\partial}{\partial x}(H\tau_{yx})\right)+\frac{1}{\rho}\left(\frac{\partial}{\partial y}(H\tau_{yy})\right) \tag{3}$$

式中：H 为水深；U、V 分别为 x、y 方向的平均流速；$q_x=HU$，$q_y=HV$；S_{ox}、S_{oy} 分别为 x、y 方向的河床坡度；S_{fx}、S_{fy} 分别为 x、y 方向的阻力坡度；τ_{xx}、τ_{xy}、τ_{yx}、τ_{yy} 分别为横向紊动应力张量。

2.3 河床阻力模型

河床阻力主要与河床剪切应力有关，以 x 方向为例，河床阻力公式如下：

$$S_{fx}=\frac{\tau_{bx}}{\rho gH}=\frac{\sqrt{U^2+V^2}}{gHC_s^2}U \tag{4}$$

式中：τ_{bx} 为 x 方向的河床剪切应力；C_s 是无量纲的谢才系数，与有效糙率高度 K_s 和水深有关。

C_s 表达式如下：

$$C_s=5.75\lg\left(12\frac{H}{k_s}\right) \tag{5}$$

对于给定的水深 H 和曼宁糙率 n 值，k_s 可由下式表示：

$$k_s=\frac{12H}{e^m} \tag{6}$$

$$m=\frac{H^{1/6}}{2.5n\sqrt{g}} \tag{7}$$

当 $\frac{H}{k_s}<\frac{e^2}{12}$ 时，式（5）用下式替代，为

$$C_s=2.5+\frac{30}{e^2}\left(\frac{H}{k_s}\right) \tag{8}$$

该式为经验公式。

2.4 横向剪力模型

采用 Buossniesq 涡流黏滞性表达式对平均深度横向紊动剪切力进行模拟，计算公式为：

$$\tau_{xy}=\nu_t\left(\frac{\partial U}{\partial y}+\frac{\partial V}{\partial x}\right) \tag{9}$$

ν_t 为涡流黏滞系数，计算公式如下：

$$\nu_t = \varepsilon_1 + \varepsilon_2 \frac{H\sqrt{U^2+V^2}}{C_s} + \varepsilon_3^2 H^2 \sqrt{2\frac{\partial U}{\partial x} + \left(\frac{\partial U}{\partial y} + \frac{\partial V}{\partial x}\right)^2 + 2\frac{\partial V}{\partial y}} \tag{10}$$

式中：ε_1、ε_2、ε_3 为模型参数，取值分别为 0、0.5、0.1。

2.5 湿/干地区处理

对于水深非常浅或有地表出露的情况，River2D 模型用地下水流动方程替代地表水流动方程，公式为：

$$\frac{\partial H}{\partial t} = \frac{T}{S}\left[\frac{\partial^2}{\partial x^2}(H+z_b) + \frac{\partial^2}{\partial y^2}(H+z_b)\right] \tag{11}$$

式中：T 为渗透系数；S 为含水率；H 为地下水位以上的水深；z_b 为地下水水位。

3 River2D 模型软件

River2D 模型主要是基于一个非恒定的瞬时模型，同时也可以用于恒定流加速收敛的状况，可使用在天然河道中的二维水深平均浅水动力学模型和鱼类栖息地模型。其显著特点是利用干湿区域解算法和近似超临界法计算河道的边界部分。

River2D 模型软件还可提供使用者将一些特定区域单独划分出来的功能，可单独对特定区域划分更为详细的网格。利用 River2D 模型可以模拟出溃坝洪水淹没范围图、淹没水深图、水流流速图，还可以单独得到各点的流速与水深。

River2D 模型软件包括四个软件包，分别是 R2D _ Bed[4]，R2D _ Ice[5]，R2D _ Mesh[6] 和 River2D。图 1 为 River2D 模型软件运行流程图。

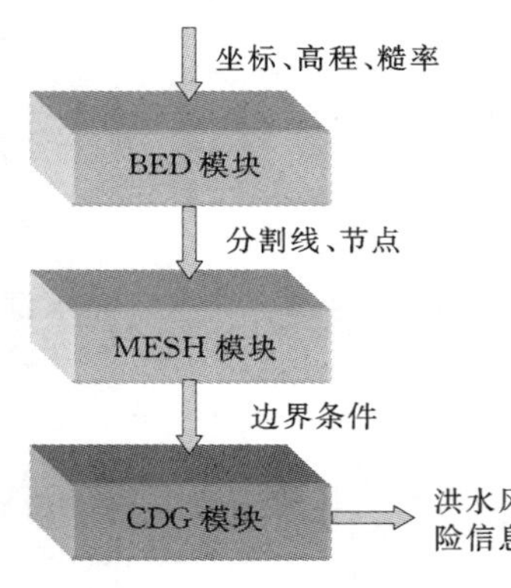

图 1 River2D 软件运行流程图

4 River2D 模型在溃坝洪水中的应用

4.1 数据准备

大余县油罗口水库大坝研究区域概况[7]：油罗口水库下游以章江河为主干流，全县国土面积 1367.63km²，县内流域面积 1360.4km²，为全县国土面积的 99.52%，而流经县内河长为 110km。

本文利用 River2D 水动力学模型模拟油罗口水库在校核洪水位下副坝发生管涌溃决后下游洪水演进过程。模拟范围：起始油罗口水库大坝坝趾，终至石下坝，全长约 12.68km。

本次模拟，模型的内边界只考虑了洪泛区地势较高的山包，未考虑所有支流和桥梁等内边界条件。初始条件：入口给定初始流量及水面高程，出口为水面高程；边界条件：入口流量为大坝溃口流量，图 2 为利用 Breach 模型模拟副坝溃决得到的溃口流量过程线，出口给定水深—单位流量关系，河岸采用动边界。河道糙率系数无实测资料，根据现场查勘，河道糙率取 0.034，陆地糙率取 0.04，城市各建筑群的糙率取 0.3。

4.2 模拟计算

此次模拟时间步长为 10s，图 5.3～图 5.5 为油罗口水库副坝在校核洪水位发生溃坝达到洪峰（$Q=10983\mathrm{m}^3/\mathrm{s}$）时的洪水风险图。

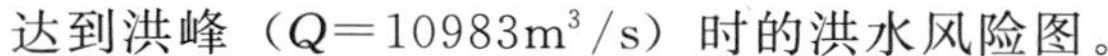

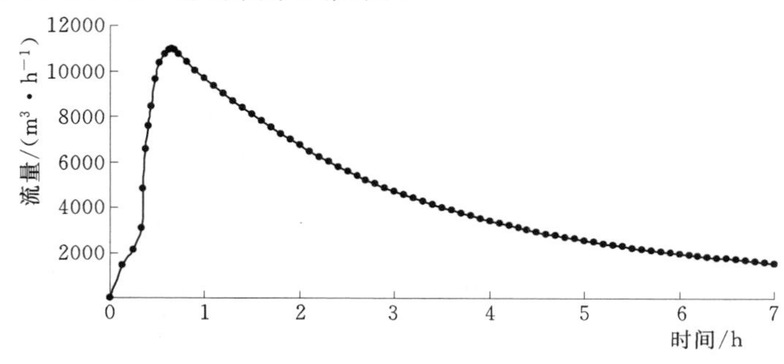

图 2 溃口流量过程线

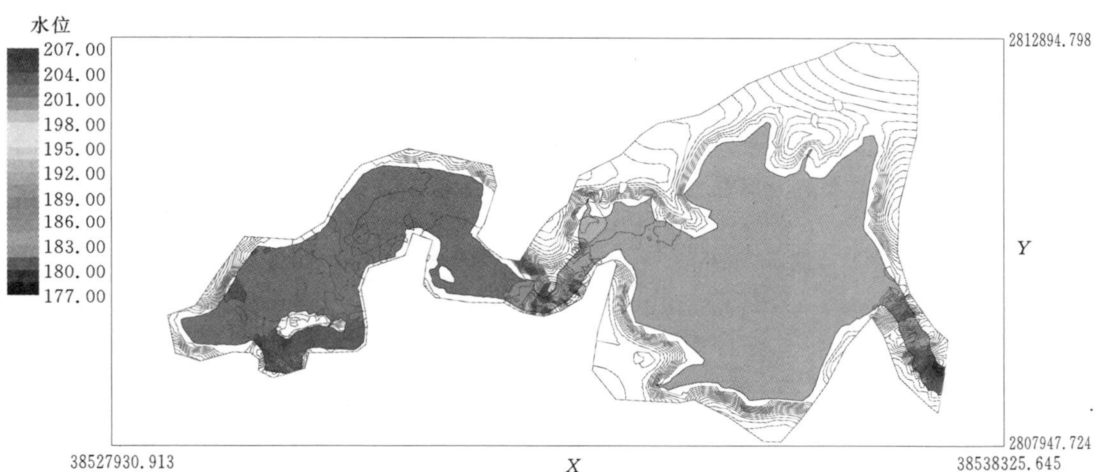

图 3 油罗口水库溃坝洪水风险图

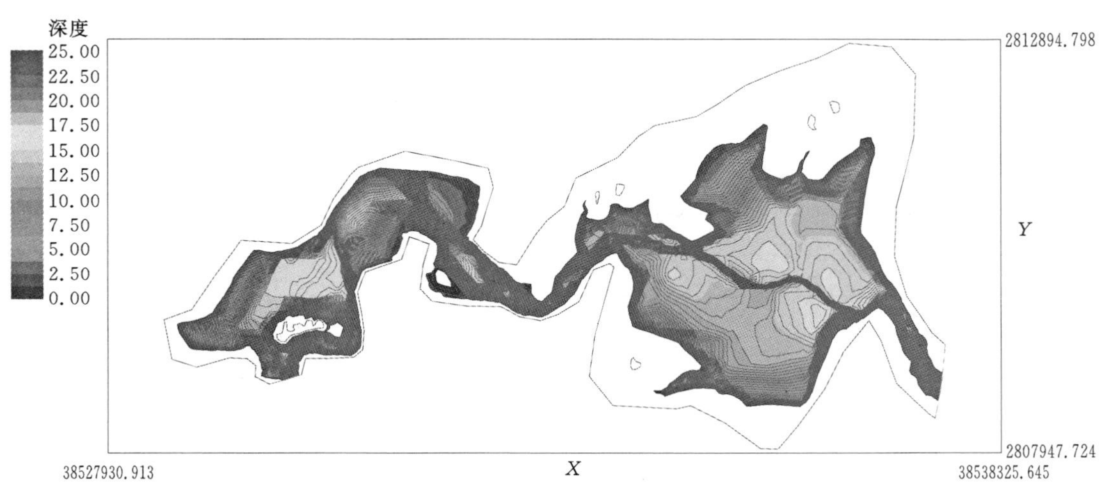

图 4 油罗口水库溃坝下游淹没区水深分布图

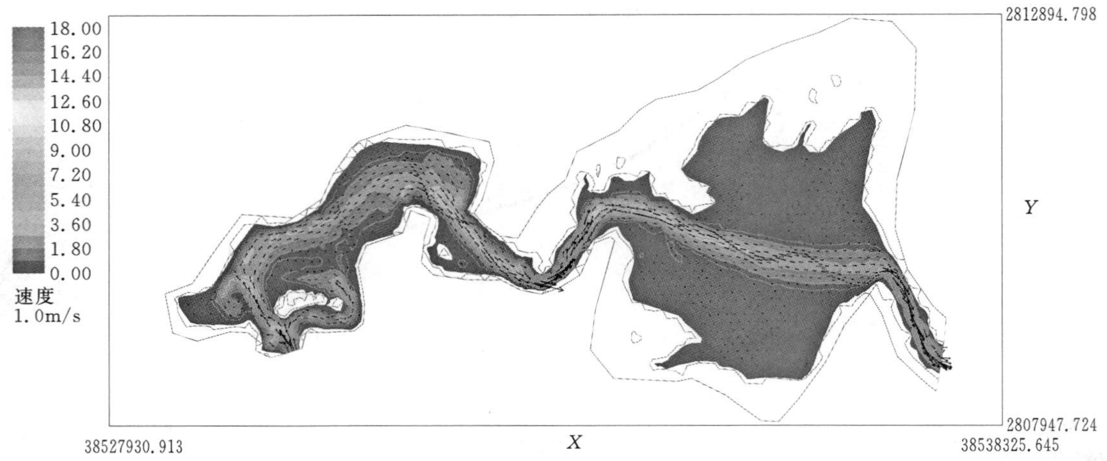

图 5　油罗口水库溃坝洪水水流速度矢量图

由流速矢量图可以看出水流流向规律，水流在河道中心较急，在河道两边趋缓。此时计算区域内行政区洪水淹没局部最大水深超过 16m。表 1 为计算区域内各行政区的洪水淹没信息统计。

表 1　溃坝洪水风险信息统计

序号	行政取划分	平均流速 /(m·s^{-1})	平均水深 /m	DV 值 /(m^2·s^{-1})	溃坝洪水严重性 SD
1	竹木村	2.62	12.01	31.50	高
2	浮江村	2.19	14.21	31.09	高
3	车里村	1.92	4.21	8.07	中
4	石桥下	1.32	14.81	19.61	高
5	北门河	1.71	14.74	25.18	高
6	余西街	2.00	14.28	28.51	高
7	余东街	1.86	14.76	27.48	高
8	建桂街	1.98	14.48	28.66	高
9	总窿口	1.99	10.45	20.82	高
10	五里山	0.64	12.71	8.10	中
11	新民村	1.53	15.15	23.13	高
12	新珠村	1.96	13.94	27.29	高
13	新余村	1.87	14.07	26.32	高
14	新安村	1.54	12.94	19.89	高
15	东山街	3.12	4.95	15.45	高

根据 DV 值判别溃坝洪水严重性程度[8]：①低严重性，$DV \leqslant 4.6\text{m}^2/\text{s}$；②中严重性，$4.6\text{m}^2/\text{s} < DV \leqslant 12\text{m}^2/\text{s}$；③高严重性 $DV > 12\text{m}^2/\text{s}$。根据溃坝洪水数值模拟成果分析得

知，油罗口水库一旦发生溃坝，其下泄洪水极其严重，对下游人民生命财产安全将造成不可估量的威胁。

5 结语

River2D模型不仅可以模拟河床地形，还能有效的模拟溃坝洪水演进过程，为大坝的安全管理提供准确的洪水风险信息，其模拟精确程度依据实际地形数据的准确度而定。以油罗口水库为例，利用River2D模型软件模拟了校核洪水下水库溃坝洪水演进过程。通过对下游淹没区淹没范围、淹没水深与流速的分析，说明River2D模型软件对溃坝洪水的数值模拟结果，对油罗口水库下游洪水的演进、事故早期预警及人员撤离等具有一定的参考价值。

参考文献：

[1] 崔玲，陈余道，蒋亚萍．River2D模型及其在漓江桂林市区段的初步应用［J］．广西水利水电．2007.

[2] 吴欢强．溃坝生命损失风险评价的关键技术研究［D］．南昌大学．2009.

[3] Steffler. P，Blackburn. J. Introduction to depth averaged modeling and user's manual ［R］. University of Alberta，2002，9.

[4] Steffler. P，Blackburn. J. Bed topography file editor user's manual ［R］．University of Alberta，2002，9.

[5] Steffler. P，Blackburn. J. Introduction to depth averaged modeling and user's manual ［R］. University of Alberta，2002，9.

[6] Steffler. P，Blackburn. J. Introduction to depth averaged modeling and user's manual ［R］. University of Alberta，2002，9.

[7] 武汉大学设计研究总院．江西省大余县油罗口水库大坝安全综合评价报告［R］．2006.

[8] 周克发，李雷，盛金保．我国溃坝生命损失评价模型初步研究［J］．安全与环境学报，2007.

震损水库应急抢险风险决策和技术

喻蔚然[1]，魏迎奇[2]

1. 江西省水利科学研究院；2. 中国水利水电科学研究院岩土所

摘　要："5.12"汶川大地震使得大量的水库出现严重的险情，威胁了下游的人民生命财产安全；因此，及时对震损水库进行应急检测，制定切实可行的应急抢险方案成为摆在各级政府和部门面前的一个重大课题。本文在综合目前现有的各种抢险技术的基础上，提出针对震损水库主要险情的应急抢险技术，同时鉴于地震灾区存在诸多风险，重点强调了风险决策的基本要求和思路。

关键词：震损水库；应急抢险；风险决策

1　研究背景

水库是国家重要的基础设施，发挥着重要的农田灌溉、发电、供水、养殖等效益。随着社会经济的发展，越来越多的人和越来越多的工矿企业聚集在水库下游。四川汶川"5.12"特大地震后，水利工程震损情况严重，许多水库甚至具有溃坝的高危险情，严重威胁下游的人民生命和财产安全。

我国是个水库大国，无法对8万余座水库进行抗震设计和处理，因此，研究切实可行的应急抢险方案并及时准确、快速地采取相应的抢险措施，尽可能减小震损破坏，是应对震损水库的有效途径。然而我国现有水库应急抢险技术的研究还不成熟，与国际先进国家相比，普遍存在不足，如紧急状况的识别、判断及评估不够准确，大多只有简单的定性评述，过于笼统。需进一步细化，并拓宽紧急事件范围，增加险情分类分级、可能危害程度等分析等。

2　"5.12"汶川大地震中震损水库基本情况[1]

地震虽然发生在四川省境内，但是周边的重庆市、陕西省、甘肃省也受到较大的影响。本文仅以四川省为例进行说明。

2.1　规模和数量

据统计，在汶川大地震中，四川省1997座水库受到不同程度的震损，占已建水库的29.7%，其中：大型4座、中型61座、小（1）型335座、小（2）型1597座。在1997座震损水库中，土石坝1885座，占总数的94.4%，拱坝91座，占总数的4.6%，其他坝

本文发表于2013年。

型21座，占总数的1.1%。

2.2 震损水库分布

震损水库分布于四川省17个市、96个县（市、区），如图1所示。

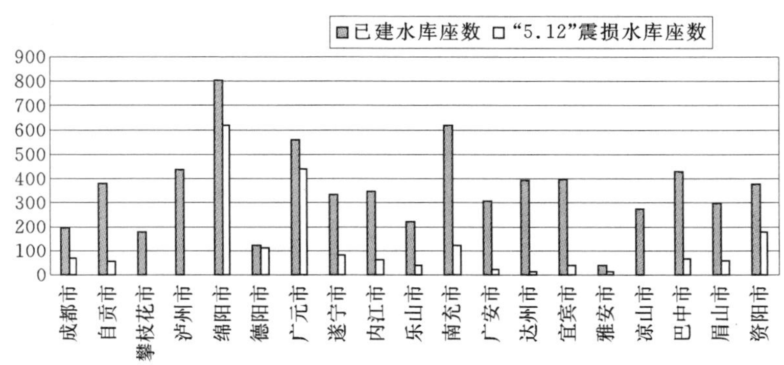

图1 四川省震损水库数量及其分布

2.3 险情类型

震损土石坝出现的险情主要类型有裂缝、滑坡、渗漏、塌陷、涵管破损等，另外还有防浪墙破损、护坡块石脱落、管理房损坏等小震损，50%以上的水库同时出现多种险情。

（1）裂缝。所有震损水库大坝及附属建筑物都出现了不同程度的裂缝，有的裂缝十分严重，成为危及大坝安全的主要因素。从核查的情况看，纵缝的数量和规模都远大于横缝，大部分纵缝都是坝顶附近的浅表裂缝，不会危及大坝安全。但是部分大坝的纵缝宽度大、条数多，最大的宽度已经达到60～70cm，有的坝在坝顶附近有4～5条纵缝。

（2）滑坡、脱坡。有部分水库已经发生了滑坡，部分水库存在滑坡迹象，有部分水库存在滑坡风险。有滑坡迹象或者存在滑坡风险的水库，多表现为衬砌断裂、护坡脱落。

（3）渗漏。地震使得坝体内部出现贯穿裂缝、坝肩和坝基基岩松动、泄放水设施结构破损等，都导致坝下游渗水量增大或出现新的漏水点。

3 应急抢险的风险决策

3.1 震损水库应急抢险的风险特点

震损水库的应急抢险与水库防洪抢险、病险水库除险加固有着明显的不同，特点鲜明，主要体现在如下几个方面：

（1）震损水库数量多，范围大。一般一场洪水可以使一条河流或者一个流域内的数座或者几十座水库发生险情，但是汶川大地震可以使得震区内1997座水库发生险情，其规模远远超过了洪水的影响。这么多的震损水库首先直接影响下游好多一些县城和乡镇，直接影响到生命安全，其次是影响到抗震救灾的整体工作。比如紫坪铺水库的抢险，不仅涉及到群众安全问题，还有救灾的正常秩序，因为下游已经发生群众的恐慌。

（2）震损水库险情多，风险高。险情多不仅体现在震损水库出现的险情数量多，而且表现为险情种类多。统计表明，50%以上的水库同时出现多种险情，险情主要类型有裂缝、滑坡、渗漏、塌陷、涵管破损等，另外还有防浪墙破损、护坡块石脱落、管理房损坏等小震损。水库的险情不光是大坝安全问题，而且关系饮水安全，甚至疫情的发生，这些事态风险是很高的。

（3）震损水库资料少，处置难。各个水库管理用房都被震坏，水文资料、工程资料大部分丢失或损毁。另外，地震造成各种条件极为恶劣，尤其信息不通，其结果是要判断一个大的险情较难，更何况各种风险因素不确定性很大，比如余震、气象等。再有就是时间非常紧迫，必须在汛期到来之前全部处理掉，否则会形成汛期洪水的叠加。以上各种状况更使得震损水库的应急处置非常艰难。

3.2 基本要求和思路

3.2.1 政治上要高度重视

发生地震，许多灾区人民家破人亡，损失惨重，正需要党和国家政府给予关怀和帮助，因此不应该也不允许发生次生灾害再次对人民群众造成生命伤害，因此对震损水库应急抢险进行风险决策时，不仅要从专业上高度重视，更要从政治上高度重视。

从政治上高度重视是党和国家的要求。国务院高度重视，始终强调这样的一个基本理论：对次生灾害不允许发生伤亡。大震之后，社会需要稳定，民心需要稳定，绝对不能再承担政治上的任何风险。因而，政治上高度重视各类风险，是一项政治任务。

从政治上高度重视也是自身的要求。震损水库风险数量多，种类多，往往由于难以照应而有所忽视，或者专业上重视，但由于险情相对较轻而重视程度不足，这都是要不得的，其反映的是政治上重视不够。只有在政治上重视了，才能真正做到专业上重视。

3.2.2 应急抢险要主动迅速

对震损水库进行应急抢险，各种各样的不确定性太多，不确定的风险太多，所以要规避这种风险的叠加，最好的办法是立足于早决策、早处理，坚持对应急事件主动从速地进行处置。一方面，在第一时间到达现场开展险情检测和抢险工作，对于震损险情明显的水库制定处置方案，尽早解除险情，稳定下游人民和社会的情绪，如汶川大地震发生当天，水利部工作组、四川省水利厅工作组在地震发生后第一时间组织专家赶赴紫坪铺水库，及时排查水库险情。同时，各地震灾区市州县也立即组织专家工作组，分头到各水库开展险情核查工作。另一方面，对于震损险情不明显的水库主动地进行相关检测，判断险情是否仍在发展，是否会有新的险情发生，是否对下游会造成大的影响，从而制定相应的方案进行处置。

3.2.3 统筹兼顾，综合风险最小

每一座震损水库都有可能存在多种风险，处置这些风险也存在多种方法和手段，这些方法和手段有的可以相容，而有的则可能相斥，也就是说，在处置某一风险时，有可能使得另一风险增大或者使得另一风险无法处置。此时，必须考虑各种方案的综合风险，把握统筹兼顾，综合风险最小的原则处置风险。综合风险最小的方案并不是说能解决所有的风险，而是为了规避最大的风险，必须要敢于承担次一级的风险，在这种情况下，看似承担了一定风险，但是综合起来，它的综合风险是最小的。

比如说，应急抢险时存在余震、滑坡、洪水等风险，这些风险是巨大的，一不小心有可能对抢险人员造成新的伤害，对抢险成功形成阻碍。如果仅考虑这些风险，则抢险不可行，施工都无法进行。但是考虑到一旦溃坝将对下游的人民造成更大的损失，则相对来说

抢险的风险是小的，是可行的。总而言之，在不是单一风险的前提下，一定要综合考虑各类风险，能够达到综合风险最小来进行处置。

3.2.4 风险应合理分担

地震造成了灾区重大的损失，或者说，灾区是地震损失的直接承担者。同时，地震也带来了许多的次生灾害风险，在消除次生灾害风险的过程中，灾区同样是风险的承担者，不能认为地震以后抢险救灾全部是政府的责任和义务。政府是有责任和义务除险，但政府是风险的消除者，政府组织抢险队伍和专业人员在工程上消除风险，这种风险或责任由抢险队伍和专业人员承担。灾区的老百姓处在这种情况下，本身就是风险的承担者，因此有主动规避的义务，地方政府有责任主动的转移老百姓，这样除险的人才能够比较踏实。所以说，复杂的风险条件下要合理地分担风险，即灾区群众应该承担主动规避风险的义务，政府应该承担解除风险的责任，二者必须有机地统一。

3.2.5 建立有效的风险决策机构和应急机制

决策机构是应急抢险的指挥中心，是震损水库应急处置的有力保证。决策指挥机构主要由抗震救灾指挥部、各工作组以及现场指挥部组成。指挥部是领导核心，负责部署、决策各项抗震救灾工作，其中包括震损水库的应急抢险，统一指挥、协调各方面的力量开展水利抗震救灾和灾后重建工作。指挥部下设综合协调组、工程抢护组、水情测报组、专家组、后勤保障组等5个工作组，指挥部办公室设在综合协调组。现场指挥部下设领导小组、抢护小组、监测小组、后勤保障小组等4个工作小组。

应急机制包括内部机制和外部机制。内部机制是在应急抢险过程中风险决策机构运转正常的工作机制，其特点是合署办公、集体协商、共同决策、地方落实；外部机制是在应急抢险过程中风险决策机构与外界联系的合作机制，其特点是广泛合作、信息共享、相互协调、形成合力。通过风险决策机制的实施，可以保证震损水库应急处置迅速、科学。

4 主要险情应急抢险技术与决策[2]

4.1 渗漏（管涌）

并非所有的渗漏都需要进行应急抢险，要进行判别。一般来说，当渗漏对大坝安全造成较大影响时才需要及时抢险，主要有以下几种类型：①存在漏洞，集中渗漏量较大，类似于管涌；②较大范围的集中渗漏区，可见明显细颗粒被带出；③大面积的散浸区；④管涌或流土；⑤有渗漏，坝坡较陡，存在滑坡的可能性；⑥坝下涵管破损，引发较大的接触渗漏。

大坝渗漏与库水有着密切的关系，进行渗漏应急抢险首要的就是降低库水位，不仅可以直接减少大坝的渗漏量，还可以减轻大坝内部的渗透压力，防止渗透破坏。采取的措施可以是拓宽溢洪道、挖深溢洪道、坝下涵管（泄洪洞）放水、抽水，对泄洪能力不足的水库，根据具体情况采取打开非常溢洪道、降低堰顶高程等措施，少量震损严重、不具备正常挡水能力的大坝已开挖临时泄洪缺口，并进行口门及下泄通道的临时防护。若险情发展迅速，已来不及等水位下降时，可先实施抢险，待险情稳定后降低水位。必须要注意的是，当放水涵洞存在严重的渗漏对大坝稳定有较大威胁时，须控制放水，注意观察水流状态。

具体采用哪种方法，要根据渗漏的类型、水库地形、抢险队伍人员素质和经验、现有抢险材料和设备等条件综合选定，不一定要选用技术最先进、装备最好的技术手段。

4.2 裂缝及其引发的滑坡

当地震造成坝体出现较大的裂缝后，大坝整体性遭到破坏，库水通过裂缝渗入坝体，不仅增大了大坝内部的渗透压力，而且降低了土体的抗剪强度，使得坝坡出现滑坡等险情；若已经产生滑坡，坝体断面变小，稳定性有所降低，当库水位在较高的情况下，易产生溃坝险情，因此，应急抢险第一项措施就是尽可能迅速降低库水位，采取的措施见本文4.1节。针对裂缝具体的应急抢险技术包括裂缝开挖、回填、封堵、覆盖，滑坡坡脚压重及上部削坡减载，贴坡反滤排水结合导渗沟等，方法简单，易操作，无需大的设备。

针对滑坡采取的具体的应急抢险技术应根据情况而定。一种情况是：震损出现的滑坡常伴随严重裂缝，采取的措施可以有：换土回填，固结灌浆，桩基加固滑动面或滑动带，增加抗滑能力；上下游坝坡压重固脚，防止新的滑动；放缓上下游坝坡，提高加固体回填土的干密度等。另一种情况是震损出现的滑坡主要是由于渗漏所致，可按渗漏抢险的措施进行处理，如挖导渗沟、铺设反滤等。

4.3 决口（漫溢）

虽然汶川地震未造成水库大坝决口溃决，但是四川省有69座水库具有溃坝的高危险情。作为最为严重的险情，其抢险决策有别于其他险情。

对于有溃坝危险但尚未决口的大坝，首先应采取的措施是尽可能降低水库水位，然后按照上述有关险情的应急抢险措施进行抢险。

对于已经库水漫溢或者已经决口的大坝，应考虑采用合适的堵口技术，尽早堵复。堵口方法有多种，传统的堵口方法主要有平堵、立堵和混合堵三种。随着科学技术的发展，新机械、新设备、新材料不断出现，施工技术、施工工艺不断创新，堵口的方法也越来越先进。近年来出现的新方法有钢木土石组合坝堵口技术、沉箱堵口技术、铁菱角堵口技术等。

采用哪种方法，要根据口门过流量、水位差、地形、地质、料物采集及抢险人员对堵口方法的熟练程度等条件综合选定，不一定要选用技术最先进、装备最好的技术手段。

决口抢险需要注意的是：①由于地震的突然性，受条件限制，大多不能当即堵口合龙，且决口初期水流压力大，流速快，堵口的难度很大，所以应考虑在水位下降到一定程度后进行堵口；②宜根据口门条件、堵口方式、易于施工、断流迅速、参就地取材、不危及对岸堤防安全等原则慎重选定坝轴线，一般有三种形式：①按原坝轴线堵口；②外堵，即坝轴线向上游侧凸出前进；③内堵，即坝轴线向下游侧凹入。

4.4 非工程措施的采用

非工程措施在应急抢险中也是必不可少的，主要有以下几个方面：

（1）启动应急预案，落实各级部门和人员的责任制。对于已经决口或即将决口的水库，加紧实施群众转移疏散，以保生命为主；对有溃坝险情和高危险情的水库，进一步细化了群众疏散转移预案，做好紧急撤离的准备。

（2）加强水库安全巡查和监测，随时了解水库大坝的危险情况，若水库已经决口，需关注水流、水位变化情况。

（3）建立会商制度，及时通报有关工作进展情况，加强实施堵口的各项工作联系。

（4）加强宣传工作，一方面及时说明情况的严重性，让大家做好相关撤离、抢险准备，另一方面注意不要引发混乱，否则反而影响抢险工作。

5 结语

水库大坝应急处置问题不再仅仅是各级政府、水行政主管部门和水库管理单位关心的工程安全问题，已成为公众关注的公共安全热点问题。震损水库的应急抢险不同于一般的防汛抢险，它风险数量多，类型多，各种不确定性因素多；因此，应急抢险的风险决策显得尤为重要。本文总结了各地应急抢险的经验教训和应急抢险的技术手段和方法，尤其是风险决策的基本思路和要求，希望有助于今后震损水库的应急抢险的成功实施。

参考文献：

[1] 四川省农田水利局．汶川地震震损水库险情排查、复查与资料收集分析研究［R］.2010.11.
[2] 刘志明．四川震损水库的特点及震害处理［J］．中国水利，2008，(14)：9-12.

Research on Simulation of Diversion Flood Routing for Kangshang Flood Storage in Poyang Lake Region

Yong Ji[a,b,c], Xinfa Xu[a], Jinbao Wan[b], Songtao Hu[a]

[a]*Jiangxi Provincial Institute of Water Sciences*

[b]*School of Environmental Science and Engineering, Nanchang University*

[c]*College of Hydraulic and Ecological Engineering, Nanchang Institute of Technology*

Abstract: Understanding the process of diversion flood routing for flood storage is very important for lake engineering purposes to manage hydraulic structures and prevent disaster from flood, and environmental purposes to maintain lake ecosystem and landscape. A two-dimensional numerical model was developed to simulate Kangshang flood storage with erodible beds and banks composed of well sorted-sandy materials. Flood flow data were estimated based on recorded flow data using standard flood frequency analysis techniques. The simulation results were compared and the model calibrated with water surface elevation records of the previous floods. Simulation results enabled prediction of maximum current speed, water depths, smbmergence scenarious and land availability under different disignated flood flows for riverbed assessment, development and management.

Keywords: Hydraulic Model; Poyang Lake; Kangshang Flood Storage; Shallow Water Quation

1 Introduction

During flood, the temporary structures created, and the activities being carried out get washed off, causing severe damage to human life and other property[1]. The reports of floods in different parts of the world for several decades reveal the same scenario of enormous economic damage and human sufferings[2]. Poyang Lake region, which plays important role in China economy, is one of the greatest grain production bases and industrial raw material suppliers. Meanwhile, it is also the national flood disaster-prone and worst affected area. At present the research on simulation of diversion flood routing for flood storage is very little in Poyang Lake region both at home and abroad[3]. Therefore, develo-

本文发表于2012年。

ping research on simulation of diversion flood routing for flood storage in Poyang lake region by the full use of the existing technical means and methods is important for minimizing the losses caused by floods, and information construction of flood controlling and the utilization of flood sources[4]. Taking full advantage of the characteristics of flood storage in Poyang lake region and on the basis of a comparative analysis of various flood routing methods, the hydrodynamic model was used for flood simulation studies of diversion flood routing for Kangshang flood storage in Poyang lake region.

2 Main Text

2.1 *two-dimension hydraulic equation*

A two-dimension model is set up to simulate the flood routing of Poyang Lake region and its diversion areas, and to quantify the flood storage and detention capacity of the wetland. As FVM can be directly used in irregular region, it's employed with unstructured mesh for area partition here. In the paper, the two-dimension shallow water equation is transformed from conservative form to discrete form. The two-dimension shallow water equation is set up by selecting water level and flow as controlling variables.

$$\begin{aligned} &\frac{\partial h}{\partial t}+\frac{\partial (hu)}{\partial x}+\frac{\partial (hv)}{\partial y}=0 \\ &\frac{\partial (hu)}{\partial t}+\frac{\partial (hu^2+gh^2/2)}{\partial x}+\frac{\partial (huv)}{\partial y}=gh(s_{0x}-s_{fx}) \\ &\frac{\partial (hv)}{\partial t}+\frac{\partial (hv^2+gh^2/2)}{\partial y}+\frac{\partial (huv)}{\partial x}=gh(s_{0y}-s_{fy}) \end{aligned} \tag{1}$$

When only the bed slope and friction are considered, the source and sink term $\bar{b}$ (q) can be written as follows:

$$\frac{\partial \bar{q}}{\partial t}+\frac{\partial \bar{f}(q)}{\partial x}+\frac{\partial \bar{g}(q)}{\partial y}=\bar{b}(q) \tag{2}$$

The expression of the limited volume after dispersement of the integral in Eq. (2) is

$$A\,\frac{\Delta q}{\Delta t}=-\sum_{j=1}^{m} T(\phi)^{-1} f(\bar{q})L^{j}+b_{*}(q) \tag{3}$$

The Osher approximate Rieman solver refer to reference [5].

2.2 *Grid Generation*

The mesh is generated using Gis to digitize the 1 : 10000 ground water measured data in 2006. The triangle mesh for hydraulic calculation is generated by the grid topology program. The principle of grid arrangement is to adopt as few grids to represent the trend of topography as possible, and to make sure that they are reasonably organized and gradually varied. In consideration of the landform, the location of the sewage drain outlet, Peclet value, Courant value, grid rate, and so on, the computational domain for the Kangshang Flood Storage is composed of a nonstructural grid with 13334 units and 6940 nodal points (Fig. 1) .

2.3 *Galculation parametes*

The Manning coefficient n is taken as 0.022. The entering boundary for the current flux in the computational domain is given and the outflow boundary fluex are $0m^3/s$. The initial water level of each unit in the computational domain takes the relative level of the lower boundary, the primary velocity of current is 0m/s.

2.4 *Simulation results*

The last time for the diversion flood routing for Kangshang flood retarding basin are 116 hours. Flow field at 1d, 3d and 5d are shown in Figs. 2, 3. and 4 respectively. It can be seen from the figures that the current flows to Kangshang safety area and Luojiaohu safety area at 4th hours. After 9 hours flooding, the current flows to Dongyuang safety area. The time of flow reaching at Luopengshang safety area and Datang safety area are 24 hours and 44 hours respectively.

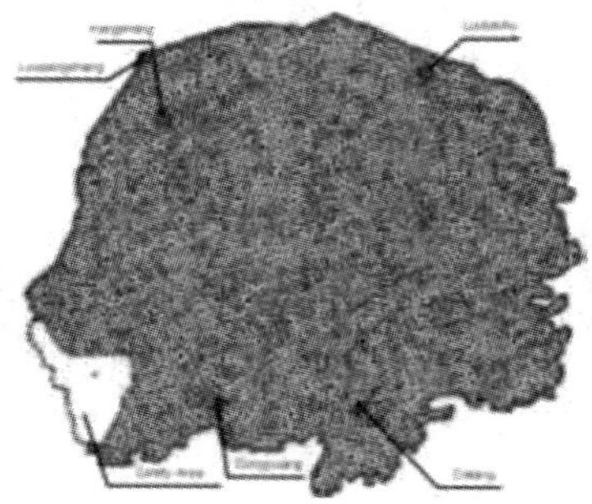

Fig. 1 Grid in Kangshang Flood Storage

Fig. 2 Velocity field in Kangshang Flood Retarding Basin at 1d

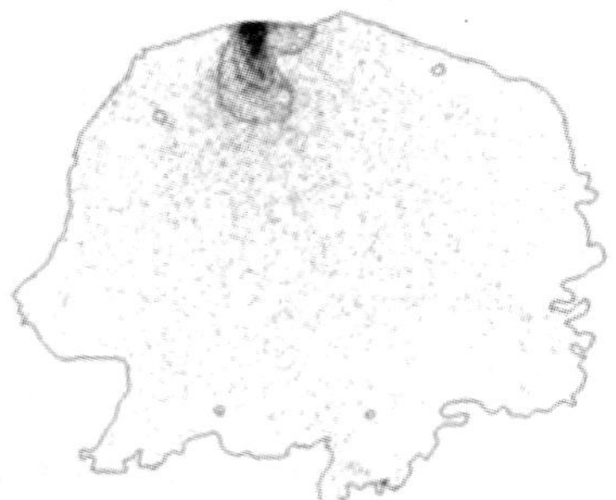

Fig. 3 Velocity field in Kangshang Flood Retarding Basin at 3d

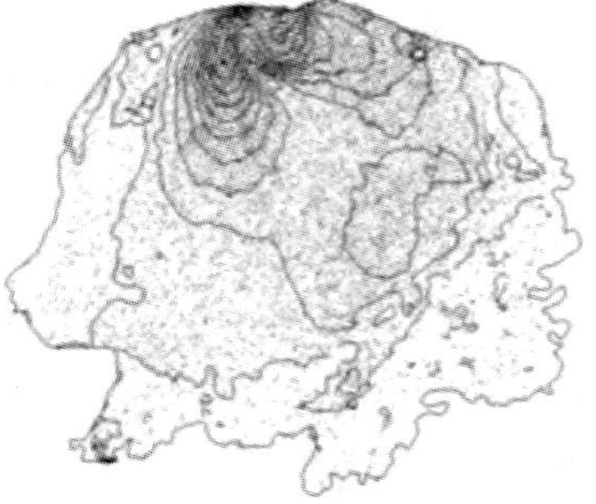

Fig. 4 Velocity field in Kangshang Flood Retarding Basin at 5d

Table 1 shows the variation of flow velocity with time in the main research areas of Kangshang flood retarding basin, and it is shown that the maximum velocity was 5.75m/s occurring at entering boundary about 2 hours after the beginning of simulation. There was no related recorded data could be used to verify the calculated velocity for the extreme difficulty of field measurement. After comparing with recent fragmentary field records, it is shown that the magnitude and variation process of the calculated velocity are reasonable. Total water depth was not affected at early periods for Datang safety area and Luopengshang safety area. Thereafter, total water depth was increased continuously from 2 days to 5 days. After 1 day of simulation, total water depth was significantly increased and reached a maximal value after 5 days exposure.

Table 1 Computed results of velocity field and total water depth at selected cells

Research Areas	1 d		2 d		3 d		4 d		5 d	
Entering boundary	5.606	2.484	5.510	4.477	3.861	5.704	2.066	8.459	0.635	9.123
Kangshang safety area	0.312	2.464	0.175	5.166	0.063	6.321	0.023	8.947	0.012	9.591
Dongyuang safety area	0.065	1.541	0.026	4.075	0.016	5.239	0.011	7.876	0.006	8.521
Datang safety area	0.000	0.000	0.001	2.768	0.007	3.932	0.005	6.570	0.005	7.215
Luojiaohu safety area	0.268	3.271	0.084	6.397	0.013	7.549	0.016	10.18	0.015	10.82
Luopengshang safety area	0.000	0.000	0.081	1.349	0.032	2.500	0.010	5.099	0.007	5.743

3 Conclusion

The results of the simulation have great practical meaning for the hazard assessment in Poyang lake region. From the simulated water depth distribution in the research area, the effects of diversion flood routing on the Kangshang flood retarding basin can be determined, and the economic effect of the diversion flood routing can thus be analyzed.

Acknowledgements

This research was supported by the Project of Jiangxi Provincial Science and Technology Department (2010BSA20300) and the Jiangxi Provinical Education Department (GJJ11246, GJJ11636). Partial support from the Ministry of Water Resources under Project No. 201001056 is gratefully acknowledged.

References

[1] Koen K, Stefan J, Vreugdenhil, DC, van der W. Effects of flooding on the recruitment, damage and mortality of riparian tree species: A field and simulation study on the Rhine floodplain, Forest Ecology and Management, 2008, 255: 3893-3903.

[2] Sun, DP Xue H, Wang PT, Lu RL, Lial XL. 2-D Numerical Simulation of Flooding Effects Caused by South-to-North Water Transfer Project. Journal of Hydrodynamics, Ser. B, 2008, 20: 662-667.

[3] Zhen L, Li F, Huang HQ, Dilly O, Liu JY, Wei YJ, Yang L, Cao XC. Households' willingness to reduce pollution threats in the Poyang Lake region, southern China. Journal of Geochemical Explo-

ration，2011，110：15-22.

[4] ROE PL. Approximate Riemann solvers，parameter vectors and difference-schemes. Journal of Computational physics，1997，135：250-258.

[5] Zhao DH，Shen HW，Tabios GQ. Finite-volume two-dimensional unsteady-flow model for river basins. Journal of Hydraulic Engineering，ASCE，1994，120：863-883.

茄坑水库溃坝机理及模式分析

喻蔚然[1,2]，傅琼华[1,2]，马秀峰[1,2]

1. 江西省水利科学研究院；2. 江西省大坝安全管理中心

摘　要：分析了茄坑水库溃坝机理，并结合渗流稳定性、抗滑稳定性等影响溃坝的因素，推断出茄坑水库溃坝的主要路径为暴雨洪水—水位突升 —渗流破坏（集中渗漏、接触渗漏）—大坝滑坡 —形成小溃口—水流下切 —冲沟、陡坎扩大加深 —两侧土体失稳坍塌—溃口增大 —溃坝，可供同类工程参考。

关键词：溃坝；机理；模式；茄坑水库

茄坑水库位于江西省上高县翰堂镇翰堂村，大坝坐落于锦江支流斜口河上游翰堂水上，于1978年8月兴建，1979年1月基本建成。水库集雨面积0.45km²，正常蓄水位59.00m（假设高程，下同）总库容38×10⁴m³，是一座以供水为主的小（2）型水库。水库枢纽建筑物包括大坝、溢洪道和坝下涵管，其中大坝为均质土坝，坝顶长70m、宽5.5m，坝顶高程62.00m，最大坝高21.5m，上游坡坡比为1∶2.90，下游坡坡比为1∶1.88，设块石贴坡。溢洪道位于大坝右坝端，为开敞式明渠，进口设公路箱涵，底高程59.00m，底宽0.68m，净高0.96m，无消能设施。坝下涵管为预制混凝土圆管，管径0.3m，管长89m，进口底高程41.00m，出口底高程40.50m，进口采用斜拉闸控制。2010年5月20～23日上高县境内连续降雨，尤其21日为大暴雨，茄坑水库水位迅速上升并达到59.90m，超出历史最高水位0.4m，溢洪道漫溢，洪水下泄后冲刷坝脚。由于溢洪道泄流有限，至22日晚水库仍高水位运行，水位为59.70m。23日早晨6：20大坝开始出现滑坡，下游土体瞬间下滑10余米，且由于大坝断面变小，上游坝体亦无法挡住洪水，逐步垮塌，至6：40大坝中部完全溃决。洪水不断下切，土体受水流冲刷影响不断被洪水带走，溃口越来越大。由于水库库容不大，至7：15抢险队到达现场时库水已基本放空。溃口位于大坝中部，顶部宽约61m，底部为12m，上游侧底高程48.00m，下游侧底高程37.50m。从溃口可看出，坝体填土、坝基覆盖层和基岩分层明显。溃口处坝体土多处开裂、结构松散，局部仍有塌方。由于洪水下切，坝下涵管完全暴露，出口段被冲毁；基岩表层强风化岩体均冲走，出露的岩体则较新鲜。为此，本文对茄坑水库溃坝事件进行了剖析，分析了溃坝机理和溃坝模式，以期为大坝安全管理和科学合理调度提供参考。

1　溃坝机理分析

目前，关于土石坝溃决的机理主要有以下两种观点：

本文发表于2012年。

（1）坝体材料受越坝或穿坝水流的冲刷，即漫顶引起外部冲刷、内部冲刷或管涌，土石坝漫顶破坏溃口发展过程主要由水流冲蚀引起的连续纵向下切和横向冲蚀及溃口边坡失稳坍塌引起的间歇横向扩展组成。管涌引起的溃决破坏进程与漫顶情况类似，只是其溃决速度较漫顶速度更快[1,2]。

（2）陡坎冲刷机理[3]，即水流在坝下游坡面上形成一个细冲沟网，并逐渐发展为一个较大的沟壑，沟壑最初包含多个阶梯状的小陡坎，并随时间不断向上游后退，同时不断扩宽，最后发展为一个大的陡坎。溃坝过程中沟壑的扩宽和陡坎向上游的发展是由陡坎下游射流冲击区周围的水流紊动和水流剪应力造成的。

结合茄坑水库溃坝的现场状况发现，其溃坝机理和发展过程包含了上述两种机理，但又有不同，具体体现在以下 4 点：①水流冲沟已形成，并逐渐向两侧、下部延伸，溃口的形成为一逐渐加大加深的过程；②由溃口 V 形断面看，两侧土体并非是光滑的表面，而是凹凸不平，表明溃口非完全因水流冲刷而成，还包括溃口不断加深加宽造成两侧土体逐块失稳而坍塌的情形；③出口部位的冲沟呈陡坎状，水流流态紊乱，进一步加剧了水流冲刷的效果；④大坝溃决非水流漫顶造成（水库水位并未超过坝顶），应与管涌破坏、坝坡失稳有关。

综合溃坝理论和溃坝实际情况，可认为茄坑水库的溃坝机理为大坝因渗漏、管涌等原因产生滑坡，加之断面减小进一步造成上游坝坡失稳、库水渗透加强，逐渐冲刷坝体，在坝体下游面形成冲沟和陡坎，并逐渐扩大、加深，向上游发展，进而在坝体和坝顶形成初始溃口。冲沟冲蚀加剧，溃口两侧土体发生间歇性失稳坍塌，同时陡坎的发展使水流流速增大，水流紊动和水流剪应力淘蚀陡坎底部，引起垂直墙面的失稳而坍塌，最终形成溃决。

2　溃坝分析

2.1　暴雨强度大

据茄坑水库附近钓田雨量站的观测记录，自 5 月 20 日起开始下雨，雨量为 19.5mm。21 日为大暴雨，雨量为 186mm，暴雨强度趋近于 50 年一遇。22 日降雨减小，雨量为 15.5mm。连续集中的大暴雨致使大坝高水位长时间运行，加大了大坝承受的荷载。

2.2　泄流能力不足

溢洪道进口为一箱涵，无压运行，其泄流能力可按明渠均匀流公式计算：

$$Q=AC\sqrt{Ri} \tag{1}$$

其中

$$C=R^{\frac{1}{6}}/n$$

式中：A 为过流断面面积，m^2；C 为谢才系数；R 为过流断面水力半径，m；i 为涵管底坡降；n 为过流断面糙率。

由式（1）可推算当水库最高运行水位达 59.90m 时，溢洪道最大泄流量为 $0.90m^3/s$，1d 最大泄流总量为 $77760m^3$（假使 24h 维持最高库水位）。据实测，21 日 8：00～22 日 8：00 雨量为 186mm，采用推理公式法推算出地表洪量约为 $120000m^3$（$P=10\%$）、$192000m^3$（$P=2\%$）。由此可看出，即使溢洪道按最大泄流量下泄洪水，也需 1.5～2.5d 才能将本次降雨洪水完全泄完。可见，溢洪道泄流能力明显不足。

2.3 渗流分析

（1）参数选取。坝体填土来源于附近山体及库区表层覆盖层，以黄色残坡积黏土为主，夹碎石，碎石粒径为2～5cm，少量大于10cm，土体结构较松散。坝基覆盖层上部为耕植土，厚约40cm，下部为残坡积黏土，较密实，厚约1～3m。下伏二迭系石灰岩。大坝典型断面见图1。通过分析大坝典型断面，将土层分为坝体填土、耕植土和残坡积黏土3层。在大坝溃口处取样4组做土的物理力学试验。根据土工试验结果及类似工程综合分析，各土层物理力学性质指标见表1。

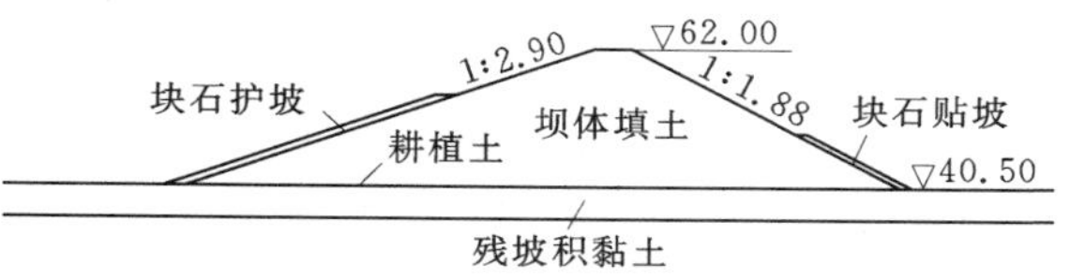

图1　大坝典型断面（单位：m）

表1　坝体土层分布及参数选取

土层	湿密度/(g·cm^{-3})	饱和密度/(g·cm^{-3})	渗透系数	黏聚力 c/kPa	内摩擦角 φ/(°)
坝体填土	1.76	1.83	4.44×10^{-4}	20.5	18.8
耕植土	1.88	1.95	5.00×10^{-4}	17.0	13.0
残坡积黏土	1.80	1.85	1.00×10^{-4}	25.0	19.5

注　渗透系数单位为cm/s。

（2）渗透稳定性。坝体填土及坝基覆盖层渗透系数均大于10^{-5}cm/s，具中等透水性，不满足规范[4]防渗要求。以溃坝当日最高运行水位59.70m作为计算工况，采用GEO-SEEP程序对大坝最大坝高断面进行渗流计算，结果见图2。由图可看出，大坝浸润线整体偏高，出逸点高程为49.52m，距离坝脚9m；出逸点位势约50%，出逸坡降达0.43，明显偏大，且出逸点高于下游块石贴坡的顶高程，渗流出口无保护，存在渗透破坏的可能。据坝坡现状和管理人员描述，大坝渗漏现象较为明显。自水库建成运行至今，坝身常年存在散浸现象，大坝未溃决部分坝坡杂草丛生，坝身非常湿润，且散浸部位较高。在坝脚沿线可见渗水，并形成一片沼泽，可见坝基存在浅层渗漏隐患。另外，从溃口处出露的坝下涵管也可看出，涵管由多根圆管拼接而成，接缝处采用水泥砂浆黏结，并在涵管外包浆砌块石。混凝土已明显老化，接缝砂浆部分脱落，浆砌块石密实度不足。由此可判断，坝下涵管存在接触渗漏，渗水经涵管接缝进入坝体，坝下渗水可能与涵管渗漏有关。综合分析认为大坝存在严重的渗漏隐患，坝体自身渗透稳定性不足。

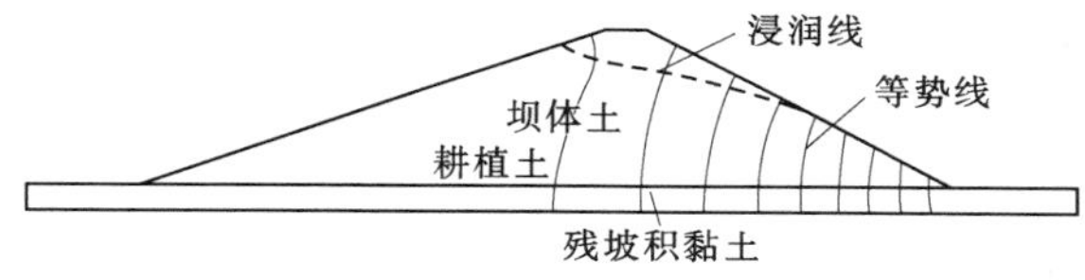

图2　当日最高运行水位下渗流计算结果

2.4 抗滑稳定性

在渗流计算基础上采用GEO-SLOPE程序对大坝下游坝坡进行抗滑稳定计算，结果见图3。由图可看出，下游坝坡抗滑稳定安全系数$K=1.031$（总应力法），小于规范[4]规定的允许值1.25，可见下游坝坡不稳定。

从溃坝当天的情形分析，由于经过3d连续暴雨，大坝处于高水位运行，坝体浸润线

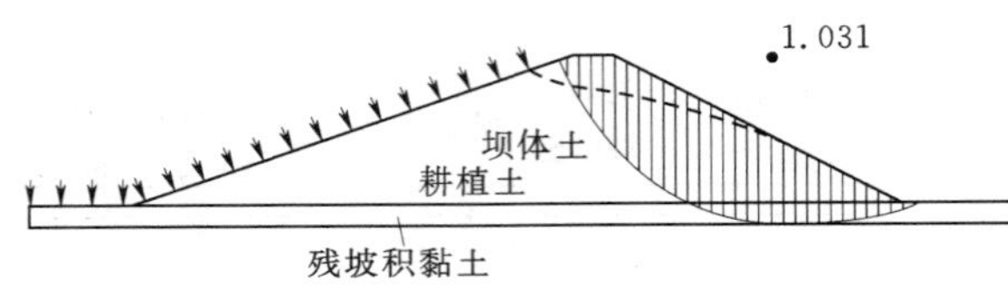

图 3　当日最高运行水位下下游坝坡抗滑稳定计算结果

居高不下，进一步降低了土体的抗剪强度；同时坝坡上杂草灌木丛生，使得雨水不能尽快流走而保存于坝体内，加剧了坝坡失稳的可能性。

2.5　监管不到位

由于该水库规模较小，因此无专管机构，多年来为村管，由人承包用于养鱼和集镇供水。承包人存在侥幸心理，未按有关部门要求降低水位运行。根据翰堂镇汛期水库巡查记录，5 月 17—20 日水库水位已达到正常蓄水位 59.00m，以致 21 日暴雨时库水位迅速上升并超出历史最高水位。同时，大坝出现险情至滑坡、溃坝时间短，且险情发展迅速，现场也无足够的砂石物料等防汛抢险物质。

3　溃坝模式分析

一般在大坝内部的薄弱环节和外部荷载共同作用下发生溃坝。由于内部薄弱环节的破坏和外部荷载的作用均存在不确定性，因此不同的荷载组合作用下，内部薄弱环节破坏不同，亦会出现不同的溃坝模式。对土石坝可能溃坝的模式主要有：

（1）漫顶溃坝。其原因是结构布置不合理、无溢洪道等泄洪设施；防洪标准低、溢洪道泄量不足或坝顶高程不足；泄洪闸门故障等。

（2）渗流破坏溃坝。其原因是坝体、坝基或坝下埋管发生渗流破坏造成水库溃坝，一般表现为集中渗流发展为管涌、流土或沿两种不同材料的接触面渗漏并带走周边土体。

（3）结构破坏溃坝。其原因是大坝整体单薄或筑坝材料较差、稳定性不足引起坝体滑坡或大坝局部结构隐患（如裂缝逐渐发展为渗漏通道、侵蚀坝体）引起局部结构破坏，进而导致整体溃坝。

（4）由于地震引起的水库大坝滑坡、坍塌、裂缝、渗漏等造成水库大坝溃坝。

根据上述分析结果，推断茄坑水库存在五种可能溃坝模式，见表 2。由表 2 推测出茄坑溃坝的主要路径为暴雨洪水—水位突升—渗流破坏（集中渗漏、接触渗漏）—大坝滑坡—形成小溃口—水流下切—冲沟、陡坎扩大加深—两侧土体失稳坍塌—溃口增大—溃坝。

表 2　　大坝溃决模式初步分析

建筑物及发生部位	破坏模式	机理	原　　因	后果
下游坝坡	滑坡	强度降低	下游坝坡陡，土体饱和	坝顶降低，漫顶溃坝
坝顶部	漫顶	超高不足	暴雨或引发洪水	漫顶溃坝
坝体	渗流破坏	内部冲蚀形成渗漏通道	填土料差，含水量大，碾压不密实，存在严重施工质量缺陷	管涌型溃坝
坝下涵管	渗流破坏	接触面冲蚀	涵管裂缝或断裂，沿管线形成渗透破坏	管涌型溃坝
坝基	渗流破坏	接触面冲蚀	常年形成的多个涌水点在洪水冲刷下形成渗漏通道	管涌型溃坝

4 结语

（1）茄坑水库溃坝是由渗流破坏和滑坡引起的大坝溃决，其溃坝模式和主要路径符合一般均质土坝溃坝特征。

（2）暴雨洪水引起库水位迅速上升是导致茄坑水库溃坝的诱发因素，大坝质量差、渗漏隐患严重、严管不到位是导致溃坝的内在因素，泄流能力不足是导致溃坝的外在因素。

参考文献：

［1］ 陈生水，钟启明，任强．土石坝漫顶破坏溃口发展数值模型研究［J］．水利水运工程学报，2009（4）：53-58.

［2］ 王占军，朱杰，袁辉，等．基于元胞自动机的土石坝溃决模拟［J］．水电能源科学，2011，29（7）：53-56.

［3］ 朱勇辉，廖鸿志，吴中如．国外土坝溃坝模拟综述［J］．长江科学院院报，2003，20（2）：26-29.

［4］ 黄河水利委员会勘测规范设计研究院．碾压式土石坝设计规范（SL 274—2001）［S］．北京：中国水利水电出版社，2002.

多层次模糊分析法在病险水库评判中的应用

洪文浩[1]，傅琼华[1]，徐镇凯[2]，胡国平[1]

1. 江西省水利科学研究院 江西省水工程安全工程技术研究中心；

2. 南昌大学 建筑工程学院

摘　要： 针对病险水库安全性态及其溃坝风险评估存在诸多的模糊性和不确定性等状况，提出基于多层次模糊分析法的病险水库溃坝风险评判模型，对病险库坝整体的安全性态进行多方面、多层次评估，为溃坝风险排序提供了一种新的途径。

关键词： 病险水库；溃坝；风险等级；多层次

对一座病险水库[1]进行综合整体安全性态评判，属于一个多目标决策问题，涉及到多方面、多层次、多因素的分析，具有一定的模糊性和不确定性，这就使得综合评判只能做出定性分析，难以做出量化评价，尤其是对一些病险水库所存在的溃坝风险往往给决策、管理者留下一个“模糊”的概念（只知道某病险水库存在一定的溃坝可能性，但它究竟处于一个什么样的等级、溃坝风险到底处于什么水准，定量分析）。本文提出的基于多层次模糊综合评价法的病险水库溃坝风险评判理论模型可以很好地解决这一问题，一方面，对病险库坝溃决风险的深入研究有一定的参考价值；另一方面，可根据溃坝风险大小，对批量的病险水库除险加固先后排序决策有一定的指导意义。

1　病险水库溃坝风险等级评判模型的构建

1.1　指标体系的建立

水库作为一个系统，由许多建筑物（主坝、副坝、溢洪道等）所组成，而每一座建筑物的工作性态将直接影响整座水库的安危，因此，通过逐层建立各影响因素指标形成一个层次结构（见图1），从而进行综合整体评判。一般情况下，建立三级因素评价指标体系，

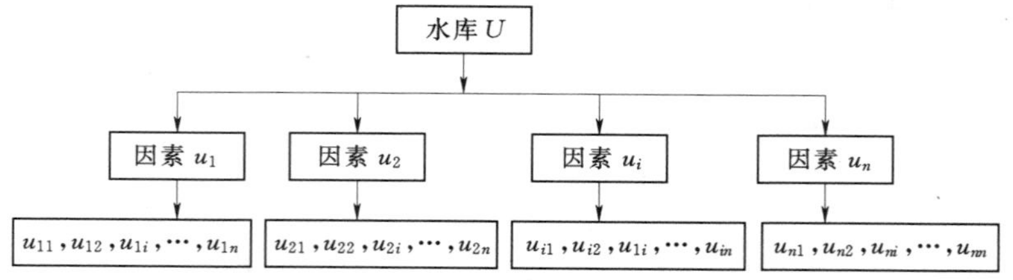

图1　病险水库溃坝风险综合评判指标体系

本文发表于2012年。

其中，建立和运用指标的目的不同，指标（U_i、u_{ij}，$i=1$，2，…，n，$j=1$，2，…，n 可根据实际情况和具体问题赋予不同指标涵义）的作用也不同。

1.2 综合评判模型的构建

病险水库溃坝风险等级综合评判模型的相关理论及实施方法如图2所示[2]。

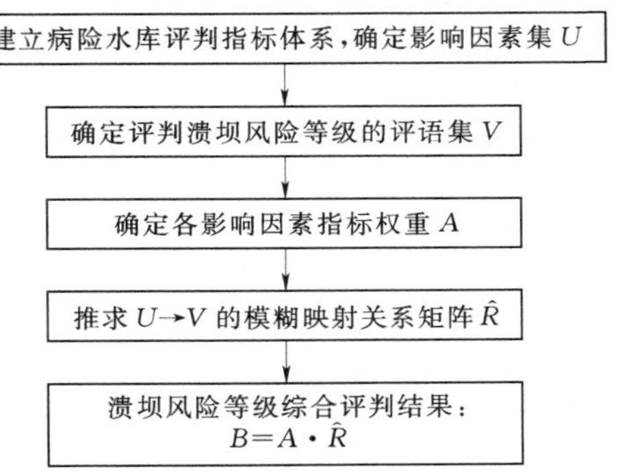

图2 病险水库溃坝风险等级综合评判实施程序及步骤

1.2.1 确定被评判病险水库的因素集

对被评判的对象进行分析，将既有水库分为 n 个组成部分，其中每个组成部分有 m 个影响因素，所有这些因素构成对水库溃坝风险等级评判指标体系，即因素集：

$$U=[U_1,U_2,U_i,\cdots,U_n],U_i=[u_{i1},u_{i2},u_{ii},\cdots,u_{im}]$$

1.2.2 确定被评判病险水库的评语集

病险水库溃坝风险评判指标的评价值的不同，往往会导致形成不同的等级，一般情况下，水库溃坝风险评判的评语集可分为4个级别，即 $V=[V_1, V_2, V_3, V_4]$，其中的含义分别代表［低风险、中等风险、高风险、超高风险］。

1.2.3 确定被评判病险水库各因素的权重

由于组成水库的各种水工建筑物及其各个因素在综合评价中所起的作用也不尽相同，综合评判的结果不仅与各因素影响大小有关，而且在很大程度上还依赖于各因素对综合评价所起的作用。因此，需要确定一个因素之间的权重分配，它通常被视为 U 上的一个模糊向量，$A=[a_1, a_2, a_i, \cdots, a_n]$，其中，$a_i$ 代表第 i 个影响因素权重的大小。权重的确定，通常有以下几种方法：Delphi 法、加权平均法、众人评估法、层次分析法。综合考虑，本文采用层次分析法来求解各影响因素的权重。

1.2.4 病险水库各指标的单因素评价

参照相关各因素、各等级的评价标准，对不同的定量和定性描述，确定不同的隶属度，则需要推求 $U\to V$ 的模糊映射关系矩阵：$\hat{R}_i=(r'_{ij})_{n\times m}$，其中 r'_{ij} 为因素 U_i 被评为 V_j 的隶属度（$j=1$，2，3，4）。隶属度的确定，需要通过借助隶属度函数来计算；隶属函数可以用来定义模糊集合[3]。在溃坝风险等级评语集 V 中对任一集合 $V_i\in F(u)$，可确定一个从 U 到 $u_{vi}(u)$ 的一个映射

$$u_{vi}:U\to u_{vi}(u),U\to u_{vi}(u)=\begin{cases}1,u\in v_i \\ u_{vi}(u), & u\text{处于中间状态} \\ 0,u\notin v_i\end{cases}$$

任一从 $U\to u_{vi}(u)$ 的映射 u_{vi} 都有唯一子集与其对应：$V_i=\{u\,|\,u_{vi}(u)=u_{vi}\}$，$V_i$ 以 $u_{vi}(u)$ 为其隶属函数，V_i 的隶属函数在 u 处的值 $u_{vi}(u)$ 即为 u 对 v_i 的隶属程度。

水库的溃坝风险等级，其隶属度的计算可以采用专家打分法来确定各风险事件的影响后果 C 的估计值及其发生概率 P 的估计值，然后将 P 与 C 的乘积代入溃坝事件对于风险等级的隶属函数，即可得到溃坝风险事件对应的风险等级隶属度[4]。

（1）确定溃坝风险事件的影响后果 C 值及风险概率 P 值。根据病险水库的工程实况，

借鉴相关参考资料及研究成果，总结得出如下溃坝风险事件的影响后果估值的转换方法（见表 1、表 2）。

表 1　　溃坝风险事件影响后果估值转换

事件严重性/级别	估值	生命损失 L/人	风险人口/人	城　　镇
特别重大（Ⅰ级）	5	$L \geqslant 50$	$>10^6$	首都、省会、直辖市
重大（Ⅱ级）	4	$50>L\geqslant 10$	$10^4 \sim 10^6$	地、县级市府或城区
较大（Ⅲ级）	3	$10>L\geqslant 3$	$10^2 \sim 10^4$	乡镇政府所在地
一般（Ⅳ级）	2	$L<3$	$<10^2$	乡村和散户

表 2　　溃坝事件发生概率估值转换

定性描述	估　　值	发生的频率范围	备　　注
罕见的	1	<0.0003	溃坝极难发生一次
偶然的	2	0.0003～0.003	溃坝不大会发生
可能的	3	0.003～0.03	溃坝有可能发生
预期的	4	0.03～0.3	溃坝不止一次发生
频繁的	5	>0.3	溃坝事件频繁发生

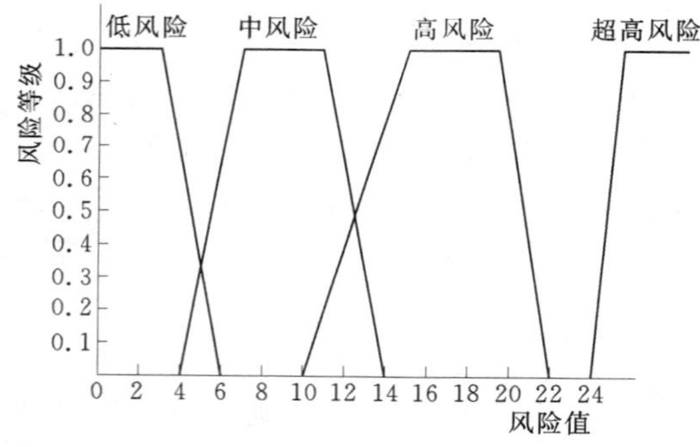

图 3　溃坝风险等级隶属函数

由表 1 与表 2 所估算的值得到 P 与 C 的乘积作为分级的指标，再根据模糊数学理论及其相关原则确定对应的隶属函数（如图 3 所示），$R'_i=[r'_{i1}, r'_{i2}, r'_{i3}, r'_{i4}]=$[低风险，中等风险，高风险，超高风险]，$(i=1, 2, \cdots, n)$。

$$r'_{i1}=\begin{cases}1, 0<v\leqslant 4\\ \dfrac{8-v}{2}, 4<v<6\\ 1, v\geqslant 6\end{cases}, r'_{i2}=\begin{cases}0, v<4, v>14\\ \dfrac{v-4}{2}, 4\leqslant v\leqslant 6\\ 1, 6<v\leqslant 10\\ \dfrac{14-v}{4}, 10\leqslant v\leqslant 14\end{cases},$$

$$r'_{i3}=\begin{cases}0, v<10, v\geqslant 22\\ \dfrac{v-10}{4}, 10\leqslant v<14\\ 1, 14\leqslant v\leqslant 20\\ \dfrac{22-v}{2}, 20<v\leqslant 22\end{cases}, r'_{i4}=\begin{cases}0, v<20\\ \dfrac{v-20}{2}, 20\leqslant v<22\\ 1, v\geqslant 22\end{cases}$$

（2）溃坝风险因素指标的量化分析。在对某水库进行溃坝风险等级综合评判时，可参照水建管〔2007〕164 号文《水库大坝安全评价导则》，将大坝安全分为 A、B、C 三级；其中，A 级为安全可靠，B 级为基本安全，但存在缺陷，C 级为不安全。因此，本文也将各项风险因素评价指标分为三级，并赋予相应的风险率评分来估值，即 A 级 0～2；B 级 2～4；C 级 4～5。溃坝风险影响后果评分值可参照相关资料结合实际工程进行大致估值。

2 病险水库溃坝风险等级综合评判结果

通过隶属函数结合相应指标推求的映射关系矩阵 $\hat{R}_i$ 与权重向量进行模糊运算，从而得到各级风险事件的隶属度向量 B_i，即 $B_i = A_i \cdot \hat{R}_i$（$i$ 表示层次数目，$i=1, 2, 3, \cdots, n$）。一般而言，有 3 个层次，通过逐层计算，最终可得到整座水库溃坝风险等级的隶属度，记 $B = A \cdot \hat{R} = (b_1, b_2, \cdots, b_m)$，若 $b_j = \max\{b_1, b_2, \cdots, b_m\}$，则由最大隶属度原则认为该评判对象相对隶属度等级为 v_j，从而可以判定该病险水库溃坝风险级别。

3 工程实例

3.1 工程简介

某水库位于章江河上游大余县境内，是一座以防洪为主，结合发电、养鱼、灌溉供水的综合利用的大（2）型水利枢纽工程。坝址控制集水面积 557km^2，水库总库容 $1.19 \times 10^8 \text{m}^3$，防洪保护人口 8 万余人，农田 $3.8 \times 10^4 \text{m}^2$，改善灌溉面积 $4.0 \times 10^4 \text{m}^2$，发电装机容量 6000kW，多年平均发电量 $2.416 \times 10^7 \text{kWh}$。该水库枢纽由大坝、副坝、溢洪道、发电引水隧洞及电站厂房等建筑物组成。

3.2 某水库溃坝风险等级综合评判

根据本文提出的理论方法结合某水库相关资料可知，该水库大坝属三类坝，属于病险水库范畴。按照相关步骤对其进行溃坝风险等级综合评判。

（1）某水库溃坝风险等级评判指标体系的建立。根据《某水库大坝安全分项评价报告集》中的计算复核及相关评述，结合水库溃坝风险等级综合评判理论建立以下指标结构体系（见表 3）。

（2）水库溃坝因素影响权重的计算。通过表 3 中所建立的指标层次，采用层次分析法[5]计算各影响因素的权值大小。借鉴专家的观点及意见，结合层次分析法将各指进行两

表 3　某水库溃坝风险等级评判指标层次

一级指标	二级指标	三级指标
某水库 U	工程质量风险 U_1	主坝 u_{11}（坝基及岸坡、坝体质量不满足要求） 副坝 u_{12}（坝基及坝体质量不满足要求） 溢洪道 u_{13}（基础处理不满足要求） 发电引水隧洞 u_{14}（施工质量不满足要求）
	工程运行管理风险 U_2	大坝运行调度不规范 u_{21} 大坝维修不及时 u_{22} 大坝安全监测设备不完善 u_{23}
	工程结构风险 U_3	主坝 u_{31}（坝体变形为 B 级，下游坝坡稳定为 C 级） 副坝 u_{32}（坝体变形为 B 级，下游坝坡稳定为 C 级） 溢洪道 u_{33}（混凝土结构强度为 C 级） 发电引水隧洞 u_{34}（混凝土结构强度为 C 级）
	渗流风险 U_4	主坝 u_{41}（坝体及坝基、坝肩） 副坝 u_{42}（坝体及坝基） 山体 u_{43}（副坝与溢洪道之间）

两比较其相对重要性，构造相应的判断矩阵。通过计算，影响因素 U_1 所对应的各指标之间的权重为 $A_1=[0.347，0.347，0.198，0.107]$。同理，可分别求出影响因素 U_2、U_3、U_4 对应各指标的权重 $A_2=[0.33，0.48，0.19]$，$A_3=[0.286，0.286，0.286，0.142]$，$A_4=[0.44，0.38，0.18]$。

目标层 U 对应的影响因素指标权重 $A=[0.20，0.33，0.30，0.17]$。

（3）溃坝风险等级评价指标隶属度的计算。参考《某水库大坝安全分项评价报告集》并结合本文提出的理论，综合分析可得工程质量、工程运行管理、工程结构、渗流性态对应的评估（映射关系）矩阵。

$$\hat{R}_1=\begin{bmatrix}0&0&1&0\\0&0&1&0\\0&0&1&0\\0&0&1&0\end{bmatrix},\hat{R}_2=\begin{bmatrix}0&1&0&0\\0&0&1&0\\1&0&0&0\end{bmatrix},\hat{R}_3=\begin{bmatrix}0&0&1&0\\0&0&1&0\\0&0&1&0\\0&0&1&0\end{bmatrix},\hat{R}_4=\begin{bmatrix}0&0&1&0\\0&0&1&0\\0&0&1&0\end{bmatrix}$$

（4）溃坝风险级别综合评判。由前面所计算得到的权重向量 $A_i(i=1，2，3，4)$ 以及风险因素评估矩阵 $\hat{R}_i(i=1，2，3，4)$，可得对应溃坝风险等级隶属度向量 $B_i(i=1，2，3，4)$，即 $B_i=A_i\cdot\hat{R}_i$。

先计算二级风险因素等级隶属度向量：

$$B_1=A_1\cdot\hat{R}_1=[0\ 0\ 1\ 0],B_2=A_2\cdot\hat{R}_2=[0.19\ 0.33\ 0.48\ 0]$$

$$B_3=A_3\cdot\hat{R}_3=[0\ 0\ 1\ 0],B_4=A_4\cdot\hat{R}_4=[0\ 0\ 1\ 0]$$

将上述隶属度向量 $B_i(i=1，2，3，4)$，组成新的评判矩阵，可得到水库大坝整体模糊综合评估矩阵 $\hat{R}$，即 $\hat{R}=[B_1\quad B_2\quad B_3\quad B_4]$，再将目标层对应的影响因素指标权重 A 作模糊运算，即 $B=A\cdot\hat{R}$，则可得到水库溃坝风险等级综合评判隶属向量，从而可判定该水库溃坝风险级别。

$$B=A\cdot\hat{R}=[0.20\quad 0.33\quad 0.30\quad 0.17]\times\begin{bmatrix}0&0&1&0\\0.19&0.33&0.48&0\\0&0&1&0\\0&0&1&0\end{bmatrix}$$

$$=\begin{bmatrix}\text{低风险}&\text{中等风险}&\text{高风险}&\text{极高风险}\\0.0627&0.1089&0.8284&0\end{bmatrix}$$

通过上述计算结果，对比水库管理人员反映实际情况及现场查勘可知，评判结果与现状工程实际基本吻合。根据模糊数学理论中的最大隶属度原则综合分析可知，该水库溃坝风险等级处于高风险状态，其中工程质量影响因素处于低风险等级，工程运行管理影响因素处于中等风险状态，工程结构失事影响因素是引起溃坝的主要风险所在。因此，必须重点对组成水库建筑物的各个部分进行重点加固，以降低或避免溃坝事件发生的风险概率。

4 结语

本文运用模糊综合评判理论对病险水库溃坝风险等级及其安全性态进行多方面、多层次评估。较客观地反映了病险水库大坝的整体安全性，将定量与定性分析相结合，为病险水库溃坝风险分析评估提供了一条新的途径，同时也为病险库坝的除险加固提供了科学依据，该理论方法的提出也为群坝的病险排序提供了新的思路。

参考文献：

[1] 彭雪辉．风险分析在我国大坝安全上的应用［D］．南京：南京水利科学研究院，2003.

[2] 朱丽楠，陈坚，张雷．模糊数学理论在泵站工程老化评价中的应用［J］．武汉大学学报（工学版），2003，36（4）：32－35.

[3] 刘云，王亮，申林方，等．大坝安全风险评价的模糊层次综合模型［J］．水科学与工程技术，2007（1）：26－29.

[4] 何锡兴，周红波，姚浩．上海某深基坑工程风险识别与模糊评估［J］．岩土工程学报，2006，28（S1）：1912－1915.

[5] 许树柏．层次分析法原理［M］．天津：天津大学出版社，1988.

康山蓄滞洪区洪水风险图研制分析

胡松涛，孙军红

江西省水利科学研究院

摘　要： 根据康山蓄滞洪区基础资料，分析不同洪水风险图编制方法，并研制了康山蓄滞洪区洪水风险图，给防汛部门实施指挥决策、抢险救灾提供依据，确保康山蓄滞洪区分洪运用前人员安全及时转移，最大限度地避免人员伤亡，减轻财产损失。

关键词： 江西；康山蓄滞洪区；洪水风险分析；风险图

长江中下游防洪实行蓄泄兼筹，以泄为主的方针。蓄滞洪区是综合防洪体系重要组成部分，在防洪紧急关头能够发挥削减洪峰、蓄滞超额洪水，是牺牲局部，保护全局，减轻洪水灾害损失的有效措施和防洪调度的重要手段[1]。根据长江流域防洪规划方案，江西省鄱阳湖蓄滞洪区分洪运用条件为当湖口水位达到 20.59m（吴淞高程 22.50m），并预报继续上涨且危及长江重点堤防安全时启用。

1　康山蓄滞洪区概况

1.1　工程概况

康山蓄滞洪区是国家重要蓄滞洪区之一，是江西省最大并首先启用的蓄滞洪区，负责分担蓄洪量 $15\times10^8m^3$。康山蓄滞洪区位于鄱阳湖东岸，上饶市余干县城西北部，赣江南支、抚河、信江三河汇流口的尾闾。蓄滞洪区北部以康山大堤与鄱阳湖分界，西抵康垦农场，东至古竹，南达瑞洪镇。区内以湖积平原洼地为主，地势南高北低。北部地形平坦开阔，湖汊密布，间或有丘陵出露，地面高程一般在 14.00～16.00m 之间，南部属浅丘地貌，地面高程一般在 13.00～23.00m，区内结合农业开发筑有分片生产堤，堤顶高程一般为 17.00～18.00m。分洪口门位置在康山大堤 18＋120～18＋500 桩号处，口门宽 380m。

康山蓄滞洪区集雨面积为 $450.3km^2$（包括信瑞联圩 $106.9km^2$），其中蓄洪面积 $312.37km^2$，总蓄洪容积 $16.58\times10^8m^3$，有效蓄洪容积 $15.92\times10^8m^3$（分蓄洪区高程—面积—容积关系见表 1），设计蓄洪水位为 20.68m（吴淞高程 22.55m），设计蓄洪量为 $16.58\times10^8m^3$，设计进洪流量为 $9630m^3/s$。

1.2　社会经济情况

康山大堤是在 1966 年当时的历史条件下，以围垦灭螺为主要目的而兴建的。2010 年区内有瑞洪镇、石口镇、三塘乡、大塘乡、康山乡、康山垦殖场、大湖管理局等 7 个乡、

本文发表于 2011 年。

表 1　　康山蓄滞洪区高程—面积—容积关系表

高程（黄海）/m	面积/km^2	总容积/$10^8 m^3$
15.00	260.00	
16.00	277.88	2.69
17.00	289.83	5.53
18.00	294.83	8.45
19.00	300.50	11.43
20.00	307.46	14.47
20.68	312.37	16.58
21.00	314.68	17.58

镇、场、局共 55 个行政村，共计 21711 户、109322 人。蓄洪水位以下需要安全转移的居民有 8439 户、36801 人，占总户数 38.9%，总人口的 33.7%。区内居民居住集中，大部分在南面丘陵岗地。

蓄滞洪区属鄱阳湖冲积平原，土地肥沃。现有耕地 13134hm^2，养殖水面 11140hm^2。区内经济以农业为主，主要是水稻、油菜作物和水产品等。据统计，年产粮食 6.2×10^4t，油料 350×10^4kg，水产品 350×10^4kg，工农业总产值 71280 万元，其中工业产值 35220 万元，农业产值 36060 万元。

2　洪水风险图研制

2.1　洪水风险图的意义及作用

洪水风险图是反映洪水风险大小和区域分布的地图，一般标明特定频率洪水发生时的淹没范围、受淹区域内的最大水深和淹没历时等信息。从国内外经验看，它是一种被广泛采用的防洪非工程措施。洪水风险图能够为防洪规划制定、运用预案编制、蓄滞洪区转移与避难方案编制、洪水保险费率制定等提供洪水风险信息。绘制康山蓄滞洪区洪水风险图，标示洪泛区内各处受灾的危险程度，准确掌握蓄滞洪区基础资料，可确保蓄滞洪区运用前人员安全及时转移，最大限度地避免和减少人口伤亡，减轻财产损失[2]。

2.2　洪水风险图的内容及编制方法

2.2.1　洪水风险图的内容

洪水风险图编制的步骤为收集整编相关资料、选择洪水风险分析方法、进行风险分析、绘制洪水风险图、编写编制说明等。洪水风险图编制说明的主要内容包括：流域水文、气象特征、排水系统和水利工程概况，历史典型洪涝灾害特点及后果；流域社会经济特征统计；分区域阐述洪水灾害的特点和性质，灾害后果和经济损失；洪水风险图的编制依据、方法和存在问题；洪水风险发生时的应急措施、洪水风险图的应用范围；其他说明。完整的洪水风险图应该是三位一体的组合，即洪水发生的频率、洪水的淹水区域分布及有关说明、洪水灾害可能造成的各类损失[3]。

2.2.2　洪水风险图的内容和编制方法

2.2.2.1　实际洪水法

该法的基本假定是流域自然地理特征保持基本不变条件下，洪水具有重现性。流域历史上已经发生过的大洪水实际淹没情况，可以作为现在和未来同类洪水重现时的淹没状态。分析历史洪水淹没实况的主要途径如下：

（1）对于近期发生的洪水，利用流域实测水文资料和灾情资料可以较为可靠地分析洪水特性及相应的淹没范围、淹没深度和淹没时间。

（2）对于缺乏资料或年代较为久远的洪水，可以通过调查考证的途径分析洪水发生时的淹没情况。

（3）通过对洪水地貌分析，初步判断洪水径流的强度、范围和水深，作为分析淹没实况的依据。

2.2.2.2 水文学和水力学方法

根据流域现状或规划中土地利用特征和工程条件，采用水文学和水力学方法分析流域洪水泛滥后的淹没状况。目前，国内外流行的水文学和水力学方法的模型较多，采用何种方法和模型应该针对流域水文地理特征、工程调度方式、资料条件以及计算精度来选择。

水文学法以洪水不确定性反映洪水风险。通过洪水风险分析、区域水文分析，获取不同洪水频率的洪量、水位等信息，然后绘制不同频率洪水的淹没范围、淹没水深，最后完成不同频率的洪水风险图。

水力学法是基于不同洪水调度或工程运行条件下，根据洪水过程中的水力学特征值变化情况，研究洪水发生时可能的威胁及危害程度。洪水风险分析一般采用一维水力学法和二维水力学法。一维水力学法主要用于推求不同频率的河道水面线；二维水力学法主要用于分析堤防、大坝溃决后的洪水演进路径、淹没区范围。

2.2.2.3 通过 GIS 反演洪水推进法

随着地理信息系统（GIS）技术的高速发展，将 GIS 技术与水文模型相结合，再根据数字高程模型 DEM 预测模拟显示洪水淹没区，并进行灾害评估，已成为 GIS 应用和水利领域一个研究热点，应用地理信息系统方法反演洪水最大淹没水深的基本原理和计算方法，对淹没区所发生的洪水进行分析。

2.3 康山蓄滞洪区洪水风险图研制

通过调查分析康山蓄滞洪区地理特征、降雨与水文特征、防洪工程与排水设施、历史洪水和洪涝灾害等的基础资料。根据康山蓄滞洪区特点及洪水风险图制作的目的与用途，综合以上情况，采用 GIS 反演洪水推进法进行康山蓄滞洪区洪水风险分析和风险图制作[4]。

2.3.1 数据准备

利用 GIS 技术和数据库技术，将模型计算结果数据通过接口转换程序，形成系统中专用的方案模型数据，并存储在系统数据库中。将数字地面高程模型（DTM）、防洪工程分布图、行政区划图合成为工作底图，把等高线、高程点、道路面、水系面等高程数据分别导入各自的 Access 数据库，在 Arcgis 中以不同图层显示各自信息。

2.3.2 数据处理

利用 GIS 的空间数据管理、分析和制图等方面的优势，通过编程实现图形导航、放大、缩小、漫游、标注、图层控制等基本功能，将基础数据库中的记录快速反映到图形对象上，然后系统利用等值线自动生成模块，形成最大淹没水深等值线图层，依据洪水风险分析提供的成果，进行危险程度分区。再在图中叠加区内各村镇转移安置情况、转移线路等数据并用不同颜色绘制成洪水风险图。康山蓄滞洪区洪水风险图是按照达到分洪水位20.68m时，分洪口门充分进洪，直至蓄满的静态水平面的淹没范围，如图1所示。

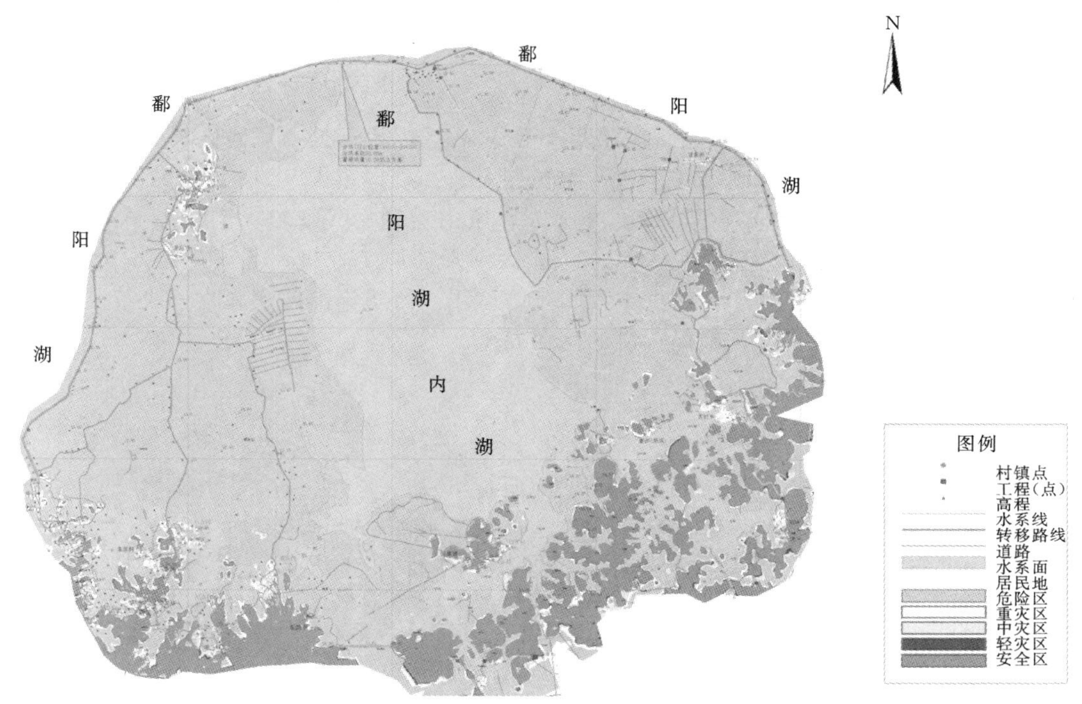

图1 康山蓄滞洪区洪水风险示意图

2.3.3 编制步骤

1）以等高线的高程字段、高程点的高程字段、水系面的高程字段、道路面的高程字段，加一个裁剪边界的图层建模，构TIN（3D分析/创建TIN）。

2）从TIN→DEM（栅格），栅格数据格网大小选2～6m（3D分析）。

3）根据预案将洪水风险分区，危险程度分5类：危险区、重灾区、中灾区、轻灾区和安全区，分区标准见表2（3D分析/重分类）。

表2　　蓄滞洪区危险程度分区表

危险分区		危险区	重灾区	中灾区	轻灾区	安全区
高程/m	黄海	＜18.68	18.68～19.68	19.68～20.18	20.18～20.68	＞20.68
	吴淞	＜20.55	20.55～21.55	21.55～22.05	22.05～22.55	＞22.55
淹没水深/m		＞2	1～2	0.5～1	0～0.5	＜0
颜色		橙	黄	绿	蓝	灰

4）配置颜色，添加其他图层。

康山蓄滞洪区洪水风险图比例尺大小是根据康山蓄滞洪区面积、洪水频率、淹没范围、资料条件及精度要求确定。在勾绘洪水淹没范围边界时，考虑了洪水的可能路径，结合地形情况，对可能淹没区域设置彩色编码区，其颜色深浅表示淹没深度的变化。风险图上标注了重要的部门和单位，如政府机关、学校、医院、居民区、村镇以及重要设施，水利工程等。

3 结语

从康山蓄滞洪区洪水风险图中可以得出，分洪后区内大部处于危险区。分洪后沿湖低洼位置可采取就近就高的原则转移。各乡需转移人数：石口镇1254户计6157人，大塘乡782户计3582人，瑞洪镇4740户计20809人，康山垦殖总场110户计572人，康山乡1531户计5618人，大堤管理局6户计18人，大湖管理局16户计45人，共计36801人。康山洪水风险图的编制保证了蓄滞洪工程的正常运用，给防汛部门实施指挥决策、抢险救灾提供依据，做到临危不乱，确保人民的生命安全。

参考文献：

[1] 中华人民共和国防洪法［S］. 1997.

[2] 国家防汛抗旱总指挥部办公室．洪水风险图编制导则（试行）［S］. 2005.

[3] 孙桂华，等．洪水风险分析制图实用指南［M］. 北京：水利电力出版社，1992.

[4] 杜国志，李开杰．洪水风险图有关问题探析［J］. 中国水利，2006（5）.

柘林水库大坝防洪安全风险分析

余　雷[1,2]　伍春平[3]　兰盈盈[4]

1. 江西省水利科学研究院；2. 河海大学水电学院；

3. 江西省赣抚平原水利工程南昌县管理站；4. 南昌工程学院水利工程系

摘　要： 水库防洪调度中存在水文、水力、防洪调度方法等不确定性因素，这些因素都将带来水库防洪安全风险。选择1954型和1998型洪水，按照各因子的分布规律，利用蒙特卡罗方法，分别研究水文、水力等因素给柘林水库防洪所带来的风险以及它们的组合风险。研究结果表明，洪水要素对大坝防洪安全影响较大，尤其是洪水线型的影响更大。考虑因素越多，风险值越大，但综合分析认为柘林水库大坝是安全的。

关键词： 防洪风险；蒙特卡罗；随机模拟；柘林水库

1　概述

在水库泄洪过程中，存在着许多人们难以预料和控制的不确定性因素，包括水文的不确定性、水力的不确定性、数据的不确定性和计算中的不确定性等[1]。其中，入库洪水（水文的不确定性）是影响调洪最高水位的主要因素[2]。对于这些不确定性对泄洪风险的影响，人们在认识上不断深入[3]，在风险计算中，从最初只考虑水文不确定性，到综合考虑水文和水力两方面的不确定性[4]，从只考虑水文不确定性中的洪峰流量的不确定性到既考虑洪峰流量的不确定性，又考虑洪水过程的不确定性。

在水库调洪过程中，不确定性因素主要有：①入库洪水过程的不确定性，包括洪峰流量、控制时段洪量及洪水过程线形状等；②出库泄洪能力的不确定性，引起泄洪风险的水力不确定性因素主要有测量误差和模型误差；③防洪限制水位的选择，由于行政性限制，汛限水位一旦确定，无法随意更改，实际操作中只能存在观测误差，但其误差较小；④库容与水位关系的不确定性，包括库容变化及量测计算误差方面因素。这些因素导致了库水位变化过程的随机性，也最终导致了与防洪风险率相关的水库调洪最高水位的随机性。大坝防洪风险是一种综合风险，由诸多影响因子组合而成。本文考虑水文和水力因素对江西柘林水库防洪安全进行风险分析。

2　研究区概况

柘林水库位于江西省武宁县中部，属亚热带季风气候，气候温和湿润，四季分明，雨水充沛。柘林水利枢纽是一座以发电为主兼有防洪、灌溉、航运、旅游和水产等综合效益

本文发表于2009年。

的大型水利水电工程，于 1975 年末建成。水库集雨面积 9340km²，占整个修水流域面积的 63.5%，总库容 79.2 亿 m³，为目前亚洲已建成土坝库容最大的水库，其中防洪库容 32 亿 m³，兴利库容 34.4 亿 m³；设计水位（1000 年一遇洪水）为 70.13m，设计洪峰流量 18250m³/s，5d 设计洪量 41.5 亿 m³，可能最大洪水的 5d 洪量 70.9 亿 m³；可能最大洪水的洪峰流量 25600m³/s，可能最大洪水的最高水位为 73.01m，设计坝顶高程 73.5m（主坝）、73.4m（副坝）。

3 入库洪水序列的模拟

3.1 洪水分析

研究区的洪水为暴雨洪水，降雨强度、时空分布不同，会产生形状不同的洪水过程线，所以洪峰流量、洪水总量和洪水过程线的变化均会对防洪带来水文风险，尤其是洪水线型对水库防洪安全产生一定的影响。洪水特征变化是洪水风险率分析的主要对象，本文洪水分析的依据为 1953—2006 年入库洪水实测资料。按典型洪水的选取原则，经过分析筛选，从坝址水文站的洪水资料中选取了 2 个较典型的洪水，分别为 1954 年洪水（峰高量大），年最大洪峰流量为 11300m³/s，5d 洪量为 26.20 亿 m³；1998 年“98626”洪水（复峰，峰型较肥，主峰靠后），最大洪峰流量为 8540m³/s，5d 洪量为 24.6 亿 m³。洪水过程线如图 1 所示。由于柘林水库库容较大，为了更有利于水库防洪安全风险分析，增加洪水的风险性，这里采用 5d 洪量的同倍比放大法。

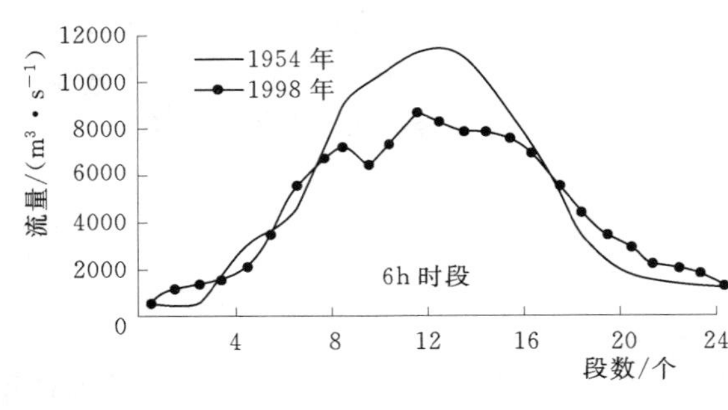

图 1 典型洪水过程线

3.2 模拟方法

近年来，国内外学者致力于这方面研究，提出了不同方法。如随机微分方程法[3]、极限状态法（或 JC 法）[5]、全概率法[6]及随机模拟法（MC 法）[7]。其中，极限状态法以最大入库洪峰流量等于溢洪道泄流能力为极限状态计算泄洪风险率，此法不考虑调洪库容的防洪作用，所以影响了其结果的合理性；全概率法受到确定随机变量联合概率分布函数的限制，不利于较复杂情况的风险分析；随机微分方程法则将整个调洪过程中水库蓄洪量（或库水位）视为一随机过程，通过求解随机微分方程，获得不同时刻库水位的随机分布，然后计算泄洪风险率。随机微分方程法假定水库蓄洪量围绕其均值过程线随机游动，且服从正态分布，分别求出其均方解和解过程的概率密度。但其求解过程复杂。本文采用的 MC 法（Monte Carlo 法），即随机模拟方法，就是用计算机模拟随机现象，根据待求随机问题的变化规律，或随机变量的概率分布，依照一个合适的概率模型进行大量的统计分析，得到分析结果，再进行分析推断，得到随机问题的求解。与其他方法相比，MC 法有以下优点：①收敛速度与问题维数无关；②受问题的条件限制影响小；③程序结构简单，计算结果比较精确。用 MC 法计算风险的关键是产生已知分布的随机变量的随机数，利用计算机进行大量模拟运算，用模拟运算的结果来估算风险。

3.3 随机数的生成

研究资料表明：所生成的洪水序列能与设计洪水特征基本一致，P—Ⅲ型模型可用于随机生成洪水[11]。用 MC 法模拟失效概率，必须要生成已知分布类型的基本随机变量的随机数，然后再按随机变量的分布类型将其转换为符合该分布类型的随机参数。具体方法和步骤如下：

（1）天然河道来水的5d洪量服从P—Ⅲ型分布[8]，根据已有资料，拟合出水库上游的5d洪量序列符合P—Ⅲ型分布（见式（1）），计算其统计参数均值、方差、变差系数和偏态系数分别为：10.2、5.71、0.56、1.68。

$$f(x)=\frac{\beta^{\alpha}}{\Gamma(\alpha)}(\overline{x}-a_0)^{\alpha-1}\mathrm{e}^{-\beta(\overline{x}-a_0)} \tag{1}$$

式中：α 为形状参数，$\alpha=\frac{4}{C_s^2}$；a_0 为位置参数，$a_0=x\left(1-\frac{2C_v}{C_s}\right)$；$\beta$ 为尺度参数；$\beta=\frac{2}{xC_vC_s}$；C_v 为变差系数；C_s 为偏态系数；$\overline{x}$ 为 x 均值。

（2）在开区间（0，1）上生成服从均匀分布的伪随机数。伪随机数产生的方法有乘同余法、混同余法等。乘同余法以其统计特性优良、周期长的特点而被广泛使用。本文所需要的（0，1）伪随机数便是采用乘同余法获得的，并进行参数检验。均匀性检验和独立性检验结果较满意。

（3）采用变换方法随机产生洪峰 Q_m 系列，利用上述洪水放大法计算洪水过程系列，根据所选用的线形，得到随机洪水数据。

4 大坝防洪安全风险分析计算

4.1 防洪安全风险模型

目前大坝防洪安全设计普遍采用传统设计方法（PFD法）确定水库的各防洪特征水位及相应的泄洪建筑物型式、尺寸、高程等[2]。国内开展的风险研究中，风险大多都是指事故概率，而不包括考虑事故的后果[9]。

大坝防洪风险率，是指在设计的坝顶高程及泄洪建筑物规模下，水库遭遇洪水时，不能安全度汛，导致大坝失事事件的概率。通常认为坝前最高水位 Z 超过坝顶高程 Z_d 是形成防洪安全风险的明确标志[9]。其极限状态方程为：

$$Z'=Z-Z_d=0 \tag{2}$$

由此确定的防洪风险率应以年最高调洪水位越过坝顶高程的频率来度量，即：

$$P_f=P\{Z>Z_d\}=P\{Z'>0\}=\int f(z)\mathrm{d}z \tag{3}$$

式中：Z'为超过坝顶高程洪水位，m；$f(z)$ 为超高洪水位 Z'的概率密度函数。

根据式（3）可得综合防洪风险计算模型为：

$$P_f=P\{Z>Z_d\}=\frac{s}{M}\times 100\% \tag{4}$$

式中：M 为随机试验总次数；s 为出现 $Z>Z_d$ 的次数。

从上述可见，影响大坝防洪风险率的主要随机因子是 Z 和 Z_d，而坝前最高库水位 Z 及相应的防洪风险率是与水库的整个调洪过程联系起来的。显然，在整个调洪过程中，涉及水文、水力、防洪调度等诸多因素，相应的防洪风险率实际上是上述诸多不确定性的综合反映。除洪水预报和大坝结构外，洪水过程、泄洪能力以及起调水位所带来的风险是目前大坝防洪安全风险分析的主要问题。本文着重研究由洪水过程、泄洪能力的不确定性而产生的防洪风险率，同时进行不同组合的综合风险率计算分析。

4.2 大坝防洪风险率计算

4.2.1 洪水风险率计算

洪水是一个十分复杂的动态随机过程，通常用洪峰流量、洪水总量和洪水过程线等洪水特征来描述。根据现有洪水资料筛选分析，选择 1954 年和 1998 年两种较为不利的典型洪水过程线，然后按照第 3 节入库洪水序列的模拟方法，得到供调洪演算的洪水序列，再根据式（4）可得洪水风险率 P_f。

4.2.2 泄洪风险率计算

泄洪风险主要由建筑物施工误差和模型误差引起的，水力不确定性因素主要包括泄洪建筑物尺寸、材料和流量系数等。国内外学者经过试验统计分析认为，这些水力不确定性因素所产生的误差绝大多数服从正态分布[10]。本研究参照水工水力模型试验结果，假定泄洪设施总实际泄流能力在设计总泄流能力的 95%～105%范围变化[10]。引入一个流量修正系数 λ_2（λ_2 为 0.95～1.05，服从正态分布），则实际泄流能力 $Q_{下泄}$ 为：

$$Q_{下泄}=\lambda_2 Q_{设计}=(0.95\sim1.05)Q_{设计} \tag{5}$$

4.2.3 综合风险率计算

本研究综合风险率计算是在综合考虑洪水过程、泄洪能力的不确定性基础上，计算调洪过程中各风险因子随机组合而产生的防洪风险。

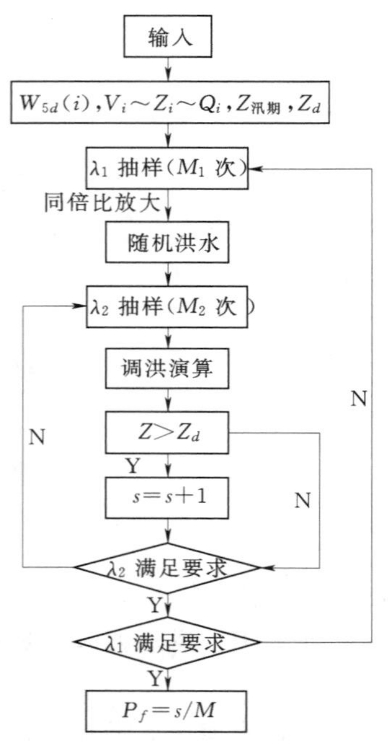

图 2　风险计算流程

对按 P－Ⅲ型总体随机生成的每场洪水过程（λ_1 抽样次数记为 M_1）按两种线型缩放，按正态分布总体随机生成泄流能力曲线（λ_2 抽样次数记为 M_2），分别进行调洪计算，求出相应的水库调洪最高水位 Z。每种洪水线型状态下获得 $M(M_1\times M_2)$ 个调洪最高水位，从中统计出 $Z>Z_d$ 次数 s，再根据式（4）计算综合风险率。计算流程见图 2。框图中 $W_{5d}(i)$ 为 5d 洪量系列；Z_d 为坝顶高程（按副坝高程 73.4m 计算）；$Z_{汛限}$ 为汛限水位；$Z(Z=Z_{max}+H)$ 为每次调洪计算所得的坝前最高水位；Z_{max} 为每次调洪计算所得水位；H 为风浪爬高。

5 计算结果及分析

根据上述研究，针对水文、水力风险因子，分别进行单因素、双因素情况下的防洪风险率计算，计算结果见表1。

表1　　柘林水库防洪安全风险分析结果

影响因素	试验次数	1954年典型			1998年典型		
		最高水位/m	$Z>Z_d$(73.4m)		最高水位/m	$Z>Z_d$(73.4m)	
			s/次	P_f/%		s/次	P_f/%
水文	10^4	73.07	0	0	73.49	312	3.12
水力	10^2	72.74	0	0	73.15	0	0
水文、水力	10^6	73.08	0	0	73.51	33826	3.38

（1）以1954年典型洪水计算，最高水位为73.07m；以1998年典型洪水计算，最高水位为73.49m，比1954年提高了0.42m，相应增加水量1.7亿m^3，占防洪库容的5.4%。结果表明洪水线型对防洪安全风险的影响是显著的。

（2）从单因素的防洪风险率计算结果分析。只考虑洪水要素时，1954年典型洪水计算得风险率为0，1998年典型洪水计算得风险率为3.12%；只考虑水力因素时，1954年典型洪水的风险率为0，1998年典型洪水的风险率为0，表明水力因素的不确定性导致坝前最高水位变化幅度不大；同时也表明，对随机生成的不同洪水序列进行了调洪计算，洪水漫顶概率有较明显的差异，入库洪水是影响调洪最高水位的主要因素。而且从双因素组合风险率计算结果分析，与单因素相比，水位有所提高，防洪风险也增加，所以考虑风险因子越多，风险值越大。

（3）对于柘林水库而言，从理论上分析，防洪风险是存在的。但由于采用最大可能洪水进行复核，将原大坝坝顶高程提高到73.4m，从柘林水库运行36a和现有55a水文资料情况分析，最大洪峰流量未超过13000m^3/s，最大5d洪量未超过30亿m^3。特别是1996年以来，由于采用预报手段和预泄方案，水库运行水位一直在68m以下（1998年最高水位67.96m），从实际运行状况分析，水库是比较安全的。

综合上述分析可以认为：①洪水特征值的不确定性会带来防洪安全风险，特别是非常不利的洪水线型能导致水力因素或其他不确定因素对防洪安全产生更大威胁，因此，在汛期防洪调度时，要密切关注雨情、水情变化，加强连续次（场）洪水、后汛洪水预报，并做好此类型洪水的预泄方案；②不确定性因子越多，综合风险率越大，所以水库控制运用中要慎重考虑洪水资源化问题，若提高汛限水位，实现洪水资源化，须配合洪水预报，实行分期汛限水位，可以减少其风险性；③防洪风险分析既重要又复杂，比如风险识别、风险分析方法、评价标准、风险与效益的耦合关系等，所以值得进一步探讨。

6　结语

风险管理是水库控制运用的发展方向。它能具体分析水文、水力因素以及防洪调度中有关参数、调度方式等对水库防洪安全所带来的影响，对水库防洪调度具有现实意义。本文综合考虑水文、水力因素的不确定性进行风险分析，认为柘林水库防洪安全风险很小，但洪水线型对水库防洪安全威胁较大，应加强洪水预报手段对非常不利的洪水情况下的调度作用。

参考文献：

[1] Yen B C. Risk in hyd rologic design of engineering project Journal of the Hydr. ASCE 1970.
[2] 梅亚东，谈广鸣．大坝防洪安全的风险分析［J］．武汉大学学报（工学版），2002，(10)．
[3] 姜树海．随机微分方程在泄洪风险分析中的运用［J］．水利学报，1994，(3)：1－9.
[4] 王长新．泄洪消能风险计算及不确定性分析［J］．新疆农业大学学报，2000，23（增）：23－25.
[5] Tung Y K，Mays L. Risk analysis for hydraulic design［J］. Proc of ASCE，J. of Hydr. Div，1980，106（HY5）：893－913.
[6] Hum berto M M. Flood safety analysis［J］. International Water Power & Dam Construction，1996，May：21－24.
[7] 丁晶，邓育仁，侯玉，等．水库防洪安全设计时设计洪水过程线法适用性的探讨［J］．水科学进展，1992，3（1）：45－52.
[8] 董胜，王腾．防洪工程项目的风险评估［J］．水利学报，2003，(9)：31－35.
[9] 肖义，郭生练，等．大坝防洪安全风险评估框架及其应用［J］．武汉大学学报（工学版），2006，(4)：19－25.
[10] 储祥元．水力不确定模型研究［J］．水利学报，1992，(5)：33－38.
[11] 庄楚强，吴亚森．应用数理统计基础［M］．广州：华南理工大学出版社，1999.

江西省大中型病险水闸风险评估

王　姣[1]，高桂青[2]

1. 江西省水利科学研究院；2. 南昌工程学院土木工程系

摘　要：根据病险水闸具体情况，依照突出重点、统筹兼顾、因地制宜、量力而行的原则，采用现场评分法，将江西省拟列入中央除险加固专项规划的病险水闸进行排序。

关键词：病险水闸；风险评价；除险加固；规划

1　项目背景

根据水利部水规办〔2009〕85号文以及病险水闸除险加固专项规划报告编制提纲要求，在水闸注册登记和安全鉴定的基础上，综合考虑大中型病险水闸的特点、险情、功能和效益，因地制宜、统筹兼顾、注重实效，合理制定除险加固方案，用3～5年时间完成全国大中型病险水闸的除险加固任务。

依据2008年江西省水闸安全状况普查：至2008年年底，全省小（1）型以上规模的水闸总数为1181座，其中大型水闸30座，中型水闸271座，小（1）型880座。经鉴定，大中型三类和四类病险水闸为164座，占大中型水闸总数的54.5%，其中三类闸90座，四类闸74座。

2　规划范围和规划水平年

规划范围为已进行注册登记，且经安全鉴定确定为三类闸或四类闸的164座大、中型水闸列入本次病险水闸除险加固规划。

规划水平年：规划的基准年为2008年，水平目标年为2012年。计划从2009年开始，用4年的时间完成列入本规划内的164座大、中型病险水闸的除险加固任务。

3　风险评估

为全面消除水闸病险隐患，权衡除险加固的轻重缓急，更合理地安排和使用有限的加固资金，规划采用了风险评估技术和加固排序方法，以水闸工程风险、失事后果风险和综合平衡因素作为分析因子，计算出病险水闸风险评估指数。

一座水闸是否安全，是否需要除险加固，不仅仅是依据工程本身的安全性，即防洪标准、结构稳定性、渗流稳定性、抗震稳定、泄输水建筑物安全及金属结构安全等是否满足

本文发表于2009年。

现行规范或标准的要求，还包括水闸失事对下游生命财产可能构成的风险，为两者的乘积。因此，进行病险水闸的风险评估，关键就是提出有关风险因子，并赋予一定的权重，计算得出风险评估指数。

3.1 风险因子的确定

水闸风险一方面是指水闸工程本身存在的病险，用 P_f 来反映，另一方面考虑水闸失事将造成的后果风险，用 L 来表示，后果包括生命损失、经济损失及对社会、生态、环境的影响。两者的乘积表示为水闸工程风险指数 R。风险评估模型如图 1 所示。

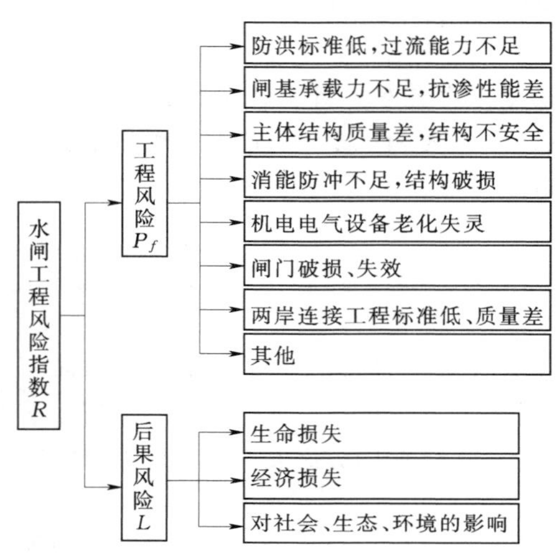

图 1　水闸工程风险评估层次分析结构模型

根据水利部 1999 年全国病险水闸调查结果，主要问题表现为以下 10 个方面[1]：①防洪标准低，不满足现行规范要求的，占大中型水闸总数 36.4%；②闸室不稳定，抗滑稳定安全系数不满足要求的占 10.0%；③渗流不稳定，闸基或墩墙后填土产生渗流破坏的占 22.3%；④抗震不满足要求或震害后没有彻底修复的占 7.2%；⑤闸室结构混凝土老化及损坏严重的占 76.4%；⑥闸下游消能防冲设施严重损坏的占 42.3%；⑦泥沙淤积问题严重的占 17.9%；⑧闸门及机电设备老化失修或严重损坏的占 76.7%；⑨观测设施缺少或损坏失效的水闸相当普遍；⑩存在其他问题（如枢纽布置不合理，铺盖、翼墙、护坡损坏，管理房失修，防汛道路损坏，缺少备用电源、交通车辆和通信设施等）的占 51%。

参照统计资料，并结合江西省病险水闸调查情况，经分析比较，采用现场评分法作为工程风险评估的手段，即确定八类工程风险作为风险因子（如图 1 所示），通过对病险水闸的类型、结构型式和存在的主要问题，将每一类工程风险细分为若干项，依据其危害程度分别赋予最大风险值，进行现场评分，每一类工程风险值为若干项专家评分总和，水闸的工程风险值 P_f 则为八类工程风险因子评分值的总和。可以用如下计算公式表示：

$$\overline{P_i} = \sum_{j=1}^{n} P_{ij}\ ;\ P_f = \sum_{i=1}^{8} P_i$$

式中：P_{ij} 为第 j 项风险因素对第 i 类工程风险因子所产生的风险值；$\overline{P_i}$ 为第 i 类工程风险因子所产生的风险值；P_f 为八类工程风险因子所产生的水闸总风险值。

江西省大中型病险水闸中 98%为以防洪、灌溉和排涝为主，经安全鉴定资料统计，每座水闸均不同程度存在多种病险隐患，其中过流能力不足的占 54.88%，闸基渗流不稳定的占 25%，主体结构不稳定的占 11.58%，而金属结构及机电电气设备老化、损坏、失效的比例为 100%，混凝土结构、消能防冲设施老化、损坏的比例超过 90%。考虑到水闸功能各异，病险隐患对工程危害程度及其产生的后果有所不同。分别确定其工程风险因子的最大风险值，结合考虑水闸规模、地理位置、重要性，以一定的比例进行组合，形成水闸工程最大风险值，见表1。

表1 水闸工程最大风险值构成表

风险因子	最大风险值	风险因子	最大风险值
防洪标准及过流能力	28	机电电气设备	13
闸基	8	闸门	15
主体结构	10	连接工程	8
消能防冲	8	其他	10

3.2 后果风险因子确定

后果风险综合评价[2]函数 L：该函数综合考虑生命损失、经济损失和对社会及环境的影响所占的权重 S_i 及其严重影响系数 F_i，采用线性加权法构成。

$$L = \sum_{1}^{3} S_i F_i = S_1 F_1 + S_2 F_2 + S_3 F_3$$

式中：S_1，S_2，S_3 分别为生命损失、经济损失和社会环境影响的权重系数；F_1，F_2，F_3 分别为生命损失、经济损失和社会环境影响的严重程度系数。

权重系数的确定：采用Saaty建议的1～9标度法（AHP法）来确定各子层因素对母层因素的权重系数。

AHP法采用1～9及其倒数作为两个元素重要性比较定量化的标度，见表2。标度1、3、5、7、9分别表示一个比另一个元素同等、稍微、明显、强烈、极端的重要程度，标度2、4、6、8分别表示上述相邻程度的中值。两个元素反过来对比为原标度的倒数。若以 $\boldsymbol{U}$ 表示总目标，u_i 表示评价元素，$u_i \in \boldsymbol{U}(i=1, 2, 3, \cdots, n)$。$U_{ij}$ 表示 u_i 对 $u_j(j=1, 2, 3, \cdots, n)$ 的相对重要性数值。由此得 $n \times n$ 阶判断矩阵 $\boldsymbol{P}$：

表2 判断矩阵标度及其含义

标度	含义
1	表示因素 u_i 与因素 u_j 比较，具有同等重要性
3	表示因素 u_i 与因素 u_j 比较，u_i 比 u_j 稍微重要
5	表示因素 u_i 与因素 u_j 比较，u_i 比 u_j 明显重要
7	表示因素 u_i 与因素 u_j 比较，u_i 比 u_j 强烈重要
9	表示因素 u_i 与因素 u_j 比较，u_i 比 u_j 极端重要
2、4、6、8	2、4、6、8分别表示相邻判断1～3、3～5、5～7、7～9的中值
倒数	表示因素 u_i 与 u_j 比较得判断 u_{ij}，因素 u_j 与 u_i 比较得判断 $u_{ji}=1/u_{ij}$

$$\boldsymbol{P}=\begin{bmatrix} u_{11} & u_{12} & \cdots & u_{1n} \\ u_{21} & u_{22} & \cdots & u_{2n} \\ & & \vdots & \\ u_{n1} & u_{n2} & \cdots & u_{nn} \end{bmatrix}$$

式中：n 为子层中相应因素的个数；u_{ij} 为各因素两两比较，对实际目标的贡献大小的赋值，且 $u_{ij}=1/u_{ji}$，$u_{ii}=1$，$u_{ij}=u_{ik}/u_{jk}$（i，j，$k=1$，2，…，n）。

本次规划在分析统计的基础上，根据功能将水闸分为以防洪为主、以灌溉为主和以排涝为主三类。对以防洪为主的水闸，生命损失与经济损失相比，重要程度比较明显，可定为 5 和 1 的关系。经济损失同社会与环境影响相比，重要程度视为稍微，可定为 3 和 2 的关系。由此，得出判断矩阵：

$$\boldsymbol{P}=\begin{bmatrix} 1 & 1/5 & 2/15 \\ 5 & 1 & 2/3 \\ 15/2 & 3/2 & 1 \end{bmatrix}$$

求得生命损失的权重系数 $S_1=0.75$，经济损失的权重系数 $S_2=0.15$，社会及环境影响权重系数 $S_3=0.10$；则后果风险综合评价函数 $L=0.75F_1+0.15F_2+0.1F_3$。

对以排涝为主的水闸，经济损失与生命损失相比，重要程度比较强烈，可定为 7 和 1 的关系。生命损失同社会与环境影响相比，重要程度视为稍微，可定为 2 和 3 的关系。由此，得出判断矩阵：

$$\boldsymbol{P}=\begin{bmatrix} 1 & 7 & 3/2 \\ 1/7 & 1 & 3/14 \\ 2/3 & 14/3 & 1 \end{bmatrix}$$

求得生命损失的权重系数 $S_1=0.11$，经济损失的权重系数 $S_2=0.74$，社会及环境影响权重系数 $S_3=0.15$；则后果风险综合评价函数 $L=0.11F_1+0.74F_2+0.15F_3$。

对以灌溉为主的水闸，经济损失与生命损失相比，极端重要程度，可定为 9 和 1 的关系。生命损失同社会与环境影响相比，重要程度视为同等，可定为 1 和 1 的关系。由此，得出判断矩阵：

$$\boldsymbol{P}=\begin{bmatrix} 1 & 9 & 1 \\ 1/9 & 1 & 1/9 \\ 1 & 9 & 1 \end{bmatrix}$$

求得生命损失的权重系数 $S_1=0.09$，经济损失的权重系数 $S_2=0.82$，社会及环境影响权重系数 $S_3=0.09$；则后果风险综合评价函数 $L=0.09F_1+0.82F_2+0.09F_3$。

影响严重程度系数参照国内外相关的研究成果选取确定，各项严重程度系数均定为 5 级。生命损失主要考虑风险人口总数、人口分布、报警时间等。本次规划暂以风险人口作为生命损失的评估标准（见表 3）。

表 3　　生命损失的赋值

影响人口/人	严重程度赋值	影响人口/人	严重程度赋值
≤1000	0.50	5000～10000	0.70～0.90
1000～5000	0.50～0.70	≥10000	0.90～1.00

经济损失包括直接经济损失和间接经济损失，本次规划暂以淹没土地田亩折算经济损

失，见表 4。

表 4 经济损失的赋值

受淹耕地/万亩	严重程度赋值	受淹耕地/万亩	严重程度赋值
≤0.5	0.50	5～10	0.80～0.90
0.5～1	0.50～0.70	≥10	0.90～1.0
1～5	0.70～0.80		

社会影响主要包括政治影响，即对国家、社会安定的不利影响；对主要交通干线和国家重要设施的破坏，给人们造成的精神痛苦及心理创伤，以及日常生活水平和生活质量的下降等；无法补救的文物古迹、艺术珍品和稀有动植物等的损失。环境影响主要包括河道形态的影响，生物及其生长栖息地的丧失，人文景观的破坏等，见表 5。

表 5 社会及环境影响的赋值

影响严重程度	主要社会影响	主要环境影响	影响程度赋值
轻微	人烟稀少，无重要交通干线和文物古迹	河道遭受轻微破坏，自然景观遭轻微破坏，一般动植物栖息地丧失	0.20
一般	影响到村庄和一般文物古迹，重要交通干线距离较远	一般河流河道遭受一定破坏，市级人文景观遭破坏，较有价值动植物栖息地丧失	0.20～0.40
较严重（中等）	影响到乡镇政府所在地、市级重点保护文物，重要交通干线距离较近	一般河流河道遭严重破坏，省级人文景观遭破坏，较珍贵动植物栖息地丧失	0.40～0.60
严重	影响县城以上政府所在地或城区、省级重点保护文物古迹、老少边地区，重要交通干线距离近	大江大河遭严重破坏，一般河流改道，国家级人文景观遭破坏，稀有动植物栖息地丧失	0.60～0.80
极其严重	影响省会城市、世界文化遗产、国家重点保护文物古迹、少数民族集居地，重要交通干线很近	大江大河改道，世界级人文景观遭破坏，（世界级）濒临灭绝的动植物栖息地丧失	0.80～1.00

3.3 风险指数

风险指数 R 表示为：

$$R=P_fL$$

即水闸工程风险因子与后果风险因子的乘积。

3.4 评估指数

鉴于各地病险水闸数量、病险程度存在较大差异，前期工作进展也有所不同，在各地申报的病险水闸加固排序的基础上，进行综合分析分组，依据优先程度的不同分为高级、中级和一般三个级别，并分别赋予一定的系数（高级 0.75～1.00、中级 0.60～0.75、一般 0.50～0.60），即为各设区市相对平衡的综合系数，将系数与水闸的风险指数的乘积即为病险水闸评估指数，作为除险加固排序的依据。

4 加固排序

根据病险水闸具体情况，依照突出重点、统筹兼顾、因地制宜、量力而行的原则，以风险评估为手段，将江西省拟列入中央除险加固专项规划的病险水闸进行排序。

对于风险指数大、位置重要和效益较好的病险水闸，特别是五河、堤防上的水闸应优先安排加固。分期实施意见如下：2009 年完成 38 座水闸的除险加固，其中大型水闸 10 座、中型水闸 28 座；2010 年完成 53 座水闸的除险加固，其中大型水闸 4 座、中型水闸 45 座；2011 年完成 49 座中型水闸的除险加固；2012 年完成 28 座中型水闸的除险加固。

参考文献：

[1] 袁庚尧，余伦创．全国病险水闸除险加固专项规划综述［J］．水利水电工程设计，2003，22（3）：6－9．

[2] 李雷．大坝风险评价与风险管理［M］．北京：中国水利水电出版社，2006．

水库大坝安全管理新理念——大坝风险管理

吴晓彬，蔡力群

江西省水利科学研究院

摘　要：大坝风险分析和管理包含可靠性理论、事故的概率分析、事件树分析技术、大坝的溃决分析、溃坝的经济分析和风险人口分析及降低风险的工程和非工程技术等，其最终目的是把大坝风险降低到使公众可接受的水平。本文介绍了大坝风险管理的内容，供大坝安全管理者参考。

关键词：大坝；安全管理；风险管理

我国现有各类水库 8.5 万余座，在防洪减灾和水资源利用中发挥了并继续发挥着重要作用。但由于我国的建坝历史跨越了不同的经济发展和技术发展阶段，工程质量参差不齐，许多工程存在建设上的先天不足和管理上的后天失调，现状安全问题突出。据统计，全国有接近 40%水库大坝存在不同程度的病险情况。至 2006 年年底，江西省共建成各类水库 9799 座，数量居全国各省第二位。据统计分析，全省共有病险水库 4000 多座，占水库总数的 41%。病险水库的大量存在严重制约了当地的经济发展，威胁着下游人民生命财产和国家公共设施的安全。

除险加固是我国目前提高水库大坝安全程度的主要措施，然而我国水库大坝除险加固的工作重点主要是工程措施，未把非工程措施放在重要位置上。如何使工程除险规模与下游经济现状和发展水平相适应，采取工程和非工程措施相结合的综合方法，把大坝风险降低到下游能够承受的程度等，就是大坝风险管理应解决的问题。

1　水库大坝安全管理存在的主要问题

（1）日常管理监管不严。对水库的管理主要采用行政手段，注重汛前检查，以安全度汛为主要目标，对日常管理监管不严。必要的管理信息发布制度不健全，社会监督有待加强。

（2）工程等级观念较强。水库大坝的除险加固基本上以工程规模或库容的大小来制定除险加固计划，安排加固资金，等级观念较强。对工程失事后果考虑较少，未按水库大坝的风险高低进行排序。

（3）安全监测设计针对性不强。大坝安全监测主要以工程规模，按照现行规范来进行安全监测设计，对工程失事模式考虑较少，安全监测设计的针对性不强。

本文发表于 2008 年。

（4）预警预报系统缺乏或不健全。水库自动化遥测和资料分析评价方面，主要根据水库规模决定是否建设自动化遥测和预警系统，很少考虑水库大坝的风险高低和下游的重要性。

综上所述，我国在水库大坝安全管理上侧重于规范化管理，风险管理理念较弱。鉴于我国经济的快速发展和大坝下游城镇的规模化发展，区域人口数量增长较快，导致水库大坝下游的经济权重日益加大，失事后果将越来越严重，风险管理的实施势在必行。

2 水库大坝风险管理

风险管理是以风险度量为理念的一种事（不利事件）前积极管理机制，是依据管理风险的法规、标准、程序和方法，识别（分析）、评估、监控和处理各种风险，实现水库大坝风险承担者所承担的风险规模与大坝结构的优化（或维修、加固规模）之间、风险与投资回报之间的平衡，是一个接受、拒绝、降低或转移风险的过程管理。它可以实现不同层次上的群坝管理，也可对单一水库大坝进行管理。其内容主要包括：确立风险准则、风险分析、风险评估、风险处理、风险监控等基本步骤。在风险管理中，一个基本的认识是：一座大坝可能达到不同的安全程度，但绝不可能没有风险。因此，安全大坝应该定义为：不会因为它的存在而对下游造成不能被接受的风险的大坝。

2.1 风险与大坝风险

风险是一个通用名词，指某一事件造成破坏或伤害的可能性或概率，风险＝伤害的程度×发生的可能性，它包含3个因素：不利的结果、发生的可能性以及现实的状态。

大坝风险是指大坝对生命、健康、财产和环境负面影响的可能性和严重性的度量，是不利事件（如溃坝、超安全泄量的泄洪等）的概率与产生后果的乘积。风险来自大坝不利事件发生的可能性和对下游所造成的损害两个方面，前者与工程安全有关，后者与下游影响区域的社会经济发展程度有关。大坝存在溃决的可能性，导致了工程安全问题；大坝溃决使得公众生命受到威胁，导致了生命安全问题；大坝溃决造成的经济损失及对经济发展产生负面影响，导致了经济安全问题；同时也将对上、下游环境产生不利影响，使得环境安全受到威胁。

2.2 大坝风险准则

风险准则是指在风险评价过程中用于确定计算定量风险分析结果是否可以接受的一种准则。任何一座大坝都有风险，只是风险程度不等，投入的增加可以减少风险，但有投入与效益（包含经济效益和社会效益等）的平衡问题，必须面对和接受某种风险，此类风险一般称为可接受风险或可容忍风险。对风险的可容忍性，英国健康和安全委员会认为：只有当减少风险是不可行的或投入的资金和减少的风险极不相称时，风险才是可容忍的；或只有当进一步降低风险是不可行的或需要采取措施，但措施的及时性、难度和付出的代价与风险的降低极不相称时，剩余风险是可容忍的。

由于国家或地区的政治、经济、历史文化背景、公众的价值观等因素的不同，目前各国现有的大坝风险准则也不尽相同，但一般应考虑3方面内容：生命风险、经济风险和环境风险。

2.3 大坝风险分析

广义的风险分析是一种识别和测算风险，开发、选择和管理方案来解决这些风险的有组织的手段。它包括风险识别、风险评估和风险管理3方面的内容。大坝风险分析是以溃坝为推动力的，考虑所有能使该大坝溃坝的模式。描述、确定大坝引起的风险的方法可以是定性的，也可以是定量的。

风险的定性分析一般采用事件树或故障树方法。列出水库大坝所存在的各种隐患，对隐患发生的可能性进行概率分析或专家判断。

定量分析是在定性分析的逻辑基础上，给出各个风险源的风险量化指标及其发生概率，再通过一定的方法合成，得到系统风险的量化值。它是基于定性风险分析基础上的数学处理过程。

风险分析的优点在于它是一个理性的、系统的过程。具体可概括如下：

（1）增强对大坝脆弱部位的了解。

（2）增强大坝安全的监测能力。

（3）增强大坝安全分析及大坝安全决策能力。

（4）将管辖内所有大坝的缺陷或隐患调查和加固、维修工作列出优先次序，可经济、有效地运用大坝加固、维修资金，即有助于除险加固工程的排序。

（5）尽量增加投资的回收。

（6）增强溃坝对下游影响的了解，尽量减少溃坝的损失。

（7）分析中不但考虑了所有荷载，也考虑了运行管理情况，处理了不确定因素。

（8）得出“该大坝的安全程度”。

2.3.1 风险识别

风险识别是风险分析的最基础环节，其目的是确定风险的来源、潜在范围、产生条件，描述其风险特征和确定哪些风险因素有可能影响大坝安全。风险识别不是一次就可以完成的事，应当在水库大坝运行过程中自始至终定期进行。一般可采用的方法有专家调查法、流程图法、经验数据法和风险调查法。

2.3.1.1 风险因素识别

通过水库大坝的风险调查，主要表现在：

（1）已建水利工程中有不少建设标准偏低，工程质量差，不能满足防洪兴利需要。

（2）长期的“重建轻管”使水库管理行业“不景气”，管理设施、技术落后，工程技术人员严重缺乏，运行管理薄弱。

（3）水库大坝安全与社会公共安全及社会经济发展间的矛盾突出。防洪安全方面：水库防洪保护范围内人口众多，城镇密集，耕地面积大；水资源利用方面：许多城镇的部分或全部生活及工业用水主要依靠水库供水；宣传教育方面：一般群众对大坝的认识仅停留在供水与抢险上；规划方面：下游发展规划中缺少大坝安全风险的分析与评估；工程管理方面：管理中缺乏公众参与及利益共享机制。

因此，水库大坝的风险识别因素可大致分为：工程风险因素、人为风险因素与环境风险因素。

（1）工程风险因素。根据对我国已溃坝资料、大型水库大坝非溃坝事故、三类坝病险原因基本情况统计，工程病险原因集中在防洪标准低、坝体质量差、管理设施与技术水平落后等。已溃坝资料统计表明，漫顶约 50.2%，坝体质量差约 34.8%，管理不当和其他形式引起的溃坝占 11.4%。另有统计资料显示，小型水库中 30m 以下的土坝为已溃坝的主体，1954—2000 年间已溃坝中 30m 以下低坝占 96.5%，土坝坝型占 97.8%，成为大坝安全管理中的难点和重点。

（2）人为风险因素。主要体现在管理体制、机制和非工程措施不完善上。管理体制落后，管理机制不健全，长期重建轻管；由于管理不到位，加剧了工程的老化和缺损；部分管理人员安全观念淡薄，风险意识不强，管理水平低下；应急预案及应急措施不能满足现实需要；水库下游的发展规划缺少水利管理部门的参与等等。

（3）环境风险因素。包括自然环境因素与社会环境因素。由于全球气候变化影响、水资源匮乏及政治因素等影响，大坝的运行环境可能发生变化。如超标准洪水，水库功能转变，战争、恐怖活动等，都可能成为影响水库大坝安全的风险因素。

2.3.1.2 破坏模式识别

溃坝模式可简单表述为荷载（水荷载、地震荷载等）→建筑物（大坝、溢洪道、涵洞等）→破坏→溃坝。经研究统计，我国乃至世界上土石坝的主要溃坝模式可分为以下几类：

（1）无泄洪设施或泄量不足、坝顶高程不足等原因引起漫顶。

（2）坝体滑坡导致的破坏。

（3）坝体、坝基或坝下埋管的渗透破坏。

（4）地震引起的大坝溃决。

（5）闸门及启闭设施失效或操作不当引起的破坏。

2.3.2 溃坝概率分析

水库大坝工程安全不仅涉及工程结构，还涉及事件本身，因而相应概率的确定相对复杂。在具体分析中，目前常用的方法有历史资料统计法与专家经验法。溃坝概率的定量分析方法多采用事件树、故障树或可靠度分析等方法。

2.3.3 溃坝后果分析

（1）溃坝影响分析。溃坝影响分析是针对某一溃坝模式和某一洪水位下，通过溃坝洪水计算及洪水演进计算，确定溃坝洪水淹没范围及损害评估。

（2）溃坝生命损失分析。溃坝生命损失是溃坝损失中的最重要指标，因为它不但会发生人员伤亡，还将造成人的心理恐慌。其影响因素涉及风险人口的数量与分布、溃坝发生的时间（白天、晚上和天气情况）、警报发布的时间与长短、水深和流速、洪水上涨速率、撤离条件等。

（3）溃坝经济损失分析。通常将溃坝经济损失分为直接损失与间接损失，计算涉及社会经济调查、溃坝洪水损失率及溃坝洪水损失增长率等。直接经济损失包括水库工程损毁所造成的经济损失和溃坝洪水淹没所造成的可用货币计量的各类损失。间接经济损失指直接经济损失以外可用货币计量的损失。

2.4 大坝风险评估

风险评估是指应用风险分析的结果来进行决策的过程，以决定已存在的风险是否可以容忍，风险控制措施是否合适，如何通过工程或非工程措施减少风险。既不是因为传统方法的安全度较小，也不是替代传统的确定性安全分析方法，而是对基于标准的传统分析方法的加强，有助于水库大坝安全管理水平的进一步提高。采用该方法的目标是尽最大可能、最大限度地降低水库大坝的风险。

风险评估包括风险排序、应对方案的制定与优化。若风险满足风险准则，则无需处理；若风险不可接受但可容忍或满足 ALARP 原则就不需要处理风险；若风险不可容忍，则必须立即对风险进行处理。

2.4.1 风险排序

风险排序涉及同一水库大坝主要风险因素或事件对溃坝概率影响的重要程度排序，及不同水库间（群坝）风险程度排序。对给定水库来说，溃坝损失后果是一定的，只需根据溃坝概率分析即可确定不同因素或事件的排序。群坝则根据水库风险比较进行排序，在溃坝概率已知的情况下，以水库溃坝后果权重确定一个综合性指标，或以几个关键指标确定。

2.4.2 应对方案

在应对方案制定上，应根据主要风险优先原则及 ALARP 原则，进行风险—费用—效益分析，综合工程措施与非工程措施进行处理。根据风险排序，优先考虑高风险水库或水库主要风险因素的处理。

2.5 风险处理

这一过程包括既定风险处理方案的编制、实施及实施后的效果评价。应在加强工程措施的同时，重视非工程措施。在风险处理方法上，主要有下述几种。

2.5.1 降低风险

包括预防及控制损失两方面。事故预防是指减少事故发生，降低大坝溃决概率，如对大坝进行除险加固处理，加强对大坝的定期检查与监控。现阶段应继续加强除险加固力度，强化防汛行政首长负责制等有效降低溃坝概率的措施。控制损失是从后果出发，减少溃坝损失后果，如建立科学合理的应急预案，完善应急救助措施，对大坝下游社会经济发展等人类活动进行风险规划与管理，降低与控制潜在的风险损失。

2.5.2 转移风险

目前，我国尚无有效的风险转移手段。国外则将保险作为转移风险的一种重要手段，同时也是业主利用保险公司的服务，检查存在隐患的一条途径。

2.5.3 回避风险

回避风险是普遍采用的一种方法，包括：完全回避，当风险过大时，通过拒绝风险的发生，如空库或降低水位运行；中途放弃，当条件变化时，中止已承担的风险。如大坝风险分析结果不满足可接受风险标准，而且降低风险措施费用与取得的效益非常不相称时，可将大坝报废退役从而回避风险。这是目前在汛期大坝安全管理中一般采用的方法，其消极方面是放弃了可能可利用的水资源。

2.5.4 承担风险

目前，水库大坝安全管理中基本上是被动性保留风险。表现在对潜在风险缺乏认识，或已知风险但受条件限制而被动自留。主动保留则是在清楚风险后果的基础上，主动承担风险的决策，如加固后满足风险准则的剩余风险等。

3 结语

（1）大坝风险分析和管理是世界先进国家近年发展起来的进一步提高大坝安全管理的先进技术。不同国家根据各自特点进行定性或定量风险分析、评价和管理，也有的部分采用这种方法，进行溃坝影响评价和风险人口分析，并据以确定风险的排序。

（2）大坝风险分析和管理是把公众安全放在十分重要的位置，其最终目的是把风险降低到使公众可接受的水平，这和传统的工程安全概念不同。

（3）大坝风险分析和管理是一项涉及到多学科的综合性技术。这项技术中包含了可靠性理论、事故的概率分析、事件树分析技术、大坝的溃决分析、溃坝的经济分析和风险人口分析，降低风险的工程和非工程技术等。

（4）我国大坝安全和管理工作需要这种技术。该项技术实质上是对已有安全管理工作的加强，适合我国水库大坝数量巨大、风险很高而加固经费又很紧缺的特点。特别是风险排序技术、应用非工程措施降低风险的技术、应用专家经验定量风险评价技术等，将大大推进我国水库大坝安全管理水平。

（5）我们在水库大坝安全管理工作中部分应用了该项技术。如《江西省病险水库除险加固规划》实施中期评估和《江西省第二期病险水库除险加固专项规划》的编制等，应用效果良好。

群坝风险评估指数排序方法的探讨

傅琼华，段智芳

江西省水利科学研究院

摘　要： 结合江西省水库的实际情况，研究了群坝的风险评估指数排序方法，综合评定了大坝工程安全风险程度因子和溃坝损失影响因子，并综合考虑水行政主管部门对水库管理因素的排序、总库容、灌溉效益、加固前期工作等情况，确定水库的综合影响因子。以这 3 大影响因子的乘积表示水库的风险评估指数的高低，风险指数越大，表明水库越危险，应优先得到除险加固。通过对江西省 200 余座不同类别的水库应用排序表明，该计算方法简单实用，可操作性强，分析结果与实际基本相符，可用于群坝除险加固排序。

关键词： 大坝险度；溃坝损失；综合因子；评估指数；风险排序

至 2002 年年底，江西省已建成水库 9268 座，为全国 1/9，其中病险水库 3488 座，占水库总数的 37.6%，其量多面广，除险加固任务重。江西省政府在《江西省病险水库除险加固实施意见》中明确提出在 2007 年前全面完成现有病险水库的除险加固任务。在当前国家对病险水库投入减少、地方投资不能足额到位的情况下，为如期完成目标，细化加固工作方案，明确责任，2005 年 4 月江西省开展了病险水库除险加固工程中期评估工作，对已加固的水库进行建设管理、工程质量、投资与效益等评价，对未实施加固的病险水库进行风险排序评估，有目标、有重点地推进加固工作。本文总结了水库群安全风险分类排队方法，建立了群坝风险评估指数排序计算方法，可用以区分轻重缓急，优先安排风险大的病险水库进行加固除险，科学合理地安排除险加固计划，为决策者提供科学合理的除险加固实施依据。

目前，在确定一座水库大坝是否是病险水库以及是否需要除险加固时，主要依据工程本身的安全性，即水库防洪标准、大坝结构稳定、抗震稳定、渗流稳定、泄水和输水建筑物安全、金属结构安全等是否满足现行规范或标准的要求。事实上，有的水库大坝虽然工程安全状况差，溃坝可能性较大，但由于溃坝产生的后果较轻，风险并不大，却优先得到加固处理；有的水库大坝虽然安全性状稍好，溃坝可能性较小，但对下游影响大，溃坝产生的后果十分严重，潜在风险很高，却迟迟得不到加固除险；还有一些水库大坝工程安全状况虽然相同或接近，但由于库容、高程、坝高以及下游经济社会发展水平、防洪保护对象等条件不同，对下游形成的风险有巨大差别，在决策时却无法定量考虑这些因素。因

本文发表于 2006 年。

此，作者对群坝风险排序方法研究时，综合分析评定大坝工程安全风险程度因子、溃坝损失影响因子，并综合考虑水行政主管部门对水库管理因素的排序、总库容、灌溉效益、加固前期工作等情况，确定水库的综合影响因子。建立群坝的风险评估指数排序的计算方法，能保证按风险大小顺序进行除险排序立项，从而使风险大的水库大坝优先得到加固除险，确保有限的除险加固经费用到最需要的地方。

1 群坝风险评估指数排序方法确定

群坝风险排序评估方法研究首先基于如下几点考虑：①满足现行规范要求；②具有科学性和先进性；③既能与国际、国内接轨，又能符合省情；④具有可操作性；⑤为江西省病险水库除险加固排序提供科学依据。为此，经过对不同类别水库工程的现场调查，引入目前国际上先进的大坝风险分析概念和分析技术，研究建立了群坝风险评估排序指数分析计算方法，其结构框图见图1。

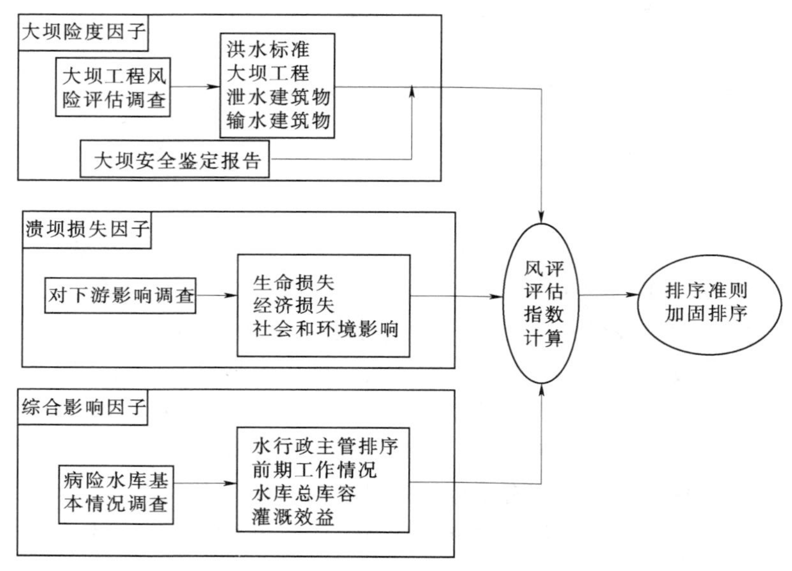

图1 计算方法

该评估方法主要包括3大影响因子：一是大坝险度因子，即大坝工程安全程度评估。通过对大坝工程安全风险程度情况调查、大坝安全鉴定报告对工程的各建筑物安全性评价分别进行赋分，从而计算出大坝险度因子系数；二是溃坝损失因子，即溃坝后果影响程度评估。通过对大坝下游情况影响调查，分析溃坝后下游生命损失、经济损失、社会和环境影响可能产生的后果，分别赋值计算，确定溃坝影响因子系数；三是综合影响因子系数。通过综合考虑水库的水行政主管部门对水库管理因素的排序、水库的总库容、灌溉效益、加固前期工作等情况综合考虑，确定水库的综合影响系数。最后对赋值量化的大坝风险度、溃坝损失、综合影响三因子系数以乘积计算，得到各水库大坝风险评估指数，按评估指数大小对群坝进行风险排序，风险指数越大，表明水库危险性越大，应优先得到除险加固。

2 影响因子的确定

2.1 大坝险度因子确定

根据大坝安全风险评价的现场评分法，综合江西省水库实际情况及以往大坝风险研究工作[1]，分析研究制定水库大坝工程险度评估赋分表及其说明。大坝险度因子值构成共分5大部分：洪水（14分）、大坝工程（32分）、泄水建筑物（18分）、输水建筑物（25分）、金属结构（11分）。当出现正常缺项时，以该项的平均分赋分。江西省水库不属地震区，一般不考虑抗震安全问题。考虑到江西省输水建筑物绝大部分是坝下涵管，工程质量差、危害性大，分值较泄水建筑物高。

表1～表5分别为大坝工程险度评估赋分计算表、大坝工程赋分表、泄水建筑物赋分表、输水建筑物赋分表和金属结构赋分表。

表1　　水库洪水影响因素赋分

风险因素	洪水（14）	
	坝顶高程不足	泄洪安全性不足
赋分	10	4

表2　　大坝工程赋分

风险因素	大坝工程风险值构成（32）															
	工程龄期/年				建筑质量				渗流态势				结构稳定			
	≥30	15～29	5～14	≤4	很差	差	一般	好	有变形、升高趋势且出露点高	有变形、升高趋势但出露点低	较稳定但出露点高	稳定且出露点低	有沉陷有滑坡	有沉陷有裂缝	有沉陷稳定	运行正常
赋分	6	3	1	0	8	6	3	0	10	8	4	0	8	6	4	0

表3　　泄水建筑物赋分

风险因素	泄水建筑物（18）										
	工程龄期/年				建筑质量				结构稳定		
	≥30	15～29	5～14	≤4	很差	差	一般	好	边坡不稳定且底板较差	边坡稳定且底板较差	边坡稳定且底板良好
赋分	6	3	1	0	6	4	2	0	6	4	0

表4　　输水建筑物赋分

风险因素	输水建筑物（25）									
	工程龄期/年				运行状况				结构稳定	
	≥20	10～19	5～9	≤4	有裂缝、渗水	有裂缝、无渗水	无裂缝、有渗水	无裂缝、无渗水	不满足要求	满足要求
赋分	10	6	3	0	10	6	4	0	5	0

表 5　　金属结构赋分

风险因素	金属结构（11）							
	运用时间/年				闸门及启闭机运行状况			
	≥20	10～19	5～9	≤4	启闭不灵	启闭尚好，供电不正常	能正常启闭，维护差	运行状况好
赋分	5	3	1	0	6	4	2	0

通过对大坝工程现场情况检查、大坝安全鉴定报告对工程的各建筑物安全性评价分别进行分析，由专家各自分别赋值，累计确定大坝工程的病险程度，从而计算出大坝险度影响因子系数。

2.2　溃坝损失因子确定

水库溃坝后果包括生命损失、经济损失及社会与环境影响。根据南京水利科学研究院研究成果[2,3]，分别对各水库生命损失、经济损失及社会与环境影响严重程度进行赋值计算。

生命损失取决于风险人口总数及其分布、警报时间、溃坝洪水强度（水深、流速及洪水上涨速率）、撤离条件及风险人口对洪水严重性的理解等，病险水库除险加固排序可暂采用 Dekay & McClelland 法确定[4,5]。经济损失包括直接经济损失和间接经济损失，由调查统计资料确定并进行折算。社会影响主要包括政治影响（即对国家、社会安定的不利影响）；给人们造成的精神痛苦及心理创伤，以及日常生活水平和生活质量的下降等；无法补救的文物古迹、艺术珍品和稀有动植物等的损失。环境影响主要包括对河道形态的影响，生物及其生长栖息地的丧失，人文景观的破坏等。将社会及环境影响的严重程度划分为 5 个级别，根据调查的影响程度进行赋值计算。

在对溃坝生命损失、经济损失及社会和环境影响分别赋值分析之后，对溃坝后果进行综合赋值分析时，涉及到各影响因素的权重问题，该权重理应由决策者经同社会科学家会商决定，研究时从技术角度上采用 Saaty[6] 建议的 1～9 标度法（AHP 法）来确定各子层因素对母层因素的权重系数。

溃坝损失因子系数[2]

$$L = \sum_{1}^{3} S_i F_i = S_1 F_1 + S_2 F_2 + S_3 F_3 \tag{1}$$

式中：S_1、S_2、S_3 分别为生命损失、经济损失和社会环境影响的权重系数，可取 0.737、0.105 和 0.158；F_1、F_2、F_3 分别为生命损失、经济损失和社会环境影响的严重程度系数。

分别对不同类别水库进行溃坝损失因子系数进行计算。

2.3　综合影响因子确定

综合影响因子系数，是通过综合考虑水库水行政主管部门对水库管理因素的排序、水库总库容、灌溉效益、前期工作情况等综合考虑，确定水库的综合影响因子系数。从风险分析的水平来看，一般分为 4 种不同的水平，即初筛（screening），初步分析（preliminary），详细分析（detail）和非常详细分析（very detail）。在详细分析和非常详细分析中，要通过可靠度理论计算来获得某一事件出现的概率，将会提出非常高的技术资料要求，往往难以达到，而且中间的某些环节是难以理论分析的。目前世界上溃坝概率分析极少采用详细分析的办法，绝大多数的风险分析是根据不同要求采用初筛或初步分析的办法。因此，在研究分析综合影响因子时，主要依靠专家经验来估算。把专家对某一事件可能出现的定性判断转化为可能出现的定量概率，围绕着事件是否发生的定性描述判断和概率分成3～5个等级来赋值。

结合江西省具体情况和国内外应用研究经验，考虑到本次评估的水库均为病险水库，存在溃坝事件发生的可能，因此，按各设区市不同水库数量，将不同类型水库分为以下各层，分别赋予不同的影响系数，见表6。按影响系数分别乘以各设区市的排序因子，得出水库的综合影响因子系数。

表6　　综合影响系数

水库类型	部补中型			省补中型			部补小（1）型				
分组层次	3			3			5				
综合影响系数	1～0.90	0.90～0.75	0.75～0.50	1～0.90	0.90～0.75	0.75～0.50	1～0.90	0.90～0.80	0.80～0.70	0.70～0.60	0.60～0.50

2.4　风险评估指数的确定

对水库的风险评估指数表示为以上3大影响因子的乘积，即

$$R=P_fLS \tag{2}$$

式中：R为病险水库风险评估指数；P_f为大坝险度因子系数；L为溃坝损失因子系数；S为综合影响因子系数。

分别对不同类别水库进行计算，得出大坝风险评估指数值，按数值大小进行分类排序。风险指数越大，表明该水库危险性大且溃坝损失影响严重，应优先得到除险加固。

3　风险评估计算排序的应用

根据以上赋分原则及系数，通过对江西省已列入除险加固规划但尚未实施的64座中央补助中型水库、50座省补助中型水库及128座中央补助重点小（1）型水库，由5位专家各自针对每座水库具体情况进行分析赋分计算，再统计以平均值计算得出病险水库的风险评估指数，按风险评估指数大小对水库群进行除险加固排序。计算分析表明，江西省中央补助类水库风险指数高于省补助类水库，而小型水库风险指数相对较低，但大部分水库风险评估指数均密集于不可容许的高风险区，表明应尽快进行加固。水行政主管部门根据本次中期评估排序结果，对江西省病险水库除险加固工作计划进行调整，促进了病险水库加固更科学有序的进行。

4　结语

应用表明，该计算方法简单实用，可操作性强，分析结果与实际基本相符，可用于群坝除险加固排序工作，能为决策层提供科学合理的加固实施依据。

参考文献：

[1]　傅琼华，苏立群．水库安全风险排序［J］．水利建设与管理，2004，(4)：46-47.

[2] 富曾慈，王仁钟，李雷，赵广和．病险水库判别标准体系研究［R］．南京：南京水利科学研究院，2004.

[3] 王仁钟，李雷，王昭升，彭雪辉．基于风险评价的病险水库除险加固排序实用方法研究［R］．南京：南京水利科学研究院，2004.

[4] Dekay M L，McClelland G H. Predicting loss of life in case of dam failure and flash flood［J］．Risk Analysis，1993，(2)．

[5] 姜树海，范子武．大坝的允许风险及其运用研究［J］．水利水运工程学报，2003，(3)：7-12.

[6] 林齐宁．决策分析［M］．北京：北京邮电大学出版社，2002.

三、工程技术推广应用

SHUIGONGANQUANYUFANGZAIJIANZAI

射水法建造混凝土防渗墙在赣抚大堤加固配套工程中的应用

杨晓华[1,2]

1. 江西省水利科学研究院；2. 河海大学水电学院

摘　要： 射水法造墙技术是建造地下连续墙的一种施工方法，通过其在赣抚大堤加固配套工程樟防1标段中的应用，介绍设备构成、工作原理、施工工艺、质量控制及质量检测等方面的内容，并提出质量控制的设想。

关键词： 射水法；造墙；防渗；赣抚大堤；应用

本文通过射水法造墙在赣抚大堤加固配套工程赣东堤樟树段1号单位工程（以下简称赣东大堤樟防1标段）的施工体会，针对一些具体情况，结合施工质量控制，对射水法建造混凝土防渗墙工艺进行一些设想与展望，供同仁参考。

1　工程概况

赣抚大堤位于江西省中部，包括赣江下游东岸大堤（简称赣东大堤）和抚河下游西岸大堤（简称抚西大堤）及其附属建筑物等，赣东大堤樟防1标段则位于其保护范围樟树市永泰镇境内，该堤防设计防洪标准为50年一遇，2级堤防。本次工程建设的主要内容为赣东大堤堤顶桩号15＋400～17＋900段的堤身堤基防渗墙处理。

1.1　工程地质条件

赣东大堤坐落在赣江下游冲击平原东岸，圩区内地形平坦，分布众多河流、沟渠、水塘等地表水系，区内地下水类型主要为孔隙性潜水，砂、砾（卵）石层为主要的含（透）水层，水量丰实，表层黏性土为隔水顶板，下伏第三系泥质粉砂岩为隔水底板。由设计单位提供的《赣抚大堤加固配套工程樟防1标段招标及合同文件》和施工单位所做的先导孔资料可知，标段内堤身填土厚5.7～8.6m，主要由黏土、壤土等组成，局部夹砂壤土、砂等，呈稍密状态。在桩号15＋600～16＋400、16＋900～17＋600处，由于砂壤土、粉细砂含量较高，填筑质量较差，易产生堤身渗漏。堤基地层岩性依次为黏土，厚1.1～5.0m、可塑状态；壤土，厚0.5～2.8m，可塑状态；砂壤土，局部分布，厚2.1m，松散—稍密状态；砾卵石，厚3.8～5.0m、松散—稍密状；基岩为泥质粉砂岩。上述的砂壤土、卵石层呈松散—稍密状态，存在堤基渗漏隐患。

本文发表于2008年。

1.2 主要设计及质量技术指标

樟防1标段范围内，不仅堤身防渗能力差，而且堤基砂壤土层、砾卵石层相对较厚，具有强透水性，每到汛期，该堤段都会出现堤身散浸、堤脚管涌、泡泉等险情，严重威胁到赣东大堤的防洪安全。

为了有效控制该堤段的渗漏险情，确保大堤能安全度汛和正常运行，根据渗透控制原理及类似工程经验，通过设计方案的比选，选用射水法造混凝土防渗墙进行堤身、堤基防渗处理。

防渗墙轴线距堤顶外边线内侧1.5m布置，处理堤线长2500m，设计造墙面积53600m^2，墙顶高程36.35m，墙底高程13.80m，墙体深入基岩面0.5m。质量标准为：墙厚不小于220mm；墙体抗压强度$R_{28}\geqslant$10MPa；墙体渗透系数$K\leqslant1.0\times10^{-6}$cm/s；墙体孔斜率不小于0.3%；允许渗透坡降$J>50$。

2 技术原理及工艺流程

2.1 射水造墙机的构成

射水造墙机主要由正反循环泵组、成槽机、混凝土浇筑机和混凝土搅拌机四个部分组成（以福建水利水电科学研究所生产的XS-Ⅲ型为例）。成槽机和混凝土浇筑机分别安装在各自的机架上，它们与正反循环泵组、混凝土搅拌机组一同用滚轮支承在同一条钢轨上，钢轨与防渗墙轴线平行，各机架移动时采用电动行走轮，固定时采用螺旋千斤顶，机器操作简单、轻便，成型器内设高压喷嘴，两侧设有侧向喷嘴，成型器底部刀具采用高强合金钢制造。XS-Ⅲ型射水造墙机的主要技术参数见表1。

表1　XS-Ⅲ型射水造墙机的主要技术参数

项目	成墙深度/m	成槽宽度/mm	水泵流量/($m^3\cdot h^{-1}$)	水泵功率/kW	冲击频率/(次·min^{-1})	冲击行程/m
技术参数	≤30.0	220～450	150～260	75.0	10～30	0.5～1.5
项目	成型器长/mm	成型器宽/mm	成型器高/mm	总功率/kW	垂直精度	平均工效/($m^2\cdot$台班$^{-1}$)
技术参数	2000	220～450	1750	180.0	1/300	50～70

2.2 工作原理

利用水泵及成型器中的射水喷嘴形成高速水流来切割破坏土层结构，水土混合回流，泥沙溢出地面（正循环）或者用砂砾泵抽吸出槽孔（反循环），同时利用卷扬机带动成型器上下往返运动，进一步破坏土体，并由成型器下沿刀具切割修整孔壁形成具有一定规格尺寸的槽孔。槽孔由一定浓度的泥浆固壁，溢出或抽吸出的泥浆与混合一起的土、砂、卵石等流入沉淀池，土、砂、卵石沉淀，泥浆水循环使用。槽孔成型后提起成型器，成槽机电动行走到隔一槽孔位置继续造孔，混凝土浇筑机就位，采用导管法水下混凝土浇筑建成混凝土槽板，并在施工中采用平接技术建成地下混凝土连续墙。

2.3 射水法造墙施工工艺流程

射水法建造混凝土防渗墙施工包括：施工准备、设备安装调试、射水建槽、清孔换浆、水下混凝土浇筑等工艺，具体工艺流程见图1。

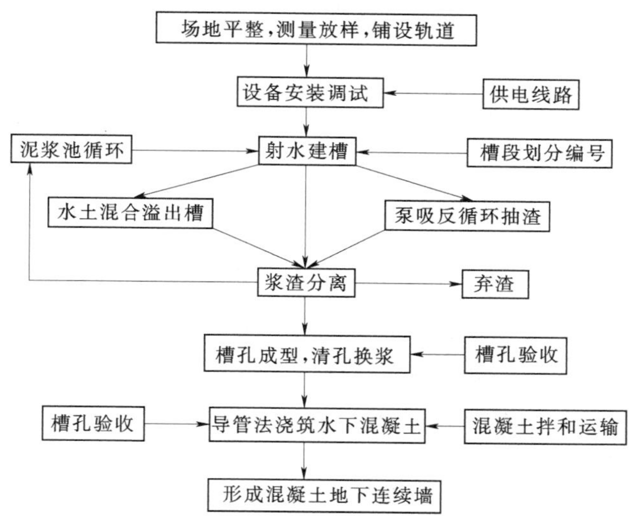

图1　射水法施工工艺流程

3　射水法施工

本工程主要施工设备采用XS-Ⅲ型射水造墙机及相应配套设备，见表2。

表2　　　主要施工设备

设备名称	规格型号	性能参数	数量
造墙机	XS-Ⅲ型	30m	3
泥浆搅拌机	JZM350型	350L	3
正反循环泵	6BS	180m³/h	3
灰浆泵	4PH60	55m³/h	3

3.1　施工准备

(1) 测量放线。在施工设备进场前，通过布设测量导线以及测量控制网，精确测量出防渗墙轴线控制点，定出防渗墙轴线。

(2) 地质先导孔。沿防渗墙轴线每100m布置一个地质勘探先导孔，用150型地质钻机钻孔取芯，探明地层分布及基岩顶面高程，以确定防渗墙底线。

3.2　造孔技术

造孔工序包括4个过程：移机定位、射水造孔、清孔换浆、移机交孔。其技术要点分述如下：

(1) 施工中采用两序孔施工法，即先施工单序孔，再施工双序孔，如此交替进行。单双序孔的施工循环间隔以48h混凝土初凝来控制，经验表明，深度为22m以内的槽孔以10～12个单序孔后返回打双序孔为好。

(2) 成槽机的两根钻杆下端与成槽器连接，上端通过高压胶管一个与正循环泵连接，一个与反循环泵连接。造孔时，正循环泵射水、反循环泵抽吸，要保证足够的射水水流压力，一般为0.4～0.6MPa。应根据不同地质条件控制成槽器的进尺速度，避免高压水喷射时间过长，孔底反射造成两侧扩孔。

(3) 垂直度的保证。造孔时的垂直度是决定能否形成地下连续墙和保证墙体搭接厚度要求的关键。本工程通过4个方面来保证机械的垂直度：①夯实作业面铺轨找平，调平机架，精校后锁固定位；②机操手提落成型器力求精良稳准，控制造墙进尺均匀；③定时检测泥浆指标，注意保护槽口以防塌槽破坏槽形；④时刻观察偏斜情况，一经发现偏斜（偏斜度大于0.3%时），立即采取纠偏措施，一般采取提起成型器自上往下修正偏斜或提出成型器焊临时扶正器修孔等方法。

(4) 入基岩面确定。按照设计要求，防渗墙底高程嵌入相对不透水层深度不小于50cm，施工中用三种方法来判断成型器是否已经达到基岩：①以感觉进尺难度及机械震动大小来初步判断；②在初步判断已到达基岩的情况下，打捞砂砾泵抽吸出来的岩块，查看其新鲜程度，判断基岩位置；③确定入基岩后，再下钻0.4～0.5m停止进尺。

(5) 清孔换浆。当造孔成槽达到设计深度，孔位、孔深、孔斜等经监理工程师验收合格后，进入清孔换浆工序。利用反循环系统进行清孔换浆，使槽孔底部清洁，泥浆达到设计指标（见表3）。然后成槽机移位，混凝土浇筑机移至孔前，下导管准备浇筑混凝土。

表3　　泥浆性能控制指标表

项目	重度 /(g·cm^{-3})	黏度 /s	含砂量 /%	失水量 /(mL·30min^{-1})	泥皮厚度/mm	pH值
新制泥浆	1.10～1.20	18～25	≤15	<30	2～4	7～9
槽内泥浆	1.10～1.20	18～25	5～8	<15	2～4	7～9
清孔泥浆	<1.10	18～20	<5	<10	2	7～9

3.3　混凝土浇筑技术

混凝土浇筑是成墙质量关键的一道工序，同样也包括4个过程：定位下管、拌和送料、浇筑监测和清场移机。其技术要点分述如下：

(1) 混凝土拌和。水泥采用普通硅酸盐水泥，按设计要求确定为32.5级，骨料选用天然砂卵石，级配选用卵石粒径不大于40mm二级配混凝土。为了获得密实的混凝土和保证拌和物有良好的黏聚性和保水性，水灰比选用0.55进行拌和。

(2) 水下浇筑。采用直升导管法浇筑，每根导管丝口上涂些黄油再进行连接，以防止导管水下漏水。第一灌混凝土浇筑前，导管内放置一个小橡胶球，以控制混凝土下降速度，并便于第一灌混凝土导入孔内能迅速将导管底部管口埋入混凝土0.3～0.5m，浇筑过程中，导管埋入混凝土的深度控制在2～4m为宜，混凝土面上升速度不小于2m/h，入孔坍落度为0.18～0.22m，扩散度为0.34～0.40m，初凝时间不小于6h。

4　混凝土防渗墙质量控制

混凝土防渗墙质量控制包括造孔成槽质量控制和浇筑成墙质量控制两部分。

4.1　造孔成槽质量控制

4.1.1　确保机具的稳定

在铺轨前整平和加固地基使其密实；减少枕木间距；前后沟开挖离铁轨不能太近，以免枕木被泥浆水泡软而下沉，影响机具稳定。

4.1.2 泥浆固壁

配制一定浓度的泥浆用于固壁，一旦发现施工中有泥浆漏失，及时在漏失出口处用麻袋装土堵漏，并同时在孔内回填黏土、粗砂或膨润土，以提高固壁泥浆浓度。

4.1.3 孔斜率

为保证槽孔的垂直精度，开钻前，用水平尺、水准仪对成槽机的平整度和导向架对地面的垂直度进行严格的检测、校核，检测合格后方可开钻，在施工中随时检测，并采用铅锤法随时检测，保证槽孔的垂直偏差不超过0.3%。

4.1.4 槽孔深度控制

槽孔嵌入基岩深度不小于50cm，施工过程中，在成型器底刃上焊接长钢齿，或者将齿尖制作成倒刺形，这样效果很好。在出渣口取到岩样后，立即在钻杆上做一标记重新进尺50cm，或观察钻杆10min左右，若无明显进尺，则表示已达到相对不透水层或基岩，可以终孔。终孔时通过钻杆量测并计算出实测孔深，然后在下完灌浆导管后，再用测绳量测得到一个孔深，两者之差即是孔底淤积厚度。此淤积厚度小于设计允许厚度10cm时，即可以浇筑，否则重新清孔换浆。

4.2 浇筑成墙质量控制

4.2.1 槽孔准备条件

清孔换浆后1h内，槽孔浇筑标准：孔底淤积厚度不大于10cm；泥浆密度不大于1.1g/cm^3；泥浆黏度不大于20s；泥浆含砂量不大于5%。清孔换浆验收合格后4h内浇筑混凝土。

4.2.2 导管准备

安装导管前，对各节导管进行检查，不使用已经变形的导管，导管的连接以丝口为主，连接前丝口涂上黄油，起润滑、止水作用，导管接好后，装入首浇隔水塞，准备灌注混凝土。

4.2.3 浇筑的连续性

混凝土浇筑必须连续，保证槽孔内的混凝土面均匀上升，其上升速度应大于2m/h，导管埋深不得小于1m，但不宜大于6m。一般每浇筑3～4斗料测一次导管埋深或混凝土面高程。浇筑过程中随时检查混凝土的坍落度和扩散度，使其分别保持在18～22cm和34～40cm之内。混凝土浇筑至孔口后，应高出设计墙顶高程50cm，以保证孔口混凝土面的质量。

5 混凝土防渗墙墙体质量检测

本工程的墙体质量检测包括施工过程中的机口混凝土取样、施工结束后开挖检查、钻孔取芯室内试验以及单元质量评定等。

5.1 机口混凝土取样

本工程共完成1216个槽孔，按质检要求，每20个槽孔为一个单元，在机口取一组试样做抗压试验，共取抗压试件61组，抗渗试件取2组，委托江西省水利厅基本建设工程质量检测中心站进行测验，测试结果见表4，从表中可以看出，混凝土抗压强度、渗透系数均满足设计要求。

表 4　　混凝土机口取样测试成果

测试项目	最大值	最小值	平均值	标准值	离散系数
抗压强度/MPa	30.5	21.7	25.2	20	0.07
渗透系数/($\times 10^{-7}$cm/s)	6.95	1.68	4.31	10	—

5.2 墙体开挖检查

施工结束后，为检查墙体搭接、墙体垂直度及厚度等情况，按照要求每 200m 进行一个检查孔的开挖。本工程共开挖 12 个检查孔，开挖尺寸 3.0m，深 2.0m，宽 1.0m。经检查，墙体外观质量好，墙面平整密实，接缝连接良好，无错位，夹泥等现象，墙体厚度均不小于 30cm，满足设计要求。

5.3 钻孔取芯检测

在完工 28d 后，根据要求，每 800m 布置一个检查孔进行钻孔取芯，在每个检查孔的中部与下部取样检测抗压强度、渗透系数、渗透比降。检查孔孔径 110mm，孔深超过墙体深度一半，且不小于 8.0m，试验结果均满足设计要求，见表 5。

表 5　　防渗墙钻孔取芯检测成果

桩号	部位/(深度，m)	抗压强度/MPa	渗透系数/($cm \cdot s^{-1}$)	渗透比降
15+500	8.5	15.4	3.41×10^{-7}	112
16+300	10.3	17.1	1.17×10^{-7}	115
17+100	15.6	13.8	1.83×10^{-7}	121
17+900	20.0	16.6	2.06×10^{-7}	138

5.4 单元质量评定

本工程项目共划分为 61 个单元工程。按照《水利水电基本建设工程单元工程质量等级评定标准》(SDJ 249—88) 的规定，经施工质量评定，单元工程质量全部合格，其中优良单元工程 58 个，优良率达 95.08%。

6 结语

(1) 射水法建造混凝土防渗墙技术近年来在赣抚大堤加固工程中应用得到了很大的发展，该技术不仅解决了均质壤土中建造连续墙问题，还解决了在砂砾石地层中建造混凝土连续墙的问题。

(2) 有必要进一步提高射水造墙机的各项性能，同时对造墙深度、造墙厚度、垂直度的保证等方面需进一步加强研究，不断提高该技术的应用范围。

(3) 发展相应的检测仪器及检测方法。尽快地将超声检测仪、水下摄像机、激光定位仪、计算机等应用于该项技术中，以便更及时准确地控制下地连续墙的施工质量。

参考文献：

[1] 李孝成，王锭一．射水法造墙技术及其在工程中的应用［J］．人民珠江，2000（1）．

[2] 黄志鹏，余强，董建军，等．射水法［M］．北京：中国水利水电出版社，2006．

[3] 彭冲．射水法建造薄型混凝土防渗墙在长江重要堤防加固工程中的应用［J/OL］．水利工程网．

[4] 吴建红，鄢忠清．黄金水库大坝塑性防渗墙施工工艺及质量控制［J］．江西水利科技，2007（4）．

薄壁抓斗法混凝土防渗墙技术在滨田水库的应用

刘　翀[1]，王太友[2]

1. 江西省水利科学研究院；2. 中国水利水电第六工程局

摘　要：混凝土防渗墙技术已广泛用于病险水库土石坝的防渗加固，本文通过该技术在江西省鄱阳县滨田水库除险加固工程中的应用，介绍了薄壁液压抓斗法混凝土防渗墙的施工技术要点及注意事项。

关键词：薄壁液压抓斗；导墙；泥浆护壁；塑性混凝土；防渗墙

薄壁液压抓斗法这一新的防渗墙施工技术出现于20世纪90年代，该技术适用于坚硬的土壤与砂砾石中成槽，成槽深度可达60m，且工程造价低，施工速度快，一台液压抓斗成槽平均工效为125m²/d，该技术已在江西省鄱阳县滨田水库除险加固工程中得到了推广运用。

1　工程概况

江西省鄱阳县滨田水库于1960年4月建成投入使用，是一座以灌溉、防洪、养殖等综合利用的大（2）型水库。

水库主坝为均质土坝，坝顶高程53.20m，坝顶长723m，坝顶宽5m，最大坝高26m。水库在多年运行后主坝存在坝身渗流稳定、坝坡稳定、坝基渗漏和坝肩绕坝渗漏等问题，严重威胁坝体安全和正常效益的发挥，被水利部列为病险库。

滨田水库除险加固工程对主坝坝体防渗选用了薄壁抓斗塑性混凝土防渗墙技术进行加固处理，由中国水利水电第六工程局承建。防渗墙轴线位于坝轴线上游0.5m处，范围0＋000～0＋723桩号，全长723m，孔口高程53.20m，墙体有效厚度0.35m，进入基岩2.5～6.2m，墙顶高程51.50m。塑性混凝土防渗墙深29m左右。本工程完成混凝土防渗墙19010m³，共耗时148d。

坝址区出露的地层岩性为前震旦系板溪群第二段浅变质岩及第四系松散堆积层。坝址区构造形迹主要表现为断层和裂隙。地层概况由上而下为：①6～29m厚的黏性壤土；②河床岩石表面呈全风化为其厚度约为2.9～5.2m；③强风化带岩石厚度约为3.1～8.5m，弱风化岩石厚度约7.5～13.0m，下伏微新岩石。

塑性混凝土防渗墙设计物理力学指标为：渗透系数$K \leqslant 5.0 \times 10^{-7}$cm/s；墙体厚度$t=$

本文发表于2005年。

0.35m；抗压强度 3MPa≤R28d≤5MPa；弹性变形模量 300MPa≤E28d≤1000MPa；允许渗透比降 $J\geqslant 80$。

2 薄壁抓斗法混凝土防渗墙成墙技术要点

2.1 造孔成槽

2.1.1 布置施工平台

抓斗施工平台设置在防渗墙轴线下游侧，原坝顶宽度只有 5m 不能满足抓斗施工所需 8m 宽的施工平台的要求。经业主同意将坝顶高程降低 1m 至 52.20m，并将防渗墙轴线向上游平移 1.75m，另外将降坝的土料填筑到坝体下游侧，以保证抓斗施工平台的宽度。在防渗墙轴线的下游侧设置平行坝轴线的排渣排水沟，断面尺寸 40cm×40cm，再按 40m 间距修建垂直防渗墙轴线的排渣排水沟，将废渣、废水排至下游坝脚，所有废渣运至弃渣场。

2.1.2 修筑导向槽

导向槽是在地层表面沿地下连续防渗墙轴线方向设置的临时构筑物。导向槽起着标定防渗墙位置、成槽导向、锁固槽口；保持泥浆液面；槽孔上部孔壁保护、外部荷载支撑的作用。导向槽的稳定是混凝土防渗墙安全施工的关键。本工程导向槽两侧墙体采用倒 L 型断面，现浇 C10 混凝土构筑，槽内净宽 45cm，顶面高于施工场地 10cm 以阻止地表水流入。其结构见图 1、图 2。

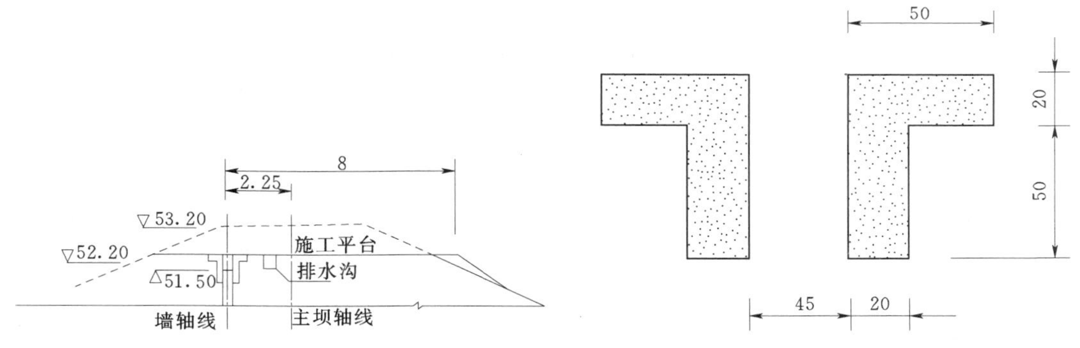

图 1 施工平台剖面图（单位：m）　　图 2 导向槽剖面图（单位：mm）

2.1.3 抓斗成槽

土石坝防渗墙开槽施工工艺主要锯槽法和挖掘法。锯槽法主要有往履射流式开槽、链斗式开槽、液压式开槽；挖掘法主要有冲击钻法、抓斗法、冲抓结合法。根据本工程地质条件及生产性试验确定采用挖掘法中抓斗法，并制定了“三抓法”的施工方案，即采用 KH—180 液压抓斗机先抓取槽段的主孔再抓取中间的副孔成槽。

2.1.3.1 槽段划分

槽段划分为Ⅰ、Ⅱ序槽段，根据设备及地质条件确定Ⅰ、Ⅱ序槽段开挖长度同为 7.5m，每个槽段分为两个主孔及一个副孔，先施工Ⅰ序槽段，后施工Ⅱ序槽段。

2.1.3.2 槽段成槽

槽段成槽采用“三抓法”，在导向槽上放样标识孔位，将抓斗对正孔位后进行垂直造孔。首先施工槽段两端2.8m的主孔，主孔完成后再抓中部1.9m的副孔。主、副孔完工即该施工槽段成槽完工，经监理确定岩层岩性，并最终确定该施工槽段成槽深度。

2.2　护壁泥浆

泥浆在造孔成槽过程中起固壁、悬浮、携渣、冷却钻具和润滑的作用，成墙后还可增加墙体的抗渗性能，本工程泥浆采用膨润土拌制，泥浆配合比为水1000kg、膨润土50kg、$Na_2CO_3$1kg；固壁泥浆性能指标密度小于1.1g/cm^{3}如漏斗黏度大于25s、含砂量小于3%。

新制泥浆经过24h膨化后，利用供浆管输送至槽孔内使用，成槽及槽段浇筑过程中回收的泥浆，经净化后可重复使用。槽孔孔口泥浆面在成槽过程中保持在导向槽顶面以下30～50cm范围内。

2.3　防渗墙体灌筑

混凝土防渗墙是在泥浆下灌注混凝土，本工程是采用刚性导管法进行墙壁体灌筑，混凝土竖向顺导管下落，利用导管隔离泥浆，使其不与混凝土接触，导管内混凝土依靠自重压挤下部管口的混凝土，并在已灌入的混凝土体内流动、扩散上升，最终置换出泥浆，保证混凝土的整体性。

2.3.1　清孔换浆

槽段终孔验收合格后进行清孔，清孔采用抓斗抓取淤泥，利用下设潜水排污泵抽浆，并及时用新鲜泥浆补充。清孔换浆结束1h后，达到下列标准：①孔底淤积厚度不大于10cm；②泥浆参数为：槽内泥浆比重不大于1.1g/cm^3，黏度不大于35s，含砂量不大于3%。清孔换浆工作可以结束。

槽段清孔换浆结束前将钢丝刷子安装在抓斗斗体上，紧贴一、二期混凝土结合面，分段上下反复提动，达到刷子上不带泥屑，孔底淤积不再增加，即接头面清洗合格。

2.3.2　槽段混凝土灌注

(1) 清孔换浆结束后，下设混凝土灌注导管，导管内径为200mm。一期槽段长度为7.5m，下设三套导管，两侧导管距槽端1～1.5m；二期槽段由于套抓接头，槽段长度为8.2m，下设三套导管，两侧导管距孔端1.0m；同时，槽段内导管间距不大于3.5m。导管底部距槽孔底板不大于25cm，当槽底高差大于25cm时将导管置于控制范围的最低处。

(2) 灌注前导管内置入可浮起的隔离塞球，灌注时先注入水泥砂浆，随即注入足够的混凝土，挤出塞球并埋住导管底端，避免混凝土与泥浆混合。

(3) 灌注过程中每30min测量一次混凝土面，每2h测量一次导管内混凝土面，根据混凝土面上升情况，决定导管的提升长度。导管在混凝土内的埋深最小不得小于1.0m，最大不得大于6.0m，在保证埋深的前提下，随着混凝土面的上升，用吊车提升导管，并将顶部的部分导管拆除。

(4) 槽孔内混凝土面上升至槽口时，采用泥浆泵抽出浓浆，并提升导管，减小埋深，增加混凝土的冲击力，直至混凝土顶面超出设计墙顶标高0.5m，即可停止浇筑，拔出导管。

2.4　槽段接头处理

相邻槽段的衔接部分即为接头，本工程采用钻凿法进行接头连接，即一期槽段浇筑完毕 12h 后，视混凝土强度进行二期槽段造孔时，将一期槽段混凝土套抓 35cm，以保证接头质量。槽段接头示意图见图 3。

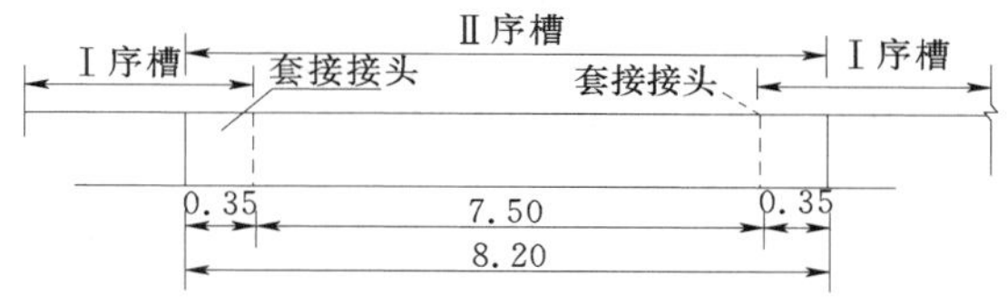

图 3　槽段接头示意图

3　薄壁抓斗法混凝土防渗墙成墙技术应用常出现的问题

3.1　坍塌、漏浆

槽段在成槽过中会出现局部坍塌和大面积坍塌，当出现局部坍塌时加大泥浆密度，出现大面积塌孔时用优质黏土（掺入 20%水泥）回填到坍塌处以上 1～2m，待沉积密实后再进行施工，同时在相应地段减小了槽段开挖长度。

槽段成槽开挖过程中，有时会出现的漏浆现象，出现漏浆现象常采用处理措施有：①平抛黏土，加大泥浆比重或抛入锯末进行堵漏；②松散地层，造孔应循序渐进，预防在先，稳中求快；③证泥浆供应强度和质量，发现漏浆及时补充；④对漏失严重的地层用速凝水泥等特殊材料处理，必要时还应对槽孔进行回填。

3.2　导管堵塞

成墙灌注混凝土过程中有时会出导管堵塞，针对导管堵塞采用捣、顿方法疏通，如果无效将导管全部拔出、冲洗、并重新下设，用泥浆泵抽净导管内泥浆后继续浇筑，同时还要核对混凝土面高程及导管长度，确认导管的埋入深度。

4　薄壁抓斗法混凝土防渗墙成墙技术应用应注意的问题

4.1　控制好抓接头的时间

抓接头时间太短混凝土没有凝固，时间太长混凝土强度太高，抓接头适宜的时间为墙体浇筑后 12h，最迟不超过 24h，抓斗抓取时斗体一侧为混凝土另一侧为土，斗体受力分布不均匀，容易造成槽孔沿轴线方向偏移，导致接头质量无法保证，同时严重影响造孔成槽进度，掌握好抓接头的时间是成槽进度快慢的关键环节。

4.2　吊装设备的配置和保养

本工程的混凝土运输采用 3 台 $1m^3$ 自卸汽车运输混凝土，1 台 12t 吊车吊卧罐的较为经济的设备组合。吊车配置偏少，在施工中吊车一旦出现机械故障，短时间内如无法修好将造成很大的经济损失及工程质量问题。

4.3　成墙混凝土灌筑应注意的问题

混凝土导管下设过程中检验螺丝紧固程度，确保导管间连接可靠。混凝土灌注具有相当高的连续性，因故中断不得超过 40min。同时，槽内混凝土上升速度不得小于 2m/h，各灌筑导管均匀放料，保证混凝土面均匀上升，使其高差不超过 0.5m。浇筑时槽口要设置盖板，防止杂物落入槽内。

扩张金属网在赣东大堤（樟树段）固岸工程中的应用

王　玢[1]，王南海[1]，王　纯[2]

1. 江西省水利科学研究所；2. 江西省水利厅

摘　要：介绍了在堤岸固脚工程中利用扩张金属网兜填充块石后，抛至堤脚深泓区域，作为抛石护脚的基础铺垫层，以起到稳固堤脚、避免抛石流失作用的一个实例。实例中采用了较巧妙的施工工艺，大大提高了工程施工效率和安全性。

关键词：扩张金属网；近岸深泓；固脚；堤防固岸工程

1　概述

扩张金属网作为一种新材料、新工艺得到了越来越广泛的应用，在水利工程中为确保质量、提高工效应用范围也在不断扩大。本文介绍应用扩张金属网兜填充块石用于江西省赣抚大堤加固配套工程中的赣东大堤（樟树段）桩号为34＋400～34＋550的150m固岸工程中的应用情况。

赣江下游东岸的赣东大堤，上起新干县，经樟树、丰城、南昌县，下至南昌市新洲闸，全长137.14km，赣东大堤从桩号31＋000～38＋000段为樟树市城区，按江西省水利规划设计院的设计要求，本项新技术推广应用工程选在桩号为34＋400～34＋550的深泓逼岸段，以扩张金属网兜填充块石后，抛至堤脚深泓区域，作为抛石固脚的基础铺垫层，以起到稳固堤脚，避免抛石流失的作用。

2　扩张金属网固岸技术应用

2.1　设计方案

赣东大堤34＋400～34＋550段，位于樟树铁路大桥的上游，这里水深流急，堤外无滩，大堤护岸固脚任务紧迫，对该段固脚工程，江西省水利规划设计院的设计方案为抛石固脚。经江西省赣抚大堤加固工程建设办公室同意，选“樟1标”的抛石固脚护岸（桩号34＋400～34＋550）断面为试验工程段，江西省水利规划设计院将原设计中用钢筋笼固脚方案改为扩张金属网兜固岸方案，网兜抛方量为1780m^3，网兜抛投断面见图1。图1说明如下：

（1）本图高程系统为黄海高程系统。

本文发表于2002年。

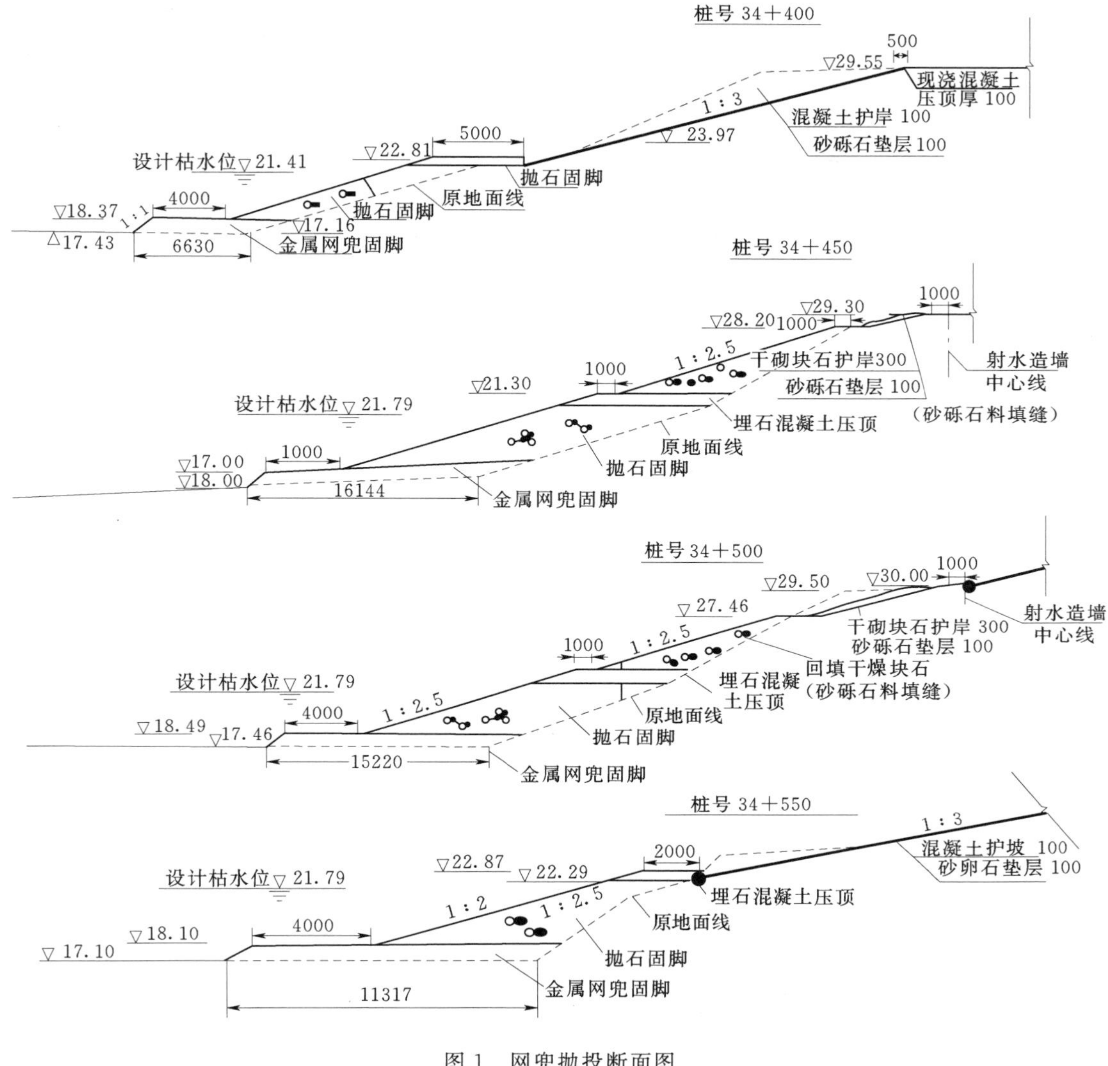

图 1　网兜抛投断面图

（2）本图高程、桩号以 m 计，其余尺寸以 mm 计。

（3）金属网兜固脚范围桩号 34+400～34+550。

（4）射水造墙中心线外侧 1m 进行 1∶3 的削坡，干砌块石护岸。

（5）抛石固脚按 0.6m 厚一层逐层抛护，技术要求按赣抚大堤抛石总说明。

（6）埋石混凝土中混凝土含量为 60%，块石含量为 40%。

（7）堤外坡回填采用干砌块石逐层堆砌，并用砂砾石填缝。

2.2　扩张金属网

本次金属网兜抛石固脚工程选用江西省新余钢铁厂生产的扩张金属网。网面宽 B 为 2m，长 L 为 4m，网孔尺寸为 $T_L \times T_B = 12\text{mm} \times 30\text{mm}$，丝梗宽 b 为 1mm，钢板厚度为 h 为 1mm，见图 2 及表 1。金属网用材的化学成分及机械性能见表 2。

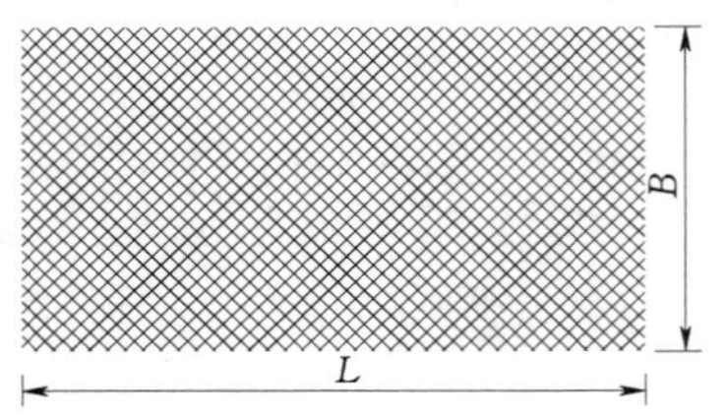

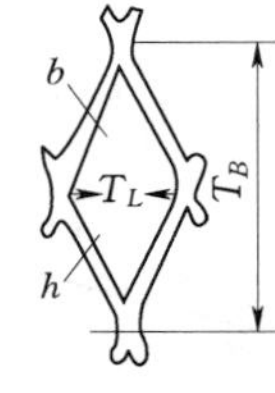

图 2 扩张金属网几何尺寸示意图

T_L—短节距；T_B—长节距；h—板厚；b—丝梗宽；B—网面宽；L—网面长

2.3 金属网兜抛投及抛设定位

本科技推广项目由江西省水利科学研究所负责技术指导，施工由九江市水建公司“樟1标”项目部组织实施，江西省水利规划设计院将原钢筋笼固脚方案经设计变更为金属网兜固脚（设计院设计变更通知书，编号：樟标〔2001〕02号），原钢筋笼固脚方量 $1059m^3$，更改后的扩张金属网兜固脚方量为 $1780m^3$，金属网规格为 2m×4m，每个网兜设计装石方量 $2.5m^3$。

表1 新余钢铁厂可供产品规格 mm

板厚	网格尺寸		标准成品尺寸（$B\times L$）
	T_L	T_B	
1	10	25	2000×4000
	12	30	
1.2	12	30	
	18	50	
1.5	18	50	
	22	60	
2	18	50	
	22	60	
	38	100	
3	24	60	
	26	100	
4	24	60	
	30	80	
	36	100	
5	32	80	（800～1200）×（1000～3500）
	40	100	
6	32	80	
	40	100	

表2 金属网用材化学成分及机械性能

板厚/mm	网格尺寸			化学成分/%				机械性能		
	T_L	T_B	C	Si	Mn	P	S	抗拉负荷/kN	抗压强度/MPa	弯曲度 α/(°)
1.0	12	30	0.7	0.27	0.65	0.028	0.018	2.0	150	180

由于条件所限，金属网兜装石固脚施工不采用钢筋笼相同的方法进行。为了提高工效，节省施工费用，保障水上施工安全，经现场兜装石抛投施工试验，以运石铁驳船弦安装翻板的方式，有效地解决了金属网兜装石、制作、定位、抛投等一系列施工工艺问题，具体做法如下：

(1) 选用 2 条装运能力为 70m³ 的铁驳船，首尾相连，经堤岸上放线定位于抛投区，见图 3。

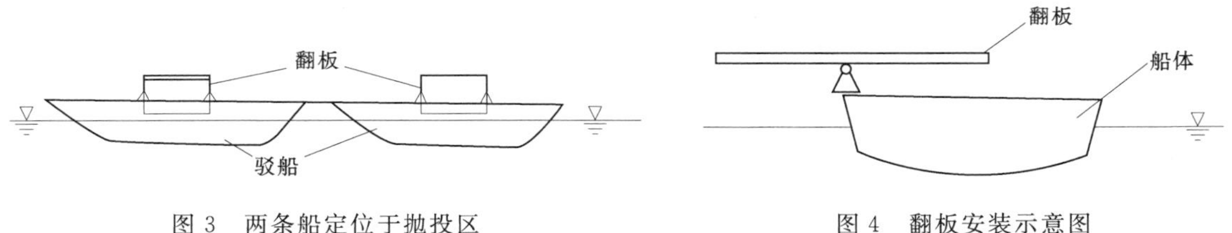

图 3　两条船定位于抛投区

图 4　翻板安装示意图

(2) 在铁驳船的近岸船侧安装一块与扩张金属网尺寸相等的铁架木质活动翻板，见图 4。

(3) 在鄱板上铺一张 2m×4m 的金属网，然后上面堆放足量的块石，见图 5。

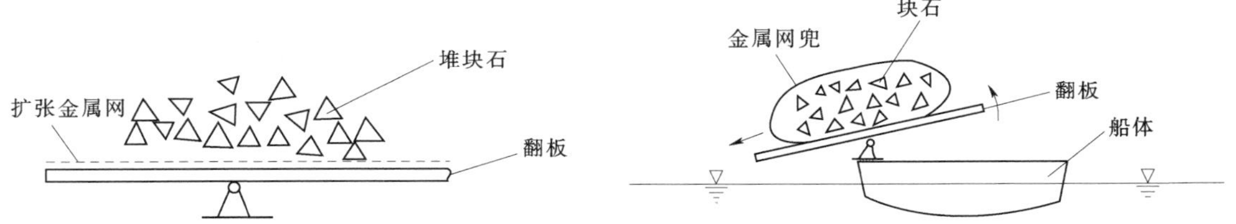

图 5　翻板上金属网装块石示意图

图 6　船体翻板上金属网装块石示意图

(4) 将金属网兜前后两边（长边方向）卷起，用铁丝穿梭封口，将网兜两侧闭合，用铁丝穿梭封口，形成一装石金属网兜。

(5) 翻板向船舷外侧转动，将金属网兜徐徐滑落于定位好的水中，坠于堤脚，见图 6。

(6) 待抛投方量达到设计要求后，适当向上或向堤线方向移定位船，再行抛投，以确保抛护均匀，达到设计要求。

2.4　施工组织

本项目施工时间从 2001 年 6 月 8—30 日，装载 70m³ 块石的 2 条铁驳船共装运抛投 25 船，每天抛运 2 船，每船装网抛投用时 6h，每船人员 10 人，船舱尺寸为 14m×5m×1m，翻板尺寸与金属网尺寸一致为 2m×4m，共用扩张金属网 1287 张，每网兜的装石方量实际为 1.36m³，共抛方量约 1750m³。

由于 2m×4m 的扩张金属网的最大理论围筒体积约 2.5m³，且该理论围筒装石量无上下盖，因此不可能每个网兜装下 2.5m³ 块石，经现场多次测量和测算，每个网兜平均装石约 1.36m³，见表 3。

表 3　　金属网兜装石方量现场测量表

项目	测次					平均
	1	2	3	4	5	
测量方量/m³	1.323	1.384	1.356	1.363	1.374	1.36
称量重量/kg	2303	2401	2388	2394	2413	239.8
按称重折算方量/m³	1.318	1.374	1.367	1.370	1.381	1.362

注　按称重折算方量计算公式：重量×[1 块石容重(1－石堆空隙率)]。

3　金属网兜抛投测量效果

2001 年 6 月 30 日，扩张属网兜抛投施工完成后，施工单位、业主和监理单位于 7 月 2 日请樟树水文站对水下抛投情况进行了水下测量，7 月 5 日提供了测量结果。测得沿堤线 150m 范围内抛投方量为 1742.5m³，见表 4。

网兜抛投断面范围基本上能够覆盖设计断面的要求，能够起到网兜作为抛石固脚基层的作用。断面范围见表 5。

表 4　　金属网兜固脚工程方量表

断面位置	34＋400	34＋425	34＋450	34＋475	34＋500	34＋525	34＋550
断面面积/m²	11.2	12.2	12.5	13.3	14.0	8.2	7.8
平均断面面积/m²	11.7	12.35	12.9	13.65	11.1	8.0	
长度/m	25	25	25	25	25	25	
方量/m³	292.5	308.75	322.5	341.25	277.5	200	
方量合计/m³	1742.5						

表 5　　金属网兜抛护断面范围比较

断面位置	34＋400	34＋450	34＋500	34＋550
设计抛护高程/m	17.43～18.37	16.10～18.50	17.49～18.50	17.10～18.10
实测抛护高程/m	17.0～19.0	17.0～18.5	17.0～19.0	16.5～21.0

从实测断面情况来看，大堤近堤岸脚已有冲沟迹象，实测堤脚高程也与原设计地形图有所差异，但实际抛护的范围从测量断面图反映，能够满足抛石固脚铺设基层的设计意图。

4　经济比较

江西省水利规划设计院在 34＋400～34＋550 段的抛石固脚基层原为钢筋笼抛石，后改为金属网兜抛石，其设计变更及实施情况见表 6。

从表中数据可知，变更设计中每网兜的装石方量不能达到 2.5m³，实际上经现场装填只能达到 1.36m³；实施情况因网兜装石方量差异较多，因此总费用上升较大；以扩张金属网兜抛石的综合单价计，明显低于钢筋笼抛石，约低 19.32%。可见金属网兜抛石比钢筋笼抛石在经济上有显著的合理性。此外，由于本项目属于小批量推广试验工程，其购置

表 6　　设计变更前后及实际工程量情况表

项目	抛石方量 /m^3	钢筋笼钢筋用量/t	钢筋笼钢筋单价 /(元・t^{-1})	金属网网兜用量/张	金属网网兜单价 /(元・张$^{-1}$)	每网兜装石方量/m^3	每网兜装总费用/元	综合单价 /(元・m^{-3})
原设计	1059	35	2500			2.5	87500	82.63
变更设计	1780			712	94.16	2.5	67042	
实际情况	1750			1287	94.16	1.36	121184	69.25

新余钢厂生产的扩张金属网的单价为 9.35 元/m^2，当购置大批量时，出厂单价会相应降低，那么，工程单价随之可降低。

5　结论

扩张金属网兜装石抛投，用于堤脚抛石固岸的基层，对解决长期以来堤脚抛石基石不稳、抛石不均、抛石流失的问题有明显的作用。而用本项目中提出的施工工艺，完全避免了像在用钢筋笼装石作抛石固岸基层施工中，所需动用电焊、船吊和机械搬运及上船、下船等费用较高、费时、安全性差、技术要求高等缺陷，其工效、经济有显著的优势。此外，由于网兜抛石较之于钢筋笼抛石更易于均匀贴布于堤脚岸坡，其做抛石基层，减小抛石底层流速更为有效，因其金属网在抛投之前已涂刷了几遍防锈防腐漆料，其抗锈蚀性能也会有所提高。

参考文献：

[1]　胡国波．如何搞好清障，疏浚及平圩行洪 [J]. 江西水利科技，1999 (1)：33 - 34.

[2]　江西省水利规划设计院．江西省信江貊皮岭分洪道工程可行性研究报告 [R]. 2000，8.

土工模袋在长江堤防崩岸治理工程中的试验应用

李良卫，张绍付

江西省水利科学研究所

摘　要：详细介绍了土工模袋在长江堤防崩岸治理工程中的施工情况及应用效果。

关键词：土工模袋；长江；堤防；崩岸

崩岸是影响江河堤防安全的心腹之患，也是水利工作者致力去攻克的难题。江西省长江堤防崩岸近年来接连不断，对沿江人民的生命财产安全构成巨大的危害。为此，对崩岸治理进行调查研究，不断探索新技术、新工艺、新材料。根据传统治理办法，即在崩岸及上下游一定范围内抛投块石，阻止崩岸的蔓延。由于抛石后改变了原河床的边界条件，水流流态随之发生急剧变化，容易出现局部漩涡，形成对块石基础的冲刷，导致块石失稳。崩岸严重的地方，汛前抛了大量的块石，汛后就发现所剩无几，只得再抛，每年都得投入大量的人力物力，却始终不能达到根治的目的，从以上不利因素分析，寻找出了一种行之有效的新方法，即土工模袋，以作长江堤防崩岸治理工程试验应用。

土工模袋是一种采用合成纤维制成的透水性新兴建筑材料，工厂加工成袋状。它的主要特点是结构整体性好，抗冲能力强，特别适用于迎流顶冲地段护岸，且外观形象好。且土工模袋起到了模板的作用，解决了水下施工时混凝土拌和物被水冲散，水泥、沙、石分离的问题。

土工模袋治理崩岸的新方法目前有拖排法和滑道法。结合长江堤防崩岸治理工程，分别选择了两个堤段进行不同方法的应用，简述如下。

1　拖排法

1.1　试验工程简况

拖排法是根据江西省水利厅与中国水利水电科学研究院合作开发的项目——“江新洲洲头北侧崩岸治理试验项目”，经现场多次试验研究得出的一种施工方法。

试验现场选在长江堤江新洲段，圩堤全长 41.36km，为一闭合圩堤。由于江新洲大堤位于长江中、下游分界处，地处分汊河段中，又近鄱阳湖出口，河道演变剧烈，江岸崩塌速度达到每年 5～10m。自 1957 年以来，年年治理，但仍然险情不断，崩岸时有发生。近年来尤为频繁、严重，因此，治理江新洲乃当务之急。

该试验段位于江新洲洲头北侧崩岸严重区，桩号 15＋500～16＋500m，全长 1000m。

本文发表于 2000 年。

流速最小 1m/s 左右，最大 2～3m/s。河底深泓最低高程－5m 左右（黄海高程，下同）。水下岸坡 1∶1.6～1∶3，局部迎水坡接近 1∶1。河床介质为粉细砂，淤积地层厚约 50m。

1.2 工程布置

根据水位的变化及工程需要，对水上部分、水位变动区、水下部分采用不同的工程措施。

水上部分为钢筋混凝土框架自锁块，下垫土工布反滤。高程 17.00～10.00m，总长度 1000m，平均宽度 23m。

水位变动区为模袋混凝土，高程 11.00～1.00m，总长度 850m，宽度为 20cm，充灌混凝土后设计厚度为 20cm。

水下部分为框格型充沙模袋软体排，高程 3.70～－5.00m，宽度为 36m。

所使用排布均采用 420g/m^2 的高强机织丙纶长丝土工布，纵向加筋带宽度为 7cm，间距为 1.1～1.2m。模袋混凝土排布为无滤点型模袋，充灌后呈哑铃状。充沙模袋软体排排布为框格型，充灌后呈现纵横交错的管状。单片模袋宽度 15m，充灌后宽度约 14m。

为保护排体的安全与稳定，分别在上游 130m、下游 20m 的范围内采取下铺土工布，上抛块石压盖措施护住两头。在充沙模袋末端抛投沙袋压脚。

自锁块四周为钢筋混凝土地梁，模袋混凝土最上端与地梁结合在一起，形成整体，共同发挥防冲护岸作用。具体布置见图 1。

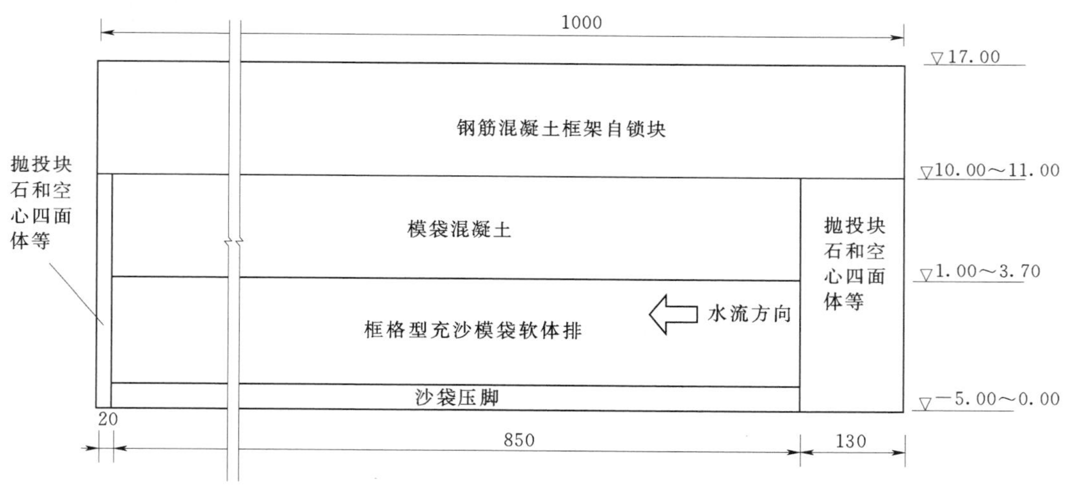

图 1　工程布置图（单位：m）

1.3 拖排施工步骤

（1）施工前准备工作。

1）理坡与排水。按 1∶2.5 进行整坡，采取上削下垫的办法进行。水下需垫坡部位最外缘抛投充沙管袋，然后从江中直接抽取河沙吹填，逐层往上施工。在水上部分已整好的堤坡上开挖盲沟，沟间距 5m，然后铺设土工布反滤排水；水位变动区从模袋幅间排出排水；水下部分渗水通过排体排出排水。

2）施工测量放样。根据施工图纸上的铺排位置，在岸上用一台经纬仪定出施工船舶的相应位置，引导施工船舶到达指定的拖排位置，然后抛锚，靠绞锚移位使拖排船位与设计船位相吻合。

（2）施工定位。在岸上安设一台经纬仪，控制好每块排体中轴线与施工船舶上2台卷扬机中线重合，确保排体准确到位。这个工作贯穿整个拖排过程，引导施工船舶的移位。

（3）钢桁架安装。拖排时为了平稳牵引，钢桁架与排体拉环连接，采用水平插销，插销采用直径45mm圆钢制作，施工船上钢丝绳通过钢桁架进行拖排作业。考虑到每块排体搭接，桁架两端上仰，仰角为30°。

（4）铺油布（或编织布）、模袋。为了减少模袋与堤坡的摩擦，避免拖拉作业时破坏已整好的堤坡，在堤坡上铺一层油布（或编织布），然后在油布上铺模袋。由于油布（或编织布）表面摩擦系数小，拖排时要严格控制好拖拉速度，避免尚未充填的模袋一并滑入水中，增加充填困难，甚至影响充填质量。

（5）模袋充沙。排体模袋在岸坡上充填。充沙采用2台沙泵轮流逐格充填，沙泵生产能力应大于$40m^3/h$。充沙后模袋饱满度达到85%以上，但不能过于饱满，充沙模袋有足够的适应地形的能力，使充沙模袋紧贴岸坡。垂直堤轴线方向充沙模袋总长度为36m，作2个单元，每个单元长18m。

（6）铺设充沙模袋。待充沙模袋第一单元充灌作业完成后，启动施工船舶上卷扬机进行拖排作业。

重复步骤（5）、步骤（6），完全充沙模袋软体排第二单元的充灌、拖拉铺排作业。

（7）模袋灌注混凝土。在岸坡上用混凝土泵进行灌注。所用混凝土为细骨料混凝土，必须有足够的流动性，可以适当掺入粉煤灰，可以适当掺入粉煤灰。垂直堤轴线方向模袋混凝土排体总长度为20m，分为2个单元，每个单元为10m。

（8）模袋混凝土排体铺设。待模袋混凝土排体第一单元混凝土灌注完成后，启动施工船舶上卷扬机进行拖排作业。

重复步骤（7），完成模袋混凝土排体第二单元的灌注，并将排体全部拖拉到位。

（9）模袋混凝土排体与钢筋混凝土地梁连接。将模袋混凝土排体最上端与钢筋混凝土地梁连接在一起，形成整体，共同发挥防冲护岸作用。

另外，组织一定的潜水员，检查排体铺设情况。

排体由下游往上游逐块铺设，相邻排体之间搭接宽度为2m，后铺排体下游边压住前铺排体上游边，见图2。

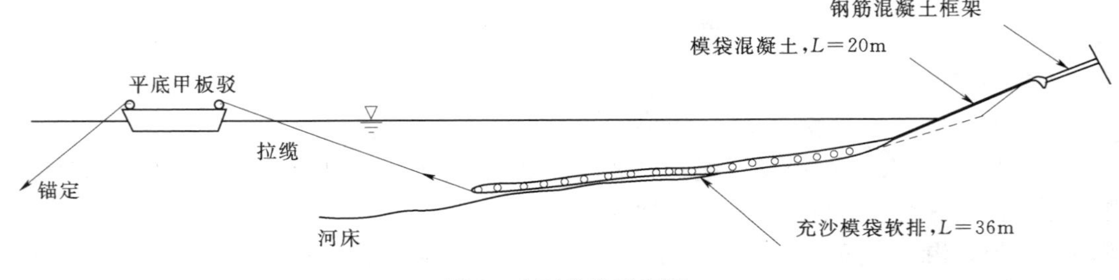

图2　水下拖排示意图

施工正常情况下，每天可以铺排 3 块，每块排体充灌后宽度约 14m，扣除排间搭接 2m，每天可完成铺排宽度约 40m，垂直堤轴线方向总长度 56m，每天可完成约 2000m^2，施工速度很快。

2 滑道法

2.1 试验工程概况

长江马湖堤模袋混凝土护坡工程采用滑道法。

长江马湖堤全长 4.189km，本试验段位于江西省九江市彭泽马湖堤 3＋230～3＋590 桩段，全长 360m。因长江水流在下三号洲尾汇合，呈约 120°急转弯，顶冲点在彭泽县城及马湖堤江岸一带形成弯道，深槽偏右岸马湖圩堤。下游小孤山和彭郎矶一对节点约束主流，使上游急骤壅水，节点上游回流紊乱，进一步加剧了这一段河床冲刷扰动，马湖堤江岸遭受上游主流顶冲和下游回流淘刷双重作用，长期处于强烈冲刷、侵蚀环境之中，致使深泓线不断南移，加上 0m 高程以下为粉细沙层，抗冲能力低，江岸淘刷日益加剧，岸坡极不稳定，江岸连年崩塌，险情不断。1996 年大崩岸，沿江近 1000m 的堤段突然滑入江中，死亡 24 人，重伤 5 人。

马湖堤段汛期最高水位约 19.00m，施工区域最大流速 1.2m/s，水深 0～33.5m；枯水期平均水位 6.62m，施工区域最大流速 0.6m/s，水深 0～29m，施工期水位 5.50～9.00m。

施工区地形平缓，坡度在 1∶3～1∶6 之间，局部有陡坡及块石。深泓底部高程－20.00m左右。

2.2 工程布置

高程 6.62m 以上为水上模袋混凝土，成型厚度为 15cm，并设有直径 10cm、间距 3m 的排水孔；高程 6.62m 以下为水下模袋混凝土，成型厚度为 20cm。模袋混凝土铺设宽度 145m 左右，其中水上部分 35m 左右，水下部分 110m 左右。混凝土设计标号 C15。单片模袋宽度 4m。陆上模袋与模袋间采用手工缝接，使充填好的模袋能够有效地形成一个整体；水下模袋与模袋采用靠贴法连接，即下一块模袋铺放在上一块模袋的连接反滤布上。

护坡模袋混凝土上缘端部采用埋固沟内回填干砌块石固定，模袋混凝土下缘端部采用铅丝石笼压脚。

2.3 滑道施工

滑道法主要施工工艺流程为：理坡、埋固沟开挖→模袋铺设→陆上充填模袋混凝土→船舶定位、滑道安装→滑板上充填模袋混凝土→水下铺设。

2.3.1 理坡

陆上理坡采取人工理坡，以突出部位削坡，凹坑部位进行回填、夯实，确保坡面平顺过渡。

水下理坡设计要求：按图纸进行坡面平整，消除表面杂物，坡面上不得有凸出块石及凹坑，基面应平顺过渡，无突变，消除向上的块石尖角，平整度要求不大于 15cm。水下坡面基本处于平缓过渡，断面没有大的突变，但有的区域原抛有块石，部分块石突起，且重叠压置，施工时，对于孤立的块石采用调离水面处理，对重叠块石回填石渣缓坡。

陆上埋固沟人工开挖，按施工图进行，弃渣用于理坡回填。

2.3.2 模袋混凝土施工

(1) 船舶移位、铺设过程定位。在浮吊船尾增抛 2 只领水锚，绞动领水锚，缓慢放松 2 根岸缆，使船往江心移动，利用船头上游的横向锚，控制船舶的轴线。在陆上测量人员的指挥下，使滑道中心线对准模袋铺设中心线。在水下潜水员的协助下，下滑充灌好的模袋，进行模袋混凝土铺设作业。

(2) 安装滑道。在浮吊船头设置一座长 36m、宽 4.1m、高 2m 的钢结构桁架，桁架始端采用铰支座，桁架上设有 4 个吊点，浮吊起重系统根据工程需要调整桁架的倾斜坡度。桁架上满铺 3m 厚铁板，作为模袋铺设、固定、填充、下滑的工作平台。

(3) 混凝土生产、输送、填充。混凝土的拌和场地选择在浮吊上进行，水泥、沙、石用船由水路运抵，沙、石用抓斗船抓吊，通过皮带机输送至配料站料斗内计量后进入搅拌机搅拌。混凝土输送采用混凝土泵进行，混凝土熟料出料后进入集料斗。再进入混凝土泵进料口，通过混凝土输送管路进入模袋充灌。

(4) 陆上模袋混凝土铺设。铺好模袋，然后充灌混凝土，从岸边水面附近进料口由低而高填充混凝土。

(5) 水下模袋混凝土铺设。陆上段混凝土填充完毕后，由岸边水面附近进料口按序向江心方向填充滑道上模袋，填充一段，铺设一段，浮吊沿模袋铺设中心线缓慢向江心移动，并同时调节滑道坡度，使灌满混凝土的模袋沿滑道下滑，准确地着落在堤坡上。

施工中应注意滑道上混凝土顶面不得低于陆上未凝固的混凝土面，以保证模袋内混凝土的密实度。

(6) 铅丝石笼固下缘。模袋混凝土充灌至末端时，将铅丝石笼固定于模袋混凝土末端，随同模袋混凝土一起下滑至水下。

重新移船定位，按上游往下游的铺设顺序，逐块充灌铺设。

施工正常情况下，每天可以铺排 2 块，每块排体充灌后宽度为 4m，每天可完成铺排宽度为 8m，垂直堤轴线方向总长度按 145m 计，每天可完成 1100 多 m^2。

3 结语

采用土工模袋治理江河崩岸险情，取得了与常规处理方法事半功倍的显著效果。在江新洲、马湖治理现场可以清楚地看到，未治理段崩岸依然非常严重，而经过治理的堤段安然无恙，治理效果非常直观、明显。事实表明，应用土工模袋治理江河崩岸是一种行之有效的好办法。

鄱阳湖二期防洪工程和赣抚大堤整治工程常用的垂直防渗措施

吴晓彬，易建州

江西省水利科学研究所

摘　要：针对江西省鄱阳湖二期防洪工程和赣抚大堤整治工程中常用的垂直防渗措施，就其基本原理、工艺流程、技术特点、质量控制和质量检测等方面予以介绍，并提出自己在施工过程中总结出来的一些经验数据和处理措施，进行探讨。

关键词：鄱阳湖二期；防洪；赣抚大堤；垂直防渗措施

1998年长江流域的特大洪水，使长江流域和鄱阳湖滨湖地区的广大人民群众生命财产、交通设施及重要城市的安全受到了严重威胁及损害。作为抵御洪水的屏障——堤防，在长时间超高水位的作用下，不断发生严重的渗漏险情，甚至溃口，暴露出了堤防防洪标准偏低，堤身隐患多、堤基渗漏等不良问题。洪水过后，江西省在今后几年将对重点江、河流域和滨湖地区堤防的关键堤段重点加固，以确保安全。为此堤防、堤基垂直防渗加固处理成为重中之重。下面就鄱阳湖二期防洪工程和赣抚大堤整治工程采用的几种垂直防渗措施作一简要介绍。

1　高压喷射灌浆法

高压喷射灌浆技术，20世纪80年代初期已在我国应用于软基础加固和垂直防渗止水工程中。从临时工程发展到永久性堤坝的基础防渗，从根本上解决了常压灌浆受地层可灌性的局限，对灌浆技术是一个极大的发展。

高压喷射灌浆法按注浆管的类型可分为单管法、二重管法和三重管法，目前较为常用的为三重管法。按喷射角度可分为定喷、摆喷和旋喷法，用于垂直防渗目的的多采用三管摆喷法。

1.1　基本原理及工艺流程

1.1.1　基本原理

利用钻机成孔至设计深度，把带有喷嘴的喷管下至设计深度，使用分别输送水、气、浆三种介质的三重注浆管，在以高压水泵等高压发生装置产生的高压水喷射流的周围，环绕一股圆筒状气流，进行高压水喷射和气流同轴喷射冲切土体，再由泥浆泵注入水泥浆液填充，喷嘴做旋转和提升运动，最后便在土、砂、砂砾石、卵石等介质中凝固为一定形状

本文发表于2000年。

的凝结体。这种施工方法具有固化地基和防渗止水的双重作用，凝结体的基本体形有圆柱形、壁板型和扇形等三种。其作用机理如下：

（1）冲切掺搅作用。强大的射流作用于土体，将直接产生冲切掺搅地层的作用。射流在有限的范围内使土体承受很大的动水压力和沿孔隙作用的水力劈裂力以及由脉动压力和连续喷射造成的土体强度疲劳等综合作用，使土体结构破坏，在射流所产生的卷吸扩散作用下，使浆液与被冲切下来的土体掺搅混合。

（2）升扬置换作用。射水冲切过程中，沿孔壁产生的升扬作用，在浆气、水气喷射时，压缩空气除了起到保护射流作用以外，能量释放过程产生的气泡将夹带孔底冲切下来的土体颗粒沿孔壁向上升扬，这样由于被灌土体部分颗粒被升扬置换出地面，同时浆液被掺搅灌进地层，使地层组成成分产生变化。

（3）充填挤压作用。射流的末端由于能量衰减，虽不能冲切土体，但能对周围土体产生挤压力。同时在喷射过程及喷射结束之后，静压灌浆仍在不停地进行，整个过程在浆柱压力作用下，对周围土体及灌入浆液将不断地产生挤压作用，使凝结体与两侧土体结合得更加密实。

（4）渗透凝结作用。喷射灌浆除了在冲切范围以内形成具有一定厚度、形状的凝结体以外，还可在冲切范围以外的周边产生浆液渗透作用，形成明显的渗透凝结层。

（5）位移袱裹作用。在冲切掺搅过程中，遇有大颗粒如卵砾石等，则随着自下而上冲切掺搅在强大的冲切振动作用下，大颗粒将产生位移而被浆袱裹，浆液亦可沿大颗粒周围绕渗直接产生袱裹充填凝结作用。

1.1.2 工艺流程

高压喷射灌浆法工艺流程见图1。

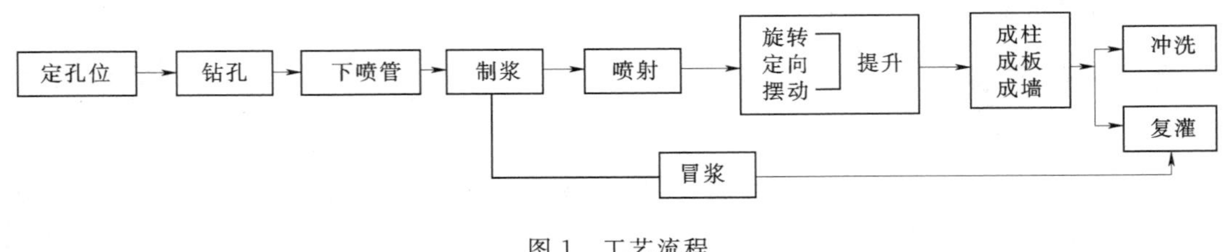

图1 工艺流程

1.2 适用范围和技术特点

1.2.1 适用范围

高压喷射灌浆技术适用于处理淤泥、淤泥质土、粉土、黏土、砂层、卵砾石层和全风化岩层等，受地层情况局限较小，适用范围很广。施工参数必须根据不同地层的最不优情况来定。处理深度可达60～70m。

1.2.2 技术特点

（1）浆液材料单一，一般采用425号普通硅酸盐水泥，原材料质量易控制。

（2）高压喷射灌浆过程为自下而上一次成墙，所成墙体连续性、整体性好。

（3）墙体具有较强的变形适应能力，凝结体由里向外强度渐低，直至与地基保持一致，对适应地基变形极为有利。

（4）墙体具有良好的防渗性能，渗水先经渗透凝结层，再进入防渗性极强的浆皮层，最后才能达到木纹状的墙体核心，然后沿相反的层次穿过防渗体，削减渗水压力作用极为明显。

（5）设备简单，摆设灵活，占地较小。

1.3 设计的一般要求

施工参数的确定应通过现场试验确定，若设计时无法或来不及进行现场试验时，则应根据工勘资料按经验数据设计。

（1）加固剂：常采用425号普通硅酸盐水泥。

（2）水压和水量：水射流的喷射压力和水量直接影响到凝结体的有效长度。增加喷射水量和水压力可增加凝结体的有效长度，但水量大将对浆液起稀释作用，使冒出的浆液增加，因此，应综合考虑选取水压和水量值（取水压不小于30MPa，水量75L/min）。

（3）喷嘴直径：喷嘴直径的大小直接影响到喷嘴出口段10～20cm范围内板墙的厚度。直径大的板墙厚，直径小则板墙薄，但直径过大会使水压降低，降低切割能力和减小喷射距离。喷嘴直径一般为1.8～2.0mm。

（4）气压和气量：压缩空气形成的气幕，可保护水束能量不过早扩散，增加喷射切割长度和升扬置换作用。因此，气压和气量的选择，也影响防渗体的材料组成和有效长度。一般情况，气压0.6～0.8MPa，气量50～80m^3/h。

（5）浆量和浆压：输浆量不小于60L/min，浆压0.1～0.5MPa，以保证浆液的充分扩散和充填。

（6）浆液比重（密度）：浆液比重（密度）不小于1.6。搅浆装置应采用搅浆筒，以保证浆液比重（密度）的稳定性和便于计量。

（7）提升及摆动速度：提升速度不仅关系到凝结体的充填均匀性及密实性，而且影响凝结体的有效长度，是确保施工质量的重要指标。大量的工程实践证明，防渗工程的卵砾石层提升速度宜控制在6～8cm/min，土、砂层应控制在8～12cm/min，摆动速度一般为10r/min。

（8）孔距、摆角和喷射形式：孔距和摆角直接影响到整个工程造价和防渗墙的搭接效果。根据以往大量防渗工程的经验，防渗工程宜采用小孔距、大摆角的摆喷形式。一般有卵砾石的地层，孔距宜不大于1.6m，摆角25°～35°。

（9）关于高压喷射灌浆参数的选定有2点需要进一步明确：①高压喷射灌浆的关键参数为水压、浆液比重（密度）和提升速度，在地层较为复杂的情况下，水压、浆液比重（密度）宜取大值，提升速度取小值；②设计的现场试验只能为设计提供一个设计极限值。因现场试验结果一般是在最优工况下得出的，最终设计参数的确定应考虑工程实施过程中的多种不利因素的影响要比现场试验结果保守一些。

1.4 质量控制

1.4.1 施工参数的控制

对高压喷射灌浆的施工参数必须严格过程控制，对已安装有仪表的应做到：①保证各压力表、流量表和电动无级调速器的可靠性和正常工作；②按照设计值定时对上述仪表进行监测；③进一步加强参数检测的自动化水平和报警手段。

1.4.2 水灰比的控制

联合搅灌机的整机化程度较高，但由于生产流程为边投料边搅拌边输送，浆液的配比难以控制，无法做到定量控制，通常根据操作者的熟练程度结合对浆液的抽检结果再行调整，水灰比的控制往往因人而异，随意性大。因此，制浆系统必须做到定量搅拌。

1.4.3 异常现象的处理

（1）喷射中，接、卸换管及事故处理后下管时，要比原停喷高度下落 50cm 以上，以使板墙的上下连贯。

（2）当出现压力下降低于设计值或压力骤增超过设计值时，都表明了灌浆管的状态有异常，或是有漏浆的地方或是接头松动甚至脱落，或是堵管堵孔等，必须立即停机检查。

（3）严禁无返浆提升。

1.5 质量检测

1.5.1 开挖检查

当凝结体具有一定强度后，即可挖坑检查，但因受开挖深度所限，常用于浅孔检查。

1.5.2 钻孔检查

当凝结体具有相当强度后，利用钻机在其上钻孔，可做如下检查：

（1）取芯观察分析，然后进行室内物理力学性能试验；

（2）全孔取芯后，观察分析凝结体的整体性和墙底、墙顶高程是否满足设计要求。

1.5.3 围井试验

在施工板墙一侧加喷 2～3 孔，与原板墙形成三边形或四边形围井，等凝结一定时间后，在井内打孔，进行压水或注水试验，其结果将直接反映该段墙体的质量。

上述几种检测手段都是局部的、随机的，工程应用的整体防渗效果才是最后的依据。

2 射水法建造地下混凝土连续墙技术

射水法建造地下混凝土连续墙技术（简称“射水造墙技术”）由福建省水电科学研究所于 1982 年开始研究，是针对我国江河堤防普遍存在并急待解决的堤基渗漏问题而提出的。经过不断地改进和完善，已发展到泵吸反循环射水造墙技术。而它的真正试验成功并得以推广，是在江西的赣东大堤上实施的。

2.1 基本原理及工艺流程

2.1.1 基本原理

利用射流泵提供造孔泥浆水射流破坏地层土体结构为主，配合下沿刀具破碎，并由下沿刀具修整孔壁，泥浆固壁，形成规格尺寸的矩形断面槽孔。被破坏的地层土体混合物由砂石泵抽出槽孔，进入沉渣池。土体混合物沉淀，泥浆水循环使用。造孔分二序，在已建成的槽孔中，采用导管法浇筑水下混凝土，建成 22cm 厚的单块混凝土槽板。序孔间采用平接技术建成连续的地下混凝土防渗墙体。

2.1.2 工艺流程

射水造墙技术工艺流程见图 2。

2.2 适用范围和技术特点

2.2.1 适用范围

泵吸反循环射水造墙技术适用于处理淤泥、淤泥质土、粉土、黏土、砂层、直径不大于100mm的卵砾石层和风化岩层等，适用范围较广。

2.2.2 技术特点

(1) 成孔可靠、直观。

(2) 侧向喷嘴清洗槽板侧面，可保证搭接质量，整体防渗性能好。

(3) 可控制沉渣厚度。

(4) 调平底盘就可控制住成孔的垂直度。

(5) 防渗墙为混凝土标量，可人工控制，防渗墙浇筑质量有保证，浇筑的混凝土墙面平整。

(6) 造墙工效较高，成本较低。

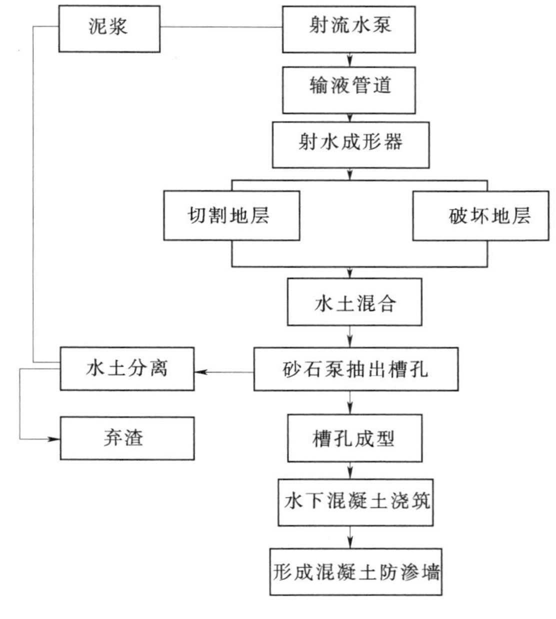

图2 工艺流程

2.3 设计的一般要求

(1) 主材：425号普通硅酸盐水泥、1～3cm卵石或碎石、中粗砂。

(2) 固壁泥浆主要性能指标：比重1.05～1.25g/cm^{3}密度，黏度18～25s，含砂量小于5%。

(3) 成槽垂直度允许偏差：1/300。

(4) 灌注混凝土前沉渣厚度：不大于15cm。

(5) 混凝土坍落度：18～22cm。

(6) 槽孔中线偏差：不大于2.0cm。

(7) 混凝土抗压等级：C15（作为防渗，C10混凝土已足够，但和易性差，难以灌注，一般选择C15混凝土）。

2.4 质量控制

2.4.1 成槽质量控制

(1) 在钢轨上的定位偏差不得大于±5mm。

(2) 泥浆固壁的泥浆质量至关重要。尤其砂、卵砾石层孔隙率大，需要由泥浆中的黏粒充填成膜，而且成槽过程中泥浆损耗大，应储备足量的优质土料，保证及时补浆，杜绝塌孔，甚至埋钻现象。一般泥浆比重在1.2g/cm^{3}以上，黏度18～25s。

(3) 保持槽孔内泥浆水水位，使其高于地下水位1.8m以上，成正压护住槽壁。成槽过程中应密切注意槽孔内水位的变化。若发现槽孔内水位骤降，应及时填入黏土，同时抽水补充。若成槽过程中，机械设备出现故障，也应随时抽水保持孔内水位。

(4) 成槽过程中随时检测底盘水平，以保证槽孔的垂直度。

(5) 造二序孔时，必须保证侧向喷嘴的畅通，使一序槽板的侧面得以冲洗干净。

2.4.2 浇筑质量控制

（1）浇筑前测定孔底沉渣厚度不超过15cm。

（2）严格按设计配合比上料、搅拌。搅拌时间90s，坍落度18～22cm，以保证水下灌注的顺畅。

（3）灌注混凝土时应保证有足够的初有量，导管下端埋入混凝土内不得小于1m。

（4）灌注混凝土过程中不得将导管下端提出混凝土面，否则会使槽板中存在泥缝，破坏槽板的整体性。

2.5 质量检测

2.5.1 开挖检查

随机抽取或在工程最重要位置的防渗墙一侧开挖一定深度，检查接缝质量，并可取样进行混凝土强度和抗渗试验。一般只能在浅层实施。

2.5.2 钻孔取样

用钻机在防渗墙顶部钻孔，取混凝土芯样进行室内试验。

2.5.3 围井试验

在已完成的防渗墙边上做围井试验，测定混凝土墙体和接缝的渗透性。由于受场地的限制，而且设备转向困难，目前一般采用较少。

3 深层搅拌桩法

深层搅拌桩法主要用于两个方面：一是用来加固软土或有软弱夹层的地基，组成复合地基，以提高软地基的承载力；二是用于深基坑的边坡支护或地下挡土结构，也同时用于深基坑的边坡防渗止水。近年来又逐渐运用于堤防的地下防渗墙工程。

国内目前的深层搅拌法按材料喷射状态分为湿法和干法两种。湿法以水泥浆为主（可加减水剂和速凝剂），干法以水泥干粉或生石灰干粉为主；目前应用较多的为湿法。按一次成桩搅拌头可分为单头、双头和多头小直径等三种。

3.1 基本原理与工艺流程

3.1.1 基本原理

利用水泥作为固化剂，通过深层搅拌机械将原位土和固化剂强制搅拌，水泥与原位土共同组成桩体。其中包含：水泥的水解和水化反应；离子交换团粒化反应；凝硬与碳酸化反应。经过一定的凝期，通过上述反应使水泥土硬结成具有良好的整体性、稳定性、弱透水性，并较之原土力学强度较高的水泥土桩。桩间相互切割搭接形成连续的水泥土防渗墙。

3.1.2 工艺流程

深层搅拌桩法工艺流程见图3。

3.2 适用范围和技术特点

3.2.1 适用范围

深层搅拌法适用于处理淤泥、淤泥质土、粉土、黏性土等软土地基，也可处理粉细砂地层，中粗砂层也能处理，但工效明显降低。处理深度不大于18m。

3.2.2 技术特点

（1）混合料采用原位土，无需开采原材料，节约了大量原材料。

（2）可以自由选择加固材料的喷入量。

（3）施工振动和噪声很小，减少了对环境的影响。

（4）引起地基的隆起较小，对周围环境影响不大。

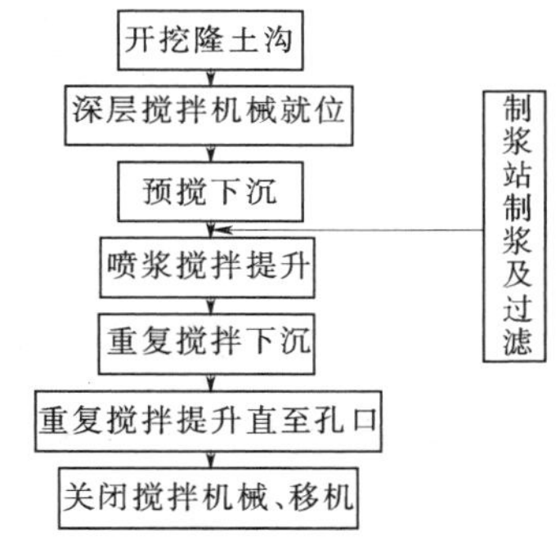

图 3　工艺流程

3.3 设计的一般要求

（1）水灰比：太高则固化慢；太低难以施工，灌浆泵、输浆管路均难以满足要求，取 0.9～1.0。

（2）加固剂：采用 425 号或 525 号普通硅酸盐水泥。

（3）桩径：0.5～0.7m，沿轴线搭接最小长度 0.15m。

（4）提升速度：宜控制在 0.6～1.2m/min。

（5）水泥掺入量：10%～18%，防渗用不宜小于 15%。

（6）喷浆量：80～90L/m。

（7）外掺剂：可根据工程需要选用具有早强、缓凝、减水等性能的材料。

（8）水泥土 90d 龄期无侧限抗压强度：粉土水泥土可达 2.5MPa，淤泥质水泥土可达 1.2～1.8MPa。

（9）水泥土的渗透系数：一般可达 10^{-6}～10^{-8}cm/s。

3.4 质量控制

除了做好设计参数和施工参数的控制外，在施工中还需进行以下几项控制：

（1）桩位控制。为确保有效搭接长度和墙体的完整性，桩位布置的偏差不宜大于 3cm。

（2）桩径控制。规定钻头直径不小于设计桩径减去 2cm。若地基为中粗砂，钻头磨损严重，应经常抽检钻头直径，及时更换叶片或钻头，保证成桩直径。

（3）桩体垂直度控制。施工中必须严格控制桩机底盘的水平，常用水平尺、水准仪或吊垂校检桩机底盘的水平和机架与底盘的垂直度。一般桩体的垂直误差控制在 0.5%范围内。在施工中若电流骤升或桩机振动剧烈，说明钻头遇障碍物，应改变桩位绕过障碍物，不宜强行钻进。

（4）桩长控制。由于钻杆本身的挠度、遇障碍物桩向的偏差和桩位放样的偏差，设计桩长一般宜控制在 18m 范围内。

（5）过程控制。为了不产生断桩，应加强施工过程的量化控制（提升、下沉速度的控制和喷浆量的控制等），强调复搅程序。

3.5 质量检测

3.5.1 开挖检查法

用于检查桩体质量和搭接质量。

3.5.2 钻孔取样法

随机选取或选取施工过程中不正常的桩体进行钻孔取样，并进行压水试验和室内测试。测试项目有渗透系数、无侧限抗压强度、弹性模量等。

3.5.3 墙体压水（注水）试验

钻孔不钻穿桩底，一般留 0.5～1.0m，进行压水试验，若止不住水改为注水试验，确定桩体渗透系数及吸水率。测试压力选用实际工程运行水头压力，测试时间宜选在水泥土 90d 龄期。

3.5.4 围井试验

该方法只能在桩底部为相对不透水层的情况下采用。

4 振动沉管劈裂灌浆法

振动沉管劈裂灌浆法是在长期施工过程中，对以前一些旧的施工机具和施工方法进行分析研究，总结经验，发现问题后，针对目前设计、施工和管理上的要求，由江西省水利科学研究所经过 3 年多的实地施工和试验而研究出来的一种简便易行、质量可靠、造价低廉的新的施工方法和施工机具。可用于江、河、湖堤防的防渗加固处理。

4.1 基本原理及工艺流程

4.1.1 基本原理

根据堤轴线附近处的小主应力面沿竖直分布的规律，沿堤轴线按一定孔距布孔，运用水力劈裂原理，通过一定的灌浆压力，有控制性地劈裂堤体，同时灌入合格的泥浆，通过浆堤互压和堤体的湿陷固结等作用，使所有与浆脉连通的裂缝、洞穴及水平砂层等隐患均得到充填、挤密，并形成竖直、连续和较为均匀的浆体帷幕，以截堵渗漏带，改善堤体应力状态，达到提高原堤体防渗性能和稳定性的目的。

4.1.2 工艺流程

振动沉管劈裂灌浆法工艺流程见图 4。

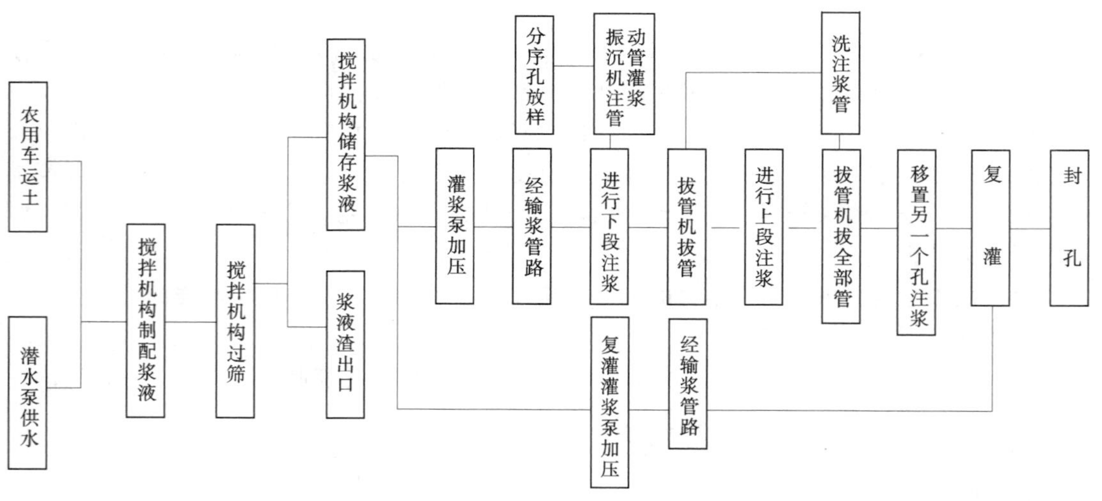

图 4 工艺流程

4.2 适用范围和技术特点

4.2.1 适用范围

该项技术适用于淤泥、淤泥质土、粉土、黏土和砂层等地层。处理深度可达 15m。

该项技术可应用于以下情况：

（1）堤坝体内存在渗水通道和软弱夹层，浸润线较高，堤坝后发生渗透变形或大面积阴湿现象。

（2）堤坝体存在沉陷裂缝和分期、分段施工结合部位质量差、碾压控制不当的残留裂纹以及易产生水力劈裂的其他情况。

（3）堤坝体与其他建筑物结合不好，如存在空隙或被淘刷等。

（4）堤坝存在生物洞穴，如蚁穴、鼠洞等。

（5）堤坝体施工普遍质量不好，干密度小、渗透系数大的松堆堤坝。

（6）堤坝体与基础的结合部位的层面渗水等。

4.2.2 技术特点

（1）止浆方式利用刚性注浆管与土体的相互挤压作用，效果好。

（2）刚性注浆管即为成孔管，省却了一道成孔工艺。

（3）采用自下而上分段灌浆，段长按设计要求灵活确定，且能够有效地针对某一部位灌浆加固。

（4）设备小巧、灵活，适用于施工场地狭窄或限高区域，动力配备小。

（5）与钻机成孔劈裂灌浆法比较，具有设备少、施工简便、施工速度快、造价低廉等优点。

（6）可对各排同序孔同时灌浆。

（7）灌入量大，单位堤长耗干土量达 $1m^3/m$。

4.3 设计的一般要求

（1）制浆土料：以粉质黏土及重粉质壤土较为适宜，其颗粒成分黏粒含量20%～50%，粉粒含量40%～70%，砂粒含量20%以下。这种比例的土料既易制成浆液，又易析水固结。

（2）起始灌浆压力：凡能自重灌浆者，孔口可不必施压；若自重灌注不吸浆，一般控制孔口压力为0.05MPa。

（3）最大孔口压力：一般为最大主应力与灌注段泥浆自重压力的差值，控制在0.1～0.3MPa。

（4）泥浆比重（密度）：一般控制在 $1.2\sim1.7g/cm^3$ 范围内。

（5）孔距、排距：孔距采用2.5～4.0m，排距1.0～2.0m。布置型式采用3排梅花形布置。

（6）成孔深度：孔深应大于隐患深度1～2m。

（7）灌浆过程：采用少灌多复、灌停相间、多孔轮灌、先稀后稠、以稠为主的方法。

（8）复灌：3～5次，间隔时间3d以上。

（9）终孔标准：以满足下列一条为终孔标准。

1）下段注浆时间45min左右，上段注浆时间35min左右。

2）由于堤顶劈裂冒浆或堤脚堤坡冒浆而停灌，间隔一定的待凝时间，再复灌，复灌3次以上即可终孔。

3）灌浆压力增幅较大。

4）由于冒浆—停灌—待凝—复灌，随着复灌次数的增加，吸浆量越来越小，每次灌注的历时越来越短，最后一灌即冒，即可终孔。

4.4 质量控制

（1）原材料控制：检测灌浆采用的主要材料（土料）和浆液的物理力学性能能否满足设计要求和有关规范规定，进而确定料场。

（2）灌浆压力控制：灌浆压力应设专人负责观测。因为从灌浆压力的大小上能直接反映出堤体质量的好坏，同时，从压力的变化上可看出灌浆泵或输浆管路工作是否正常，并可及时对灌浆压力进行修正。控制灌浆压力以内劈外不劈为原则。

（3）裂缝控制：一旦堤顶出现裂缝，其宽度应控制在1～3cm内，长度在20m内，每次停灌后12h内能回弹闭合为宜。

（4）冒浆控制：①堤坡冒浆，表明灌注的浆液通过堤内的渗漏通道出露于堤表，发现后，应立即停灌待凝，一般停1～2d，再灌注浓浆即可；②孔间冒浆，这是由于堤体内部劈裂发展到堤顶的结果，因为没有必要将堤顶大量劈裂，所以发现堤顶劈裂后，即可停灌。

（5）串浆控制：因浆液劈裂堤体或堤内原有裂缝使相邻孔连通，灌浆压力过大，致使浆液从另一孔冒出或喷出，发现串浆后，应首先降低灌浆压力或暂停灌注，也可采取两孔同时灌注的方法处理。

（6）吃浆量控制：对吃浆量大的孔，可采用少灌多复的办法，直至灌饱灌好。若原堤体土料含水量较低，而又是可压缩性土料，可选用稀浆，借以达到加速堤体湿陷的目的；若原堤体土料含水量较高，或是不易压缩的土料，可选用浓浆。

4.5 质量检测

4.5.1 探井检查

（1）检查内容：①分层检查浆脉的分布、厚度，并可绘制浆脉的分布图；②分层对浆体和浆体两侧不同距离的堤体土取样做含水量和干密度试验，绘制含水量和干密度沿深度的变化曲线，并与加固前作对比。

（2）开挖时间：因浆液必须得到充分的析水固结，开挖时间要在停灌半年后实施，不宜过早。

4.5.2 钻机取样检查

主要是分析灌浆后堤体物理力学性能的改善情况。钻孔布置在灌浆轴线的两侧0.2～0.5m处。

5 结语

（1）高压喷射灌浆技术，不仅适应黏土、亚黏土、砂层、砾石层，而且在处理卵石层、局部含块石的地层中均可获得良好的效果，其处理深度可达60～70m，是其他方法无法比拟的，但单价较高。

（2）射水造墙技术在江西省堤防处理直径不大于100mm卵石层已获得成功，处理最大深度已达25m，但不宜处理含块石较多的地层，单价适中。

（3）深层搅拌桩技术已很成熟，但搭接头较多，放样须仔细、精确；由于是回转式钻进，考虑到钻杆的挠度，深度不宜超过18m，建议用来处理堤身；考虑搭接要求，建议桩径不宜小于50cm。含块石地层不适用，单价适中。

（4）振动沉管劈裂灌浆法理论成熟、机具新颖，在鄱阳湖地区试验已获成功。可用于处理堤身及其与堤基接触带的渗漏问题，单价低廉。

新型护岸技术——四面六边透水框架群在长江护岸工程中的应用

王南海，张文捷，王　玢

江西省水利科学研究所

摘　要：介绍一种新型护岸技术—— 四面六边透水框架群护岸。通过室内试验和工程现场试验，证实四面六边透水框架群有很好的消能、减速、缓冲、落淤作用，它可以改变江岸冲势为缓冲或淤势，保护了江岸，可用于防止条崩的发生，亦可用于窝崩口的治理，是一项值得推广的护岸新技术。此技术已成功地应用于长江护岸。

关键词：四面六边透水框架；长江；护岸工程；新技术

1　护岸方法简述

长江中下游由于河道不稳定产生的崩岸，严重威胁长江大堤的防洪安全，威胁人民生命财产的安全和国民经济的发展。因此，长江护岸工程代代相传不断进行，在实践中形成了多种护岸技术与护岸形式。

长江中下游护岸工程按其平面布置形式分类，可分为平顺护岸、矶头护岸和丁坝护岸。按护岸机理分类，可分为实体抗冲护岸和减速不冲护岸。在此，重点研究不同机理的护岸工程技术。

实体抗冲护岸是以实体抗冲材料顶住江水的冲刷保护岸坡，或强迫水流转向绕过一段江岸，以保护岸坡，例如一般的抛石、砌石护岸，丁坝或矶头，沉排，土工枕，混凝土模袋等均属此类。但实体护岸往往存在基础被淘刷影响工程自身的稳定问题，例如抛石护岸以后，水流横向冲刷受阻转为下切，淘刷块石基础，坡脚河床被刷深，造成块石滑动翻滚、坍塌，以致护岸工程本身失稳。丁坝也是这样，以丁坝强行使水流转向，水流在丁坝前后各形成一个回流区，在坝头形成螺旋流，将基础沙石不断旋出流失，淘刷基础，形成巨大的冲刷坑，导致护岸工程自身失稳和破坏。

另一类护岸方法，即局部改变水流的流态，降低岸边流速，将其降到不冲流速以下甚至可以起到落淤的效果，使岸边的冲淤态势发生改变，改岸边的冲势为缓冲或淤势，以护住河岸，稳住岸坡。例如杩杈、挂柳、钢筋混凝土网格排等即属此类。钢筋混凝土网格排护岸效果好，但水下基础施工较复杂，并且需要足够的埋深，才能保住基础不被冲刷，网格排才不致破坏。水利部西北水利科学研究所韩法云观教授在整治多沙游荡性河道实验研

本文发表于 1999 年。

究中提出采用预制钢筋混凝土四面六边透水框架群治河护岸，见图 1。经模型研究证明，这种四面六边透水框架群减速落淤效果非常理想。

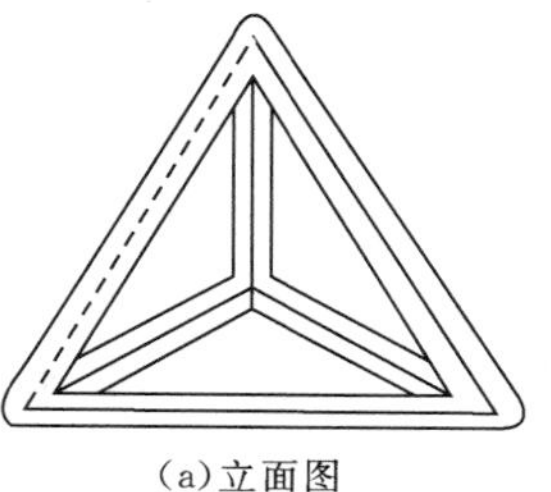
(a)立面图

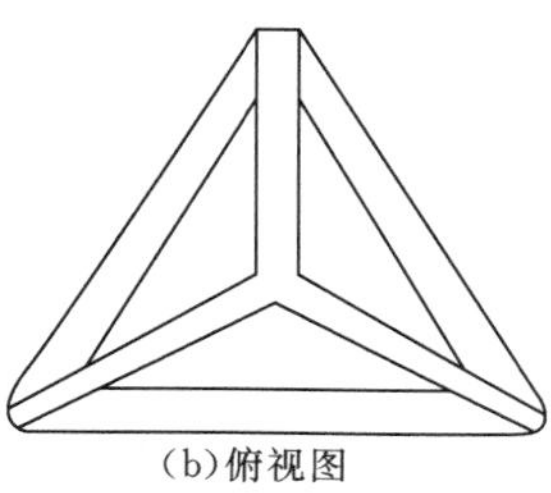
(b)俯视图

图 1　四面六边透水框架

长江堤防是两岸人民的生命线。条崩、窝崩是对堤防安全的最大威胁。条崩的原因是由于岸脚淘刷导致岸坡失稳坍塌。若采用四面六边透水框架作为护岸材料，减小近岸流速，甚至落淤造滩，就固住了岸坡，这是一种可行的有效的新型护岸措施。但长江不同于黄河，长江常年水深流量大，枯水水深常大于 10m，为了将四面六边透水框架应用于长江护岸，我们选择了 2 段长江堤岸进行室内试验和工程试验，并进行了原型观测，证明了四面六边透水框架群用于长江护岸非常有效，是一项值得推广的新型护岸技术。

2　四面六边透水框架群护岸室内试验简介

2.1　模型试验内容

（1）研究比较块石、混凝土四面体、四面六边透水框架的护岸效果。

（2）选择九江长江大堤益公堤东升段为试验段，东升段堤脚被冲刷而崩塌形成陡岸，堤外已无河滩，现将四面六边透水框架用于堤脚保护，故进行了该段的水工模型试验。试验研究四面六边透水框架群的布置方式、尺寸大小、抛投方式、减速效果等。

2.2　模型设计与制作

2.2.1　试验在室内一长 6m 宽 1.1m 的水槽中进行，动床模型用电木屑平铺于槽中。试验前用静水淹没使其密实，起动流速为 7.1cm/s，块石粒径为 2～3cm，混凝土四面体棱长 3cm，四面六边透水框架棱长 4cm，在同一流量条件下进行水力学试验研究，见图 2。

四面六边透水框架

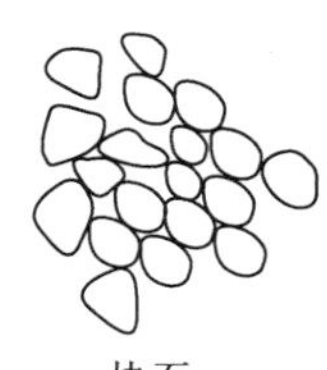
混凝土四面体

块石

图 2　室内试验护岸材料

2.2.2　东升段局部水工模型

东升段长江江岸断面见图 3。按重力相似设计模型，模型比尺为 30，见图 4。主要是模拟崩坍的陡岸及岸边河床，模型长度 6m，槽宽 1.9m。四面六边框架用直径 3mm 铅丝焊成大小 2 种类型，大框架边长 5cm，小框架边长 4cm，在试验段每隔 50m 设一测速断面，共有 6 个施测断面，即$^{\#}1$～$^{\#}6$断面。

2.3　试验结果

2.3.1　水槽试验

以块石、混凝土四面体、四面六边透水框架分别构成水中矶头，可见水流顶冲石块及混凝土四面体，即在石块前产生激烈的漩滚，并将电木屑不断带出，带出后即被水流冲走，很快在块石前和旁侧形成巨大的冲刷坑，块石翻滚，混凝土四面体下挫，导致失稳。

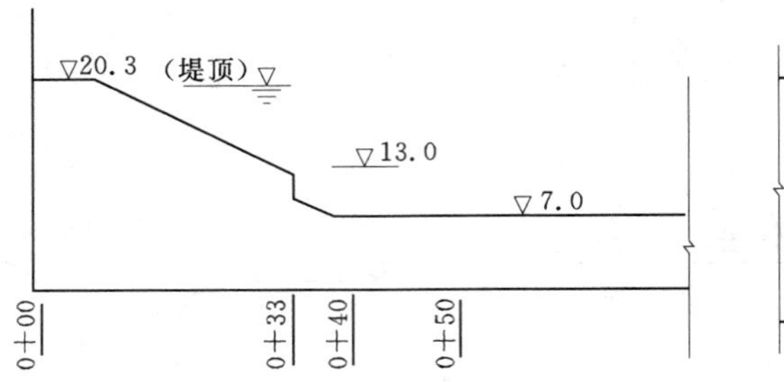

图 3　东升段长江江岸断面图

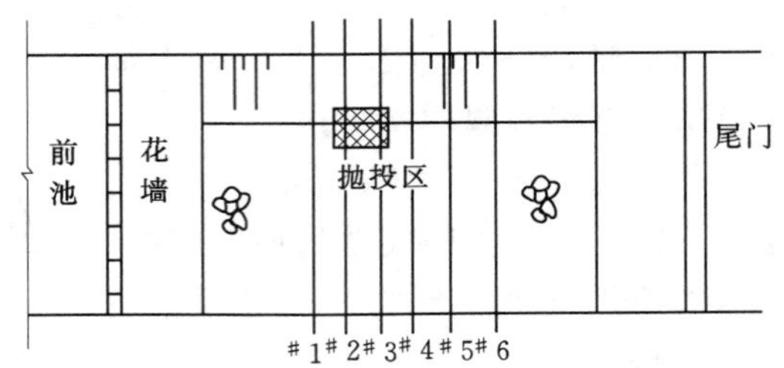

图 4　模型平面图

而四面六边透水框架群构筑矶头时，水流平顺通过框架群，不带走电木屑，不扰动床沙。

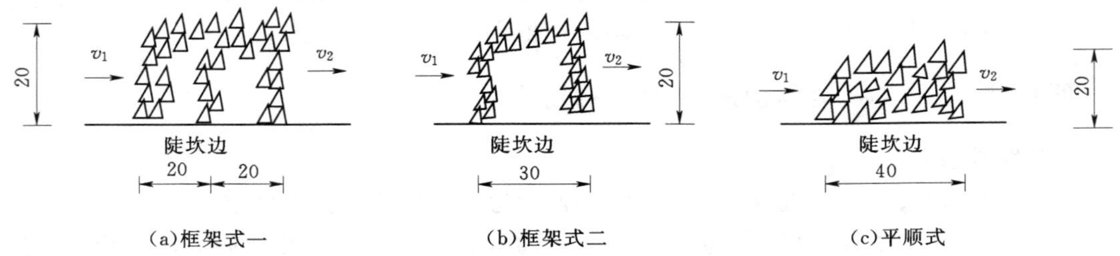

图 5　护岸型式（单位：cm）

2.3.2　东升段模拟试验

按图 5 三种型式放水试验，观测其前后临坎流速的变化，测得结果见表 1。

表 1　框架型和平顺型流速比较

位置		原河床流速 /(m·s^{-1})	框架式堆高流速 /(m·s^{-1})	流速缩减百分数 /%	平顺式堆高流速 /(m·s^{-1})	流速缩减百分数 /%
#2	底	9.1	4.5	51	4.9	46
	坎上	14.0	11.5	18	10.6	24
#3	底	8.9	2.1	76	2.6	71
	坎上	15.4	11.8	23	5.3	66
#4	底	11.2	4.4	61	4.6	59
	坎上	14.3	14.2	0.7	12.8	11
#5	底	10.3	7.5	27	8.2	20
	坎上	15.5	12.5	19	11.5	26
#6	底	12.3	9.9	20	10.7	13
	坎上	17.5	12.5	29	14.7	16

注　表中数据系在长江水位 16m 时测得。

四面六边透水框架堆放位置在#2～#3断面之间，由表1可见，平顺型较框架型布置流速减速率大，但框架式布置流速影响范围更远些。考虑施工时现场抛投形成框架型的难度较大，因而放弃框架型布置而选用平顺型布置。即在采用平顺型护岸方案下进行各工况的抛投试验。

抛投方法试验研究：分别用大小框架，在水位16m、14m、11m时进行室内抛投试验。试验结果证明单个抛投框架易重叠，造成浪费，框架群堆高所需的框架数多，也造成浪费，而3个一串、4个一串抛投，容易形成稳定的框架群，能有效地保护陡坎部位，而且架空率高。框架群堆积所占体积与四面体实体的体积比为4～6倍。

间隔抛投试验：在高、中水位情况下，即水位为19m及16m时，将框架进行间隔抛投试验，即抛投保护一段，间隔一定距离后再抛一段。共进行了3个组次的试验，抛投段为40cm（原型12m），间隔距离分别为30cm（原型9m），20cm（原型6m），15cm（原型4.5m）。对间隔段陡坎进行了3个部位的流速测量，即前抛投段的末端，间隔段的中端、后抛投段的首端，见图6。结果发现间隔段陡坎底部流速没有增加或增加极小，而坎上部流速顺流向增加很快，间隔距离越大，流速增加越多。间隔15cm所测流速见表2。

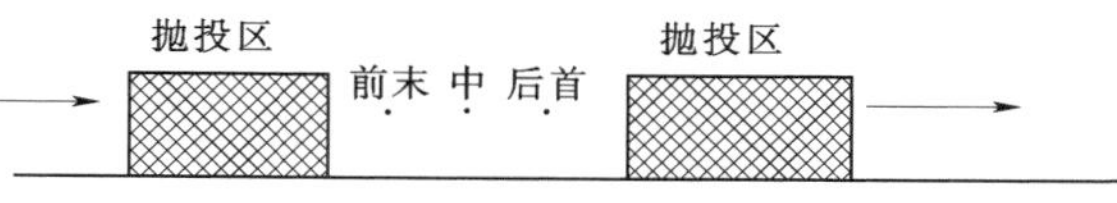

图6　间隔抛投测速点示意图

表2　　间隔抛投流速缩减表　　cm/s

位置	无抛投	间隔抛投		
	#2～#3中间	前抛投末端	中端	后抛投段首端
底	6.75	2.0	1.6	1.6
坎上	11.05	2.5	3.66	4.70

2.4　室内试验小结

采用四面六边透水框架群保护陡坎，有明显的分散水流、消杀能量、降低流速的作用，过程是渐变的，因此不会对河床产生扰动淘刷。在不同的水位组合下，四面六边透水框架群保护陡坎边的流速降低30%～70%。

四面六边透水框架群以平顺布置护岸为好。为保证抛投有效率，应3个一串或4个一串抛投为宜。施工时，为形成稳定的框架群，底层应抛投两排以上，才能达到堆高的目的。可以以流速控制，采取适当间距，间隔抛投。

3　工程实例及效果

3.1　九江市长江东升堤试验工程

该试验工程于1996年4月实施，长70m。预制钢筋混凝土杆件70m^3，杆长1.0m，断面10cm×10cm，钢筋为Φ12螺纹筋，运至工地，现场焊接成1160个钢筋混凝土四面六边透水框架，出露钢筋涂以沥青防腐，见图7。用钢管搭成滑杆，滑至施工小船上，小船用3根锚索定位，在船上用铅丝将3个框架扎成一串，并在框架上系白色泡塑块以标志抛投位置，抛投长度70m，每排抛6个，高2层，见图8。

图 7　钢筋混凝土四面六边透水框架抛投现场

1996 年 6 月，由长江水利委员会九江水文站实测抛投区前后流速，实测减速率为 47％～75％。1996 年 12 月上旬，长江水位降落，抛投区出露水面，明显看到抛投区淤积得很好。由淤积前后实测断面比较，经过一个汛期后，抛投区淤积为 0.8～1.5m^3，与钢筋混凝土杆件体积之比为 18。与抛投块石护岸比较，1m^3 钢筋混凝土预制杆造价 1025 元，18m^3 块石价约为 1250 元，效果比抛石更好，成本略低于抛石，淤积状况见图 9。

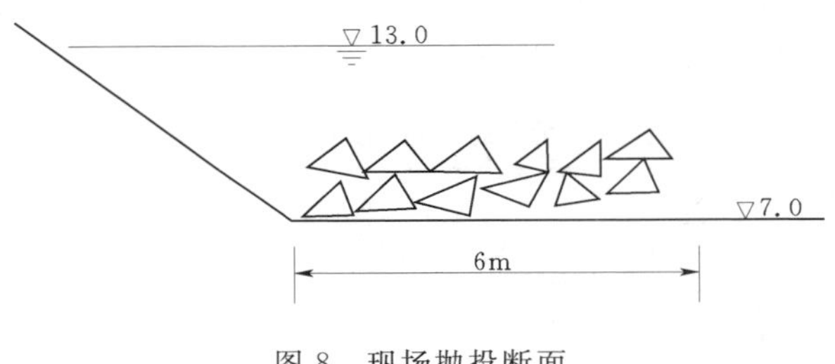

图 8　现场抛投断面

图 9　抛投区淤积全貌

3.2　长江彭泽县金鸡岭段固岸试验工程（1996 年 6 月实施）

由于庐山区东升堤抛投四面六边透水框架群减速落淤效果很好，证明利用四面六边透水框架群护岸固脚是可行和有效的。因此，选择了长江的另一段，水深很大、江岸濒危的彭泽县金鸡岭段作为固岸试验工程，该段河床高程为－18.00m。彭泽县长江马湖堤 1996 年 1 月 8 日发生特大窝崩，崩岸长 960m，宽约 200m。金鸡岭段紧邻马湖上游段，据

1997 年 1 月实测资料，深泓线又比 1993 年加深了 2～3m，并向岸边靠近了数米至几十米不等，已采取紧急抛石固岸措施，紧邻抛石段上游选了一段河岸用四面六边透水框架保护，见图 10。

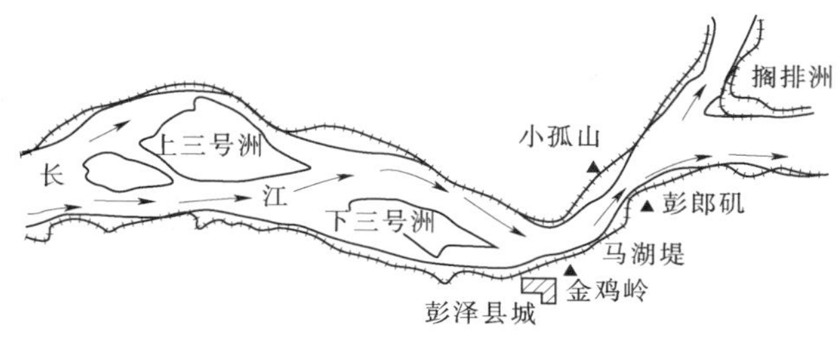

图 10　彭泽县长江段示意图

因钢筋混凝土杆件造价较高，在彭泽改用当地材料制作四面六边透水框架，用将毛竹截成 1.2m 长，捅去中隔灌进河沙，两端用混凝土封堵，用铅丝绑扎成毛竹四面六边透水框架，见图 11。因为影响减速效果的是几何形状及布置方式，因此，毛竹材料的框架与钢筋混凝土材料框架的减速效果应是一样的。

图 11　毛竹四面六边透水框架

图 12　毛竹框架投抛现场

1997 年 6 月在现场预制了 2000 个毛竹框架，用定位船控制抛投位置，每隔 2.5m 平行移动一次，3 个框架绑扎在一起投入江中，保护河岸面积 20m×50m，竹框架造价仅为钢筋混凝土框架造价的 1/3，见图 12。

1997 年 1 月和 1997 年 10 月，分别由长江水利委员会九江水文站对彭泽县长江江岸进行了抛护前后的水下地形测量，结果见图 13～图 15，图中清楚地表明了四面六边透水框架群的减速促淤效果。从剖面图看，已把抛投前 1∶2 的陡岸改变为 1∶5 的缓坡，有效地护住了岸坡，淤积物由颗分实验表明为细砂、极细砂及粉粒，潜水员下水探摸，也表明框架抛投均匀，淤积较好。由此可见，在水深很大、坡度很陡的江岸，四面六边透水框架群依然有效。

4　结语

四面六边透水框架群是一种减速促淤的新型护岸措施，四面六边透水框架自身稳定性好、透水，适合任何地形变化，不存在基础被冲刷的问题，框架群减速率可达 30%～

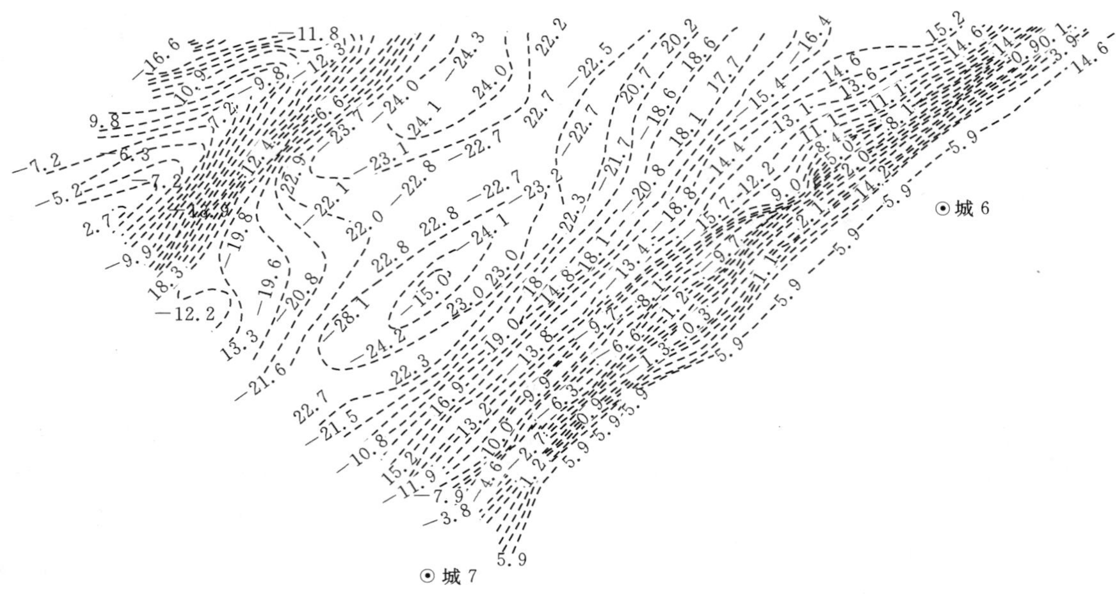

图 13　彭泽县金鸡岭段水下地形图（抛投前，1997 年 1 月测）

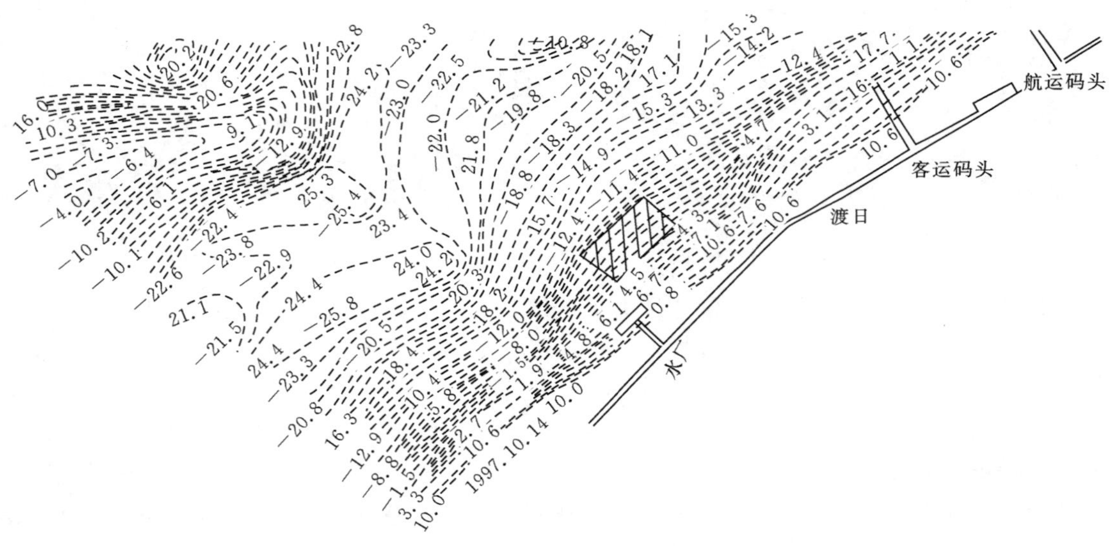

图 14　彭泽县金鸡岭段水下地形图（抛投后，1997 年 10 月测）

70%。经实验室研究和在长江九江段东升堤和彭泽县金鸡岭堤 2 段进行工程试验，证实四面六边透水框架群有稳定、减速、促淤、护岸的良好功能，即使在长江河床较深（－18.0m）且受冲崩塌的地段，也能产生淤积护岸的效果。框架材料可因地制宜，采用钢筋混凝土、毛竹或其他材料均可，造价低且效果比块石护岸还好，可以将局部冲势改变为缓冲或淤势，可以用来防治崩岸的发生，是一种值得大力推广的护岸技术。

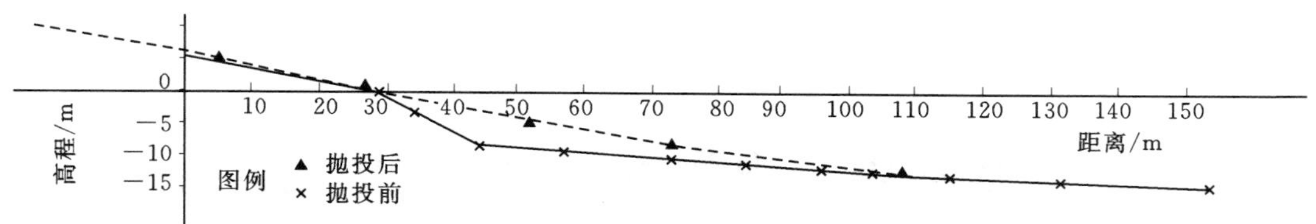

图 15　彭泽县长江金鸡岭段水下断面图

参考文献：

[1]　长江水利委员会．长江中下游护岸工程 40 年［C］//长江中下游护岸工程论文集，第四集．1989.

[2]　徐国宾，张辉哲，等．多沙河流河道整治新型工程措施试验研究［J］．西北水资源与水工程，1994 (3)．

[3]　龚成久．新型护岸工程——钢筋混凝土井柱网格排［C］//第二届中日河工坝工技术讨论会论文集. 1985，10.

土工织物在中徐电排站险情治理中的应用

刘祖斌[1]，黄　真[2]

1. 江西省水利科学研究所；2. 南昌大学

摘　要：中徐电排站由于坐落在不良的地基上，致使历年来在引水渠部位出现了泡泉、跌窝等险情。在险情的治理过程中成功地使用了土工织物，使险情基本上得到了治理。文章总结了应用土工织物作排水反滤体的设计及施工经验。

关键词：险情治理；土工织物；反滤体

1　基本情况

中徐电排站是南新联圩上一座重要的排灌工程，位于江西省南昌县南新乡境内赣江北岸南新联圩上，距乡政府约3.0km。电排站建于1971年，装机4台520kW，由于在没有掌握工程地质条件的情况下，任意选择站址，致使电排站处于不利的地基上（细砂层上），机房底板高程较低（吴淞高程13.20m），从20世纪70年代初建成后，就发生了前池泡泉，引水渠两侧发生跌窝等险情，80年代初两次对该站进行了初步整治，前池泡泉在以后的汛期中未发现，但引水渠两侧走沙跌窝现象并未减轻，在1994年汛期，引水渠两侧又发生多处跌窝，跌窝直径0.5～1.5m不等，窝深达3.0m之多，严重危及机房的稳定及大堤的安全。

2　险情原因分析和治理方法

根据地质勘察报告，中徐电排站的主要地层由人工填土层、粉质黏土层、黏土层、重壤土层、中细砂层和含砾粗砂层组成。其中粉质黏土层层厚1.0～6.2m，层底高程12.00～14.90m，该土层物理力学性质复杂，软硬不均匀，处于可塑到软塑状态，渗透性能变化较大，砂粒含量37%～42%，且底部分布有厚约10～15cm的细、粉砂夹层为薄弱环节。细砂层的允许水力坡降$J=0.13\sim0.15$；砾砂层及中细砂层的允许水力坡降为$J=0.3\sim0.35$；黏土层的允许水力坡降为$J=0.6\sim0.7$。而洪水期外河水位最高为22.82m，相应堤内引水渠水位为16.30m，水头差$\Delta H=6.52$m，堤外水闸进口至泡泉地段的最小距离（渗径）$L=100$m，则实际平均水力坡降为$J=\Delta H/L=6.52/100=0.0652$，由此可见各土层的允许坡降大于实际水力坡降，发生管涌或流土的可能性很小。因此产生泡泉及跌窝的主要原因：人工填土与下覆的粉质黏土的接触带比较薄弱，由室内渗透试验结果可

本文发表于1998年。

知，两种土层的渗透系数的比值达 40 多倍，特别是粉质黏土层底部存在厚 10～15cm 的灰黑色粉细砂夹层，此层极易形成渗漏通道。另外中细砂、砾砂层上覆的黏土层本来就比较薄，层厚约 0.7～1.3m，加之人工开挖等因素，渠底黏土层厚度减薄，抗渗能力降低，局部可能产生破坏，渗透水沿引水渠底板及两侧护坡进入填土层，在填土较松散的部位溢出地表形成泡泉。

根据险情产生的原因和产生的部位，以及泡泉出口处土料的颗粒分析，决定采用导渗减压的治理措施。具体为：在引水渠两侧浆砌石挡土墙后直立铺设土工织物反滤，并在挡土墙底部用土工织物反卷内填砂卵石做成排水沟，并在挡土墙端部（排水沟出口处）采用块石护坡并做贴坡排水，注意渗流出口保护。其目的是为了做好挡土墙后的反滤排水，并将墙后集水通过排水沟顺利排入引水渠内。由于浆砌石挡土墙质量差，施工时出现倒塌等困难，未能达到设计要求，为确保机房附近渠道两侧不再出现险情，在机房后 60m 的范围内浇注混凝土底板及挡土墙，同时向后伸延 40m，用土工织物排水反滤，以形成严封畅排的整体，达到除险加固的目的。经 1994 年、1995 年两年的整治，1996 年汛期，内外水头差达到 6.32m 时，引水渠两岸没有出现跌窝泡泉，但在混凝土挡土墙和贴坡反滤体的交界处出现浑水流出，有 2 处较大泡泉，局部贴坡块石倒塌和混凝土挡土墙发生垂直裂缝等险情。1996 年冬又在出险处铺设土工织物，做好混凝土与土工织物排水反滤体接触带的施工，注意连成整体，以减少混凝土底板渗水压力，保护挡土墙后土颗粒不被渗水带走。1997 年汛期，外河水位为 21.58m 时（内外水位差为 5.58m），整治过的引水渠两岸没有出现泡泉、跌窝等险情，电排站的险情基本得到了治理，工程亦正常发挥了其效益。由此可看出，土工织物在电排站的险情治理中发挥了很大作用。

3　土工织物排水反滤体的设计与施工

3.1　土工织物的特点

作为水利工程的排水反滤体，主要应满足反滤功能要求（水力学要求）、强度要求（力学性能要求）、耐久性要求，而土工织物又恰好能满足这些方面的要求，因此，土工织物的出现为反滤排水工程提供了一种崭新的材料。国内外的大量工程实践表明，用土工织物作反滤层除了具有粒状材料反滤层同样的功能之外，还有其独特的优点：①工程要求的反滤排水特性和尺寸可以在工厂生产过程中控制，易于满足使用要求；②具有整体性和足够的强度，可以抵御外力作用；③施工简单，易于铺设，缩短施工时间，反滤层质量容易得到保证；④运输方便，可以大大节约运力；⑤造价相对低廉，节省工程投资，特别是缺少砂砾料地区更为明显。正因为这样，土工织物代替砂砾料反滤已成为国内外岩土工程中应用最为广泛的一种新材料，在水利工程中也如此。

3.2　土工织物排水反滤体的设计

土工织物作为反滤材料，应遵循两个原则：保土准则和渗透准则。保土准则规定土工织物有效孔径的上限，渗透准则规定土工织物有效孔径的下限，这样既保证土工织物具有排水减压作用又保证土的细颗粒不会流失，以免发生渗透破坏。因此，按古典粒状反滤料设计准则推导的公式来进行设计。

从地勘资料可知，粉质黏土渗透系数 $K=5.57\times10^{-5}$cm/s，泡泉出口处土样的颗粒试验结果为：$d_{85}=0.21\text{mm}$，$d_{50}=0.159\text{mm}$，$d_{15}=0.008\text{mm}$，按土工织物设计原则（O_e为土工织物的有效孔径）：

保土性要求：$O_e<d_{85}=0.21\text{mm}$

透水性要求：$O_e>d_{15}=0.008\text{mm}$

渗透性要求：$K_g>(5\sim10)K_s=(2.89\sim5.77)\times10^{-4}\text{cm/s}$。

故选用湖南维尼轮厂的维丙 300g/m^2 的土工织物，其性能参数见表 1。

表 1　　性　能　参　数

原料	物理指标			水力特性		
	幅宽/m	重量/(g·m^{-2})	厚度/mm	垂直渗透系数/(cm·s^{-1})	孔隙率/%	有效孔径/mm
维丙	2～4	300	4.13	7.0×10^{-2}	97.3	0.09～0.1

根据工程特点，所设计的土工织物反滤结构见图 2。

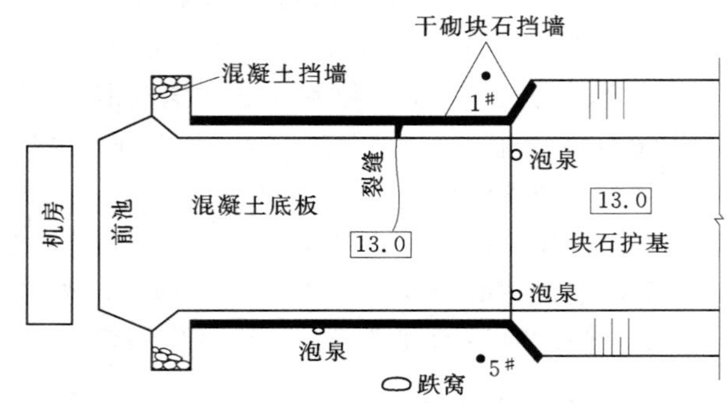

图 1　中徐电排站平面布置图

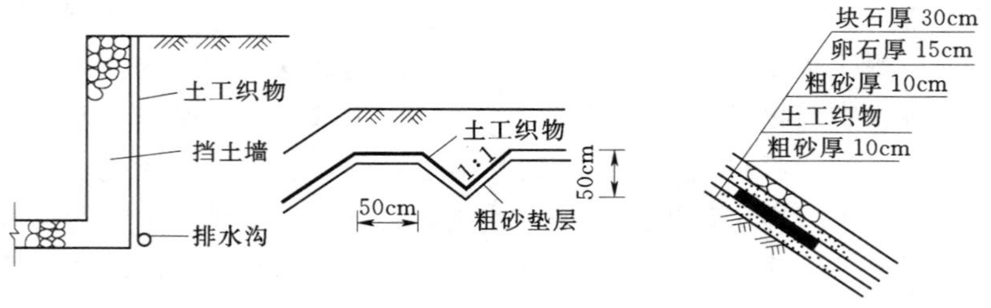

图 2　土工织物反滤结构图

3.3 土工织物排水反滤体的施工

整个引水渠险情整治工程分三个枯水季完成，土工织物的铺设也分为三次施工，但三次使用土工织物的部位及铺设工艺都不相同。在 1994 年初次尝试使用土工织物，主要是铺设在引水渠两侧浆砌石挡土墙之后，采用的是直立铺设方法，土工织物是沿着浆砌石挡土墙由下逐渐向上铺设，并在墙脚用土工织物反卷包裹砂砾石成一纵向排水沟。其目的是为了保护挡土墙后的土颗粒不被渗水带出挡土墙，减轻挡土墙及底板的渗透水压力。但由于施工困难等原因，铺设深度未达到设计高程，其效果不明显。1995 年再次使用土工织物，主要是铺设在引水渠混凝土挡土墙之后的排水反滤段，采用的是平铺方法，将土工织物铺设在渠底及两岸护坡，以保护混凝土挡土墙之后渠道不发生险情，目的是做到严封之后的畅排，以消除险情。1996年又用土工织物，主要是由于混凝土底板及挡土墙与土工织物排水反滤体接触带出现泡泉，土工织物铺设采用的是局部补洞法，铺设难度较大，用土工织物裹起混凝土底板及直立挡土墙后铺在排水反滤体上成为一扭曲面，特别应注意织物与混凝土墙、织物与原织物、织物与堤岸坡的接触带及周边缝处理。

总结三次使用土工织物，虽然每次铺设的部位不一样，施工工艺也不相同，但其目的都是排水反滤。由此可以看出土工织物的应用关键在于铺设施工，根据不同的目的要求、不同铺设部位而采用不同的铺设方法，以取得最佳的应用效果。同时也充分体现了土工织物的优越性，其具有很强的适应性、施工方便、造价低廉等，值得在水利工程中广泛推广应用。

4 效果与结语

中徐电排站经过1994年、1995年、1996年三次使用土工织物进行险情治理，在1997年汛期外河水位为21.58m时，引水渠内经整治过的部位没有发生跌窝、泡泉等险情。同时引水渠两岸测压管监测资料亦表明：当外河水位高达21.58m时，渠内水位为15.89m，内外水位差为5.68m，距机房约60m处的5号测压管（西岸）管水位为18.25m，1号测压管（东岸）管水位为17.65m。即测压管附近的剩余水头为2.36～1.76m，渗透水压力是比较大的，通过土工织物排水反滤体后，渗透压力得到了消散，证明引水渠混凝土挡土墙后的土工织物排水暗沟及土工织物的排水反滤体的排水减压效果明显。

通过中徐电排站应用土工织物整治险情的效果，进一步说明土工织物在排水工程中的应用能充分发挥其排水反滤功能，并具有适应性强，施工方便等特点，同时也证明土工织物应用于堤坝的泡泉治理具有其良好的效果，是水利工程病险治理的一种很好的新材料新技术，具有广泛的推广应用价值。

堤基劈裂灌浆防渗新技术的推广应用

刘晓容，王南海

江西省水利科学研究所

摘　要： 堤基劈裂灌浆是在总结“土坝坝体劈裂灌浆”经验的基础上发展起来的一项防渗新技术。经过近几年来的试验研究，在山东、湖北等省的一些工程应用上已获得成功。为了在江西省推广这一技术，选择了瑞昌市赛湖堤乌鱼爪典型泡泉险段进行推广应用的试验研究。结果表明，应用堤基劈裂灌浆技术处理土沙夹层、粉细层及级配好的细沙层的基础渗漏，防渗效果好、造价低，具有显著的技术经济效益。

关键词： 堤基；劈裂灌浆；防渗技术；推广应用；效益

1　概述

堤基劈裂灌浆防渗技术是近年来发展起来的一项新技术，它打破了长期以来认为中细砂层不可灌的传统观念，在一定范围内以新的灌浆理论解决了粉细砂层及土砂夹层堤坝透水地基的渗漏问题。该项技术是在总结“土坝坝体劈裂灌浆”的经验，分析堤坝地基应力场的规律的基础上，利用水力劈裂的基本原理，对堤坝透水地基进行系统的劈裂灌浆试验后提出来的，经过近几年来的试验研究，在工程应用上已获得成功。近年来，山东、湖北等省，利用这一技术形成的垂直防渗帷幕制止了不少堤基的地基渗漏、泡泉等渗透破坏，取得了显著的成效。江西省亦在1983年首次引进劈裂灌浆技术处理坝体渗漏获得成功，但在堤坝地基劈裂灌浆方面尚属空白。

江西省现有大小圩堤4000多条，堤线长7000多km，保护耕地面积54.87万hm^2（823万亩），占全省耕地面积的21.6%，保护人口705万，是江西省国民经济正常发展和人民生命财产安全的重要保障。江西省堤防较多，但达到标准的堤防少，其中一个主要问题是许多堤基存在险性隐患（约占总险情的80%）。每年洪水时，堤基险象环生，泡泉十分严重。据统计，1995年汛期全省重点圩堤出现泡泉险情780多处，万亩以上的圩堤决口8处，其中就有4处是因泡泉险情引发的。

针对全省圩堤的通病，选择了瑞昌市赛湖堤乌鱼爪典型泡泉险段进行堤基劈裂灌浆防渗新技术推广应用的试验研究，试图在长江、鄱阳湖圩堤治理中推广堤基劈裂灌浆除险加固技术，确保大江、大河、大湖安全度汛。

本文发表于1998年。

2 堤坝地基劈裂灌浆机理

堤坝地基劈裂灌浆是在总结土坝坝体劈裂灌浆的基础上发展起来的。

坝体劈裂灌浆的理论基础是水力劈裂原理，所谓水力劈裂就是在向钻孔内压水或灌浆时，作用到孔壁上的径向压力将引起孔的扩张，使孔壁土体承受劈裂拉应力，而当这些应力超过土体的抗拉强度时，就会在土体内产生一些裂缝，这样产生的裂缝称为水力劈裂。因为土坝坝体在轴线方向上是连续延伸的，作为一无限连续的散粒体，在此方向上传到孔壁上的水平应力 σ_2 较大，而在其垂直方向上，为坝的上下游临空坡面。传到孔壁上的水平主应力 σ_3 最小（详见土坝三向应力示意图，如图 1 所示），因此沿坝轴线形成的对称应力面为小主应力面，土的抗拉强度很小，比较容易沿该小应力面劈开坝体，产生沿坝轴线方向的纵向裂缝。由于在每个钻孔截面上均是如此，所以可使每个钻孔产生的劈缝上下、左右一致，形成一个完整的，与水流方向垂直的直立平面，也就是说水力劈裂造成的裂缝方向是沿坝轴线方向延伸的。

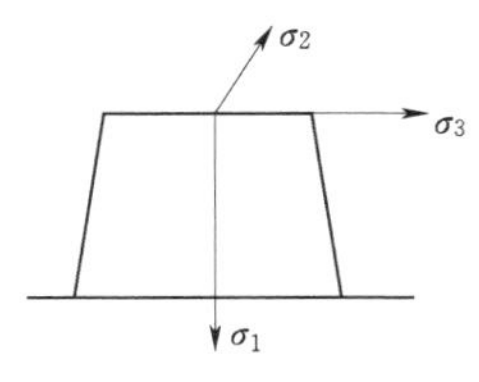

图 1 土坝三向应力示意图

经过近几年的试验研究和分析，堤坝基础的大小主应力与坝体应力相似是由天然地基的自重力和荷载产生的附加应力两部分组成，在堤坝地基中点处的小主应力面也是铅直的，并与堤坝坝体轴线处的小主应力面重合，而且由荷载引起的堤坝地基应力场内满足 $\sigma_1 > \sigma_2 > \sigma_3$ 的条件，为堤坝地基的定向劈裂提供了可能性和有利条件。

堤基劈裂灌浆就是沿堤轴线布置钻孔，钻至地基一定深度，以浆液为载体，施加一定的灌浆压力，有控制地劈裂堤基，同时注入适量的浆液，在堤基中形成一道以缝中结石为主体的密实、竖直、连续的浆体防渗帷幕，而且能挤密浆体两侧的土体和充填原有的空穴和裂缝，从而达到堤基的渗透稳定和变形稳定。

堤基劈裂灌浆技术具有设备简单，容易施工、效果好、工程造价低等优点，主要适用于基础附加应力场主要影响范围内的土砂夹层、粉细砂层及级配好的细砂层地基的渗透变形，通过实施上述专门的堤基劈裂灌浆工艺，能解决堤基的泡泉险情。

3 试验堤段的选取及渗漏原因分析

赛湖堤位于瑞昌市城东郊九瑞公路南侧，全长 15km，赛湖堤乌鱼爪段位于赛湖农场原六分场与七分场之间，该堤建于 20 世纪 50 年代，经数次加固加高逐步达到现在的规模，堤顶宽一般 4～5m，堤身高 6.5m，内外边坡为 1：2.5～1：3.0 不等，堤顶高程为 22.15～22.65m（黄海高程），堤基为原河滩，由于当年筑堤时未对堤基作彻底的处理，所以历年汛期堤基都发生泡泉险情，且曾数次溃决成灾，1995 年夏季洪水期间，该地段泡泉群突发，最大的泡泉出口达 10cm×50cm 之大，内堤脚涌出大量泥沙达 $5m^3$ 之多，严重地威胁着大堤的安全。

为了查清泡泉原因，消除隐患，1995 年冬对此段堤进行了地质勘探和电法探测，初步查明堤身、堤基土层分布大致为：堤顶以下 0～5.0m 为黄色黏土，5.0～7.5m 为黑色黏土层，7.5～10.5m 为含沙量较大的细泥沙层，10.5m 以下为黑色黏土层。据分析，堤基中这一含沙量较大的细泥沙层是产生泡泉，渗透变形，危及圩堤安全的主要原因。

4 劈裂灌浆设计

4.1 浆液选择

劈裂灌浆所用浆液，堤基采用黏土水泥浆，根据灌浆试验，水泥含量采用浆液干土重的10%，堤身采用纯黏土浆。

用于灌浆的黏土应先经过泥浆试验，选择黏料含量25%～35%，粉粒含量30%～50%，沙料含量20%左右的重粉质壤土或粉质黏土。经过试验，选用赛湖大桥的土料，其泥浆试验结果见表1。

表1　　试验项目

料场名称	密度 $/(g \cdot cm^{-3})$	失水量 $/(mL \cdot 30min^{-1})$	胶体率 /%	稳定性	泥皮厚 /mm
赛湖大桥（下层）	1.3	24.4	97	0.022	10
	1.4	22	98	0.015	8
	1.5	13	98.5	0.004	10

该土料制成的泥浆具有良好的流动性、稳定性和较强的胶结性及良好的析水固结性，适合于堤坝灌浆。

4.2 灌浆帷幕设计

4.2.1 帷幕深度设计

根据该段堤的地质勘探和电法探测资料，在堤顶以下7.0～10.5m之间是一含沙量较大的细泥沙层，10.5m以下为一相对不透水的黑色黏土层，据分析堤基中这一含沙量较大的细泥沙层是产生泡泉，渗透变形的透水层，为了防止堤基的渗透变形，封堵整个透水细泥沙层，采用接地式帷幕，深入相对不透水层2m，设计孔深12.5m。

4.2.2 帷幕厚度设计

浆体帷幕厚度按下式确定

$$\delta = H/J_{允}$$

式中：H 为水头，$H=6$m；$J_{允}$ 为浆体的允许渗透坡降，$J_{允}=J_{破}/K$；$J_{破}$ 为浆体有反滤条件下的渗透破坏坡降，经试验水泥含量为10%的黏土水泥浆结石的渗透破坏坡降为200；K 为安全系数，一般取2。

$$J_{允}=J_{破}/K=200/2=100$$

所以 $\delta=0.06$m。

从以上设计可知，所构造的黏土水泥浆的浆体防渗帷幕厚度只要大于6cm，即可满足防渗要求。

4.3 灌浆压力设计

灌浆压力的大小与堤坝断面尺寸，堤基地质，上、下游水位等诸多因素有关，不能简单地用一个数值来控制，通过灌浆试验，在本施工段所采用的灌浆压力以能保证在堤基中劈开一道裂缝，使浆液能沿裂缝渗入来控制，一般为50～300kPa。

4.4 孔位、孔距的设计

孔位：沿堤轴线布置钻孔，以便能沿堤轴线形成铅直连续的浆体帷幕。

孔距：为了保证堤基以下防渗帷幕的连续性，钻孔不分序，根据灌浆试验，孔距采用4m。

5 劈裂灌浆施工

赛湖堤乌鱼爪泡泉险段劈裂采用钻孔不分序（根据相同灌浆方法下试验结果表明，不分序所形成的浆脉衔接好），孔距4m的布置方式，本次灌浆试验共钻孔200个，形成帷幕深12m，设计堤基构造防渗帷幕面积约9600m^2。

灌浆试验具体施工步骤如下：

（1）堤顶钻孔时，先用91mm旋转钻具钻孔至细泥沙层面后，下108mm套管至细泥沙层面上0.1m，再用95mm冲击钻锥造孔至设计孔深（设计孔深12.5m），每一造孔验收后，立即下注浆管至设计高程，并在孔口安装封浆器。

（2）用于灌浆的黏土料，先经过泥浆试验，采用赛湖大桥料场的土料黏粒含量25%～35%，粉粒含量30%～50%，用于灌浆的浆液必须满足设计对灌浆浆液浓度的要求，不满足时应适时改变，使其达到设计要求。

（3）每次开灌时必须遵循稀浆开路的原则，堤身灌黏土浆，稀泥浆的密度采用1.2～1.3g/cm^3，灌注200L左右后改为浓泥浆灌注，浓泥浆的容重要求在1.45～1.55g/cm^3之间。

（4）单孔灌浆时严格执行“少灌多复”的原则，根据试验段情况确定乌鱼爪段基础灌注黏土与水泥的混合浆液，水泥掺和量采用相应浆液干土重的10%，施工时水泥掺量参照表2施加，两者必须充分搅拌后才可用于灌浆。堤基灌浆每孔大致分3次进行，第一次灌注2.0m^3左右，第二次灌注1.5m^3左右，第三次灌注1.0m^3左右，每次实际灌注量视单孔的可灌性而定，以堤顶不开裂为原则，堤顶开裂即结束。

表2　各种泥浆密度水泥掺量表

原浆液密度/(g·cm^{-3})	1.44	1.45	1.46	1.47	1.48	1.49
掺加水泥量/kg	6.83	6.96	7.13	7.29	7.42	7.58
混合浆液密度/(g·cm^{-3})	1.48	1.49	1.50	1.51	1.52	1.53
原浆液密度/(g·cm^{-3})	1.50	1.51	1.52	1.53	1.54	1.55
掺加水泥量/kg	7.75	7.88	8.04	8.21	8.33	8.50
混合浆液密度/(g·cm^{-3})	1.54	1.55	1.56	1.57	1.58	1.59

（5）灌浆压力以孔口压力表读数为准，所用的灌浆压力要能保证劈开一道裂缝，使浆液沿裂缝渗入。在灌浆过程中定期记录（一般10～20min记录一次）灌浆时控制吸浆速度在100L/6min以上。

（6）灌浆顺序采用间孔灌注，连续施工的方法，造孔和灌浆均不分序，相邻两灌次之间的间歇时间必须满足5d。

（7）每孔灌浆结束标准按满足下述要求之一来控制：

①每孔复灌次数达到4次以上；②堤顶纵向裂缝反复冒浆；③每孔灌浆历时不少于60min或每孔灌进干土不少于$2m^3$；④根据灌入的干料换算成浆体帷幕厚度并满足设计要求，$\delta \geqslant 6cm$。

$$\delta = \frac{W}{H_{cp} L r_d}$$

式中：W为灌进计算堤基中的干土料总重，g；L为计算灌浆段的长度，cm；H_{cp}为计算灌浆段帷幕的平均高度，cm；r_d为干料土密度，g/cm^3。

(8) 灌浆结束后拔出工作管和套管，向孔中自流注入浓泥浆，反复多次，直至浆液面不下降为止，然后再填黏土夯实整平即可。

(9) 在灌浆中遇到一些特殊情况时作了如下处理：

1) 串浆：灌浆时若发现相邻或隔孔串浆现象，须仔细检查后无严重后果的串浆孔采用封堵的办法处理。

2) 套管壁冒浆：若发现套管壁冒浆时，应及时停灌，然后将套管周围约50cm左右的土挖除，深度视具体情况决定，然后回填黏土夯实即可继续灌浆。

3) 不进浆：灌浆开始后遇到不进浆的现象时，应及时查明是否属工作管堵塞或灌浆泵工作不正常，及时排除故障，为了防止灌浆管堵塞，最好每次灌浆后及时拔起。

6 劈裂灌浆试验段效果检查

(1) 从防汛情况来看，此试验段处理前，是出名的险段，历年汛期都发生多处泡泉，特别是1995年夏季洪水期，水位为21.78m时，试验段内泡泉群突发，最大泡泉出口达10cm×50cm之大，内堤脚涌出大量泥沙达$5m^3$之多。1995年冬采用劈裂灌浆处理后，1996年汛期水位达高程21.50m，在原出现泡泉群的地方再也看不到任何泡泉迹象，该处堤脚干燥，经受了大洪水的考验。

(2) 从堤基测压管资料分析。为了加强对乌鱼爪险段劈裂灌浆后，堤内浸润线的观测，对灌浆效果进行检查，在险段范围内选择了2个典型断面在浆体帷幕前后各安装了2根测压管，根据原赛湖堤乌鱼爪段勘探工程钻孔柱状图将测压管孔底高程定在12.7m的相对透水的泥沙层。测压管于1997年3月25日开始安装，5月初全部安装完毕，6月24日开始观测，因今年降雨均匀，外河水位变化幅度不大，最高水位只达到17.86m，但从现已观测到的测压管水位资料来看（见表3），仍具有一定的规律性，当外河水位低于17.0m时，2个断面的测压管水位，浆体帷幕前后相差约0.4～0.6m；当外河水位高于17.0m时，2个断面的测压管水位，浆体帷幕前后相差约0.6～1.3m（因内塘水位一直保持在15.2～15.4m之间，使幕后水位无法降下），说明劈裂灌浆所形成的浆体帷幕的防渗效果是明显的。且当外河水位变化时，管内水位的变化有一个滞后现象，可以认为，劈裂灌浆在堤轴线处形成浆体帷幕的同时，对浆体两侧的土体也具有挤密和充填的作用。

另外，原内塘水位随外河水位变化而变化，外河枯水时，塘内干枯，无法养鱼，但经1996年灌浆治理后，内塘水位基本不随外河水位变化，又恢复了养鱼（全年基本保持水位不变）。

(3) 探井检查及取样试验情况。为了检查劈裂灌浆试验段堤内是否形成竖直连续的防

表 3　　测压管水位观测记录　　m

观测日期		外河水位	测压管水位				内塘水位
			帷幕前		帷幕后		
月	日		1#	3#	2#	4#	
6	24	17.676	16.542	16.576	16.100	16.168	15.245
6	24	17.706	16.932	17.156	16.160	16.278	15.250
6	25	16.156	16.732	16.831	16.370	16.358	15.245
6	26	15.836	16.572	16.586	16.265	16.308	15.235
6	27	15.686	16.477	16.381	16.225	16.206	15.235
6	28	15.676	16.372	16.306	16.165	16.193	15.230
6	29	15.686	16.352	16.321	16.100	16.153	15.230
6	30	15.696	16.302	16.311	16.070	16.128	15.230
7	1	15.686	16.257	16.276	16.010	16.098	15.230
7	6	15.676	16.287	16.266	15.995	16.098	15.230
7	13	15.966	16.667	16.693	16.235	16.363	15.280
7	15	16.376	16.752	16.771	16.140	16.378	15.280
7	17	16.930	16.792	16.881	16.270	16.373	15.280
7	25	17.10	16.922	17.031	16.170	16.373	15.33
8	4	17.52	17.267	17.401	16.240	16.533	15.35
8	8	17.68	17.442	17.586	16.33	16.648	15.37
8	11	17.83	17.602	17.736	16.36	16.668	15.36
8	15	17.86	17.612	17.756	16.36	16.678	15.36

注　表中 1#、2# 管为同一断面，3#、4# 管为同一断面。

渗帷幕，在施工结束一年多后，于 10 月 21 日我们在堤上桩号 11+084 处开挖了一探井，直径 80cm，深度 7.4m（因再往下挖有水渗出），从开挖的探井来看，劈裂灌浆堤身成墙效果明显，探井沿堤轴线两侧均有一浆体帷幕，浆体厚度 8～12cm 不等，在探井底面有一条分叉后不合拢的浆缝，宽 8～12cm。

为了检查灌浆后堤身和浆体帷幕的防渗效果，我们在深 3.5m 至 7.0m 的探井壁上取了 2 个土样，在探井底部的浆体帷幕中取了一组试样进行渗透试验，试验结果见表 4（即灌浆前后对比资料）。

表 4　　灌浆前后土样渗透试验对比结果表

深度/m	土层	灌浆前			灌浆后		
		W /%	r_d /(g·cm^{-3})	K /(cm·s^{-1})	W /%	r_d /(g·cm^{-3})	K /(cm·s^{-1})
0～3.5	黄色黏土	29.4	1.51	9.52×10^{-5}	27.7	1.51	1.31×10^{-5}
3.5～7.0	灰黑色黏土	28.4	1.53	2.28×10^{-6}	26.8	1.55	2.67×10^{-7}
7.5～8.5	泥沙夹层	20.4	1.69	1.16×10^{-4}	18.7	1.71	4.5×10^{-7}

从以上试验成果可看出，经过堤基劈裂灌浆后，不论堤基、堤身土体的渗透系数明显减小，增强了防渗性能；土体的干容重增大，提高了土的密实度。

7 劈裂灌浆的经济分析与评价

7.1 堤基劈裂灌浆的经济分析

该劈裂灌浆工程由瑞昌市水电建筑工程公司承担施工任务，江西省水利科研所技术指导，建设单位和主管部门负责验收，总投资55.6万元，构筑防渗面积9600m^2。其单价分析见表5（直接费用）。

表5　　堤基劈裂灌浆单价分析表

序号	项目名称	单位	数量	单价/(元·100m^{-1})	复价/(元·100m^{-1})
一	人工费	工日	168	9.4	1579.20
二	材料费	元			10264.43
1	水	m^3	171	0.93	159.03
2	黏土	t	68.05	20	1361
3	水泥	t	20.82	420	8744.40
三	机械费	元			9229.09
1	灌浆泵	台班	9.40	111.52 (电费按1.0元/kWh计)	1048.29
2	灰浆搅拌机	台班	8.44	55.14	465.38
3	泥浆搅拌机	台班	11.23	108.23	1215.42
4	钻机成孔下套管	m	100	65.0	6500
四	其他费用10%				2107.25
合计		23179.78			
折合57.95元/m^2					

7.2 与其他处理措施比较

以前江西省长江堤和鄱阳湖堤在治理堤基渗漏泡泉险段时，多采用水平防渗，据1985年编写的《鄱阳湖区重点圩堤泡泉险情及治理调查研究成果报告》附录资料，在治理赣东大堤西河的南木—江木泡泉险段时一般采用在堤后加50～100m长的压浸台，在压浸台后设减压井的方法，来解决堤基的渗透稳定问题，既浪费土地，工程造价又高，按3m厚50～100m长压浸台计算，每治理1m泡泉险段，仅需土石方就达150～300m^3，约需资金3000～6000元。若采用其他垂直防渗措施，套井回填和锥探灌浆均无法处理基础渗漏，只有采用高压喷射灌浆才能解决基础渗漏问题，但高压喷射水泥浆，造价高，据工程实践，每平米需花费资金（直接费）250元。

综上所述，应用堤基劈裂灌浆技术处理土沙夹层，粉细沙层及级配好的细沙层的堤基渗漏，防渗效果好，造价低，约为水平防渗的1/10，高喷的1/4，具有显著的技术经济效益。若能在全省适合于劈裂灌浆的堤段上进行推广，对堤基基础渗漏的治理有着不可估量的意义。

参考文献：

[1] 白永年，吴士宁，等．土坝加固［M］．北京：水利电力出版社，1992.
[2] 张景秀．坝基防渗与灌浆技术［M］．北京：水利电力出版社，1992.

弹性聚氨酯密封止水材料在我省水利工程中的应用

管升明，刘建成

江西省水利科学研究所

摘　要： 弹性聚氨酯是一种介于一般橡胶与塑料之间的高分子聚合材料。它既具有塑料的高强度，又具有橡胶的高弹性，且有优良的耐油、耐水、耐磨损、耐辐射特性，施工简便。该材料已在江西省内外许多水利工程的止水伸缩缝和钢筋混凝土裂缝修补中得到应用，取得了理想的效果，有着广泛推广应用的前景。

关键词： 弹性聚氨酯；止水材料；水利工程；应用

聚氨酯是一种高分子材料，早在1937年就为德国人拜耳等制出。1950年弹性聚氨酯问世，1965年以后，德国、日本、美国陆续把聚氨酯用作矿山、水工等建筑物的化灌材料。20世纪70年代初期，国内许多单位也相继研制出多种聚氨酯化灌浆材。1975年，中国科学院广州化学研究所首先使用弹性聚氨酯处理新丰江水电站主变平台混凝土裂缝漏水，并获得成功。进入80年代，聚氨酯新材料在我国有了全面迅速的发展，并且开始工业化生产。北京水利水电科学研究院，长江科学研究院、华东勘测设计院科研所等单位相继研制出各种不同性能的弹性聚氨酯密封胶，在水利工程中应用，取得了较好的经济效益和社会效益。为了更好地解决江西省水工建筑物伸缩缝和变形缝的止水补强处理，江西省水利科学研究所于1989年开始进行了弹性聚氨酯密封止水材料的室内性能试验研究和现场应用，取得较好的效果。

1　弹性聚氨酯的性能

弹性聚氨酯是分子结构中含有氨基甲酸酯基［—NH(CO)O］的橡胶状高分子材料的总称，它是一种介于一般橡胶与塑料之间的高分子聚合材料，既具有塑料的高强度，又具有橡胶的高弹性，它的强度高，延伸率大，硬度范围广，且具有优异的耐油、耐水、耐磨损、耐辐射等特性。据有关资料介绍，弹性聚氨酯的通性有以下几点：

（1）抗撕裂强度高，最高可达61.7kN/m。

（2）抗拉强度高，最高可达40MPa。

（3）弹性好，伸长率在100%～1000%之间。

（4）硬度范围宽，最低为邵尔氏硬度A10，一般都在A45～A95之间。

本文发表于1994年。

(5) 耐磨性好，这一点非常突出，比天然橡胶高9倍，比丁苯橡胶高1～2倍。

(6) 耐热性好，耐干热可达80～100℃，个别可达150℃。

(7) 耐老化、耐臭氧性能亦佳；但在紫外线照射下易褪色。

(8) 缺点是易水解，尤其是聚酯型38℃以上湿空气中就发生水解，聚醚型比较稳定。

合成预聚体所选的原材料不同，其力学性能有所不同，江西省水利科学研究所选用甲苯二异氰酸酯，聚醚（N220）及蓖麻油合成预聚体，再用二元胺摩卡与预聚体进行交联反应制得弹性体，在嵌缝时视具体部位可以加入适量的水泥、橡胶粉或膨润土等（见表1），其力学性能见表2。潮湿黏接面涂水下环氧基液对黏接强度的影响见表3。

表1　　各种配方材料表

材料名称	配方编号				
	1#	2#	3#	4#	5#
合成预聚体	100	100	100	100	100
水泥（525#普通硅酸盐水泥）	50	100	150	200	250
橡胶粉（细度系数0.971）	50	100	150	200	250

表2　　各种配方材料的力学性能表

配方编号	拉断强度/MPa	伸长率/%	永久变形/%	邵氏硬度	压缩50%抗压强度/MPa	黏接强度/MPa		
	自然养护	自然养护	自然养护	自然养护	自然养护	干黏自然养护	干黏水中养护	湿黏水下养护
0#	3.0	125	0	61	5.12	2.26	—	0.16
1#	2.3	127	3	71	—	2.44	—	0.17
2#	1.6	96	3	65	—	1.78	1.48	1.14
3#	1.4	98	2	57	—	1.78	—	0.59
4#	1.5	107	2	57	—	1.41	—	0.37
5#	1.6	81	2	65	—	1.71	—	0.49

注　0#配方为不加填料的基本配方。

表3　　潮湿黏接面涂水下环氧基液对黏接强度的影响

处理方法	潮湿水下黏接强度/MPa	备　注
未涂水下环氧基液	0.2	本试验的配方为0#配方中加入400份水泥150份橡胶粉
涂水下环氧基液	1.98	

2　施工方法

用弹性聚氨酯密封止水材料处理伸缩缝或死缝与环氧砂浆贴橡皮相比，前者操作简单、省工、省料。施工顺序可简单表示如下：基层清理→衬条填塞→嵌缝料拌制→嵌填→

养护。

基层清理：对欲处理的混凝土伸缩缝应进行全面清理，缝内的油毡、沥青等应清除至符合设计的深度要求为止，用高压水、钢丝刷等洗刷干净并晾干，对欲处理的死缝应凿成3～5cm宽，深约4cm的U型槽，并冲洗干净晾干。

填塞衬条：衬条一般嵌填在接缝的底部，其作用是调节接缝的深度和防止弹性聚氨酯密封止水材料与接缝底部黏接，衬条一般可用聚乙烯、聚氯乙烯、硬质软木和纤维板等。使用方法可参考图1和图2。

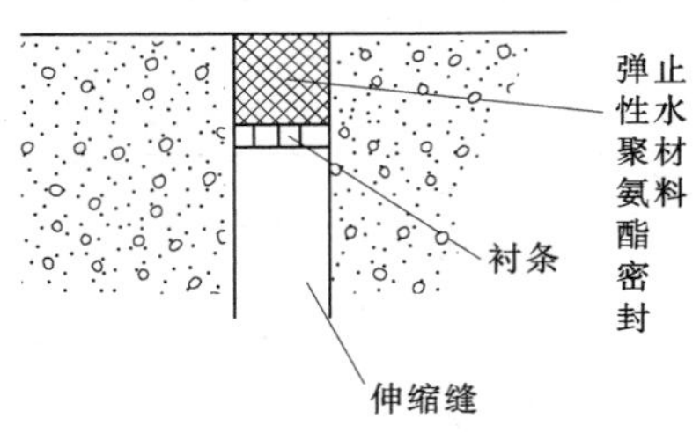

图1　用弹性聚氨酯密封止水材料处理伸缩缝

图2　用弹性聚氨酯密封止水材料处理裂缝

配料：根据不同的部位选择合适的配方进行配料。严格掌握配比和称量，充分搅拌至颜色均匀为止，一般拌和时间不少于10min，一次拌和量控制在1kg左右，最多不要超过3kg。

嵌填：先在缝的两侧黏接面上涂刷一层弹性聚氨酯基液（如果潮湿最好涂刷一层稀薄的环氧基液）。再用油灰刀将配制好的料嵌入缝中，使之比原混凝土面略高一点，盖上塑料薄膜立模加压。

养护：嵌填后在空气中养护24h即可拆膜，如两侧有被挤出的料可以用刮皮刀修整。能直接受日光照射处要用干硬水泥砂浆粉拌一薄层加以保护，防止老化。

对有动水的伸缩缝或裂缝进行处理要在凿开清洗后先用快硬水泥砂浆堵水，埋设导水管把水导走，再按上述方法处理，最后封堵好导水管。

3　工程应用实例

自1990年以来，江西省水利科学研究所先后在3个工程处理中应用了弹性聚氨酯密封止水材料，现简述如下：

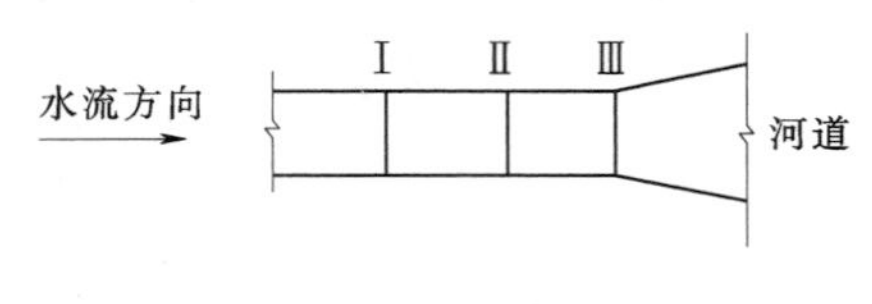

图3　渡槽接缝示意图

（1）南城县毛家瑶渡槽接缝处理。该渡槽全长332m，共33节，半圆形断面，半径50cm，过流量0.7m/s。此工程处理系课题现场试验，因此只选取了出口处的3条接缝（见图3）进行处理，Ⅰ缝宽10cm，Ⅱ、Ⅲ缝宽为5cm，原为塑料油膏止水，运行2年后老化，漏水相当严重。1990年4月，对Ⅰ缝用江西省水利科学研究所的3#配方材料嵌填，Ⅱ、Ⅲ缝用2#配方材料嵌填，并分别用水泥砂浆护面，据南城县水电局的同志反映，运行至今黏结、弹性完好如初，效果良好。

（2）泰和县万合倒虹管裂缝处理。该倒虹管横穿赣江老河床，全长780m，内径

1.5m，壁厚 22cm，设计过流量 4.0m^3/s。由于基础不均匀沉降及分缝过长（最长达 32m 一节）等原因，产生了许多环向裂缝。1990 年底，根据间距等情况，选取了 14 条缝用弹性聚氨酯密封止水材料进行处理，至今运行良好。

（3）余干县信瑞联圩四岔电排站压力水道和输水涵管裂缝处理。该电排站地处鄱阳湖信瑞联圩信丰堤段，起着抵御外湖高水位倒灌和排内涝的双重作用。压力水箱为矩形断面，尺寸 2m×2m，长 15m，输水涵管为方形断面，1.5m×1.5m，长 42m。由于该工程在新建和以后几次扩建时，对基础未作妥善处理，并分缝过长，以致产生不均匀沉降和断裂缝，共计 21 条。最大的不均匀沉降缝高低相差约 10cm，最大拉裂宽度约 8cm。1993 年 3 月，针对该工程特大沉降、拉裂缝的具体情况，选用弹性聚氨酯这一基本止水材料，调整配方，加入适量的膨润土作充填料，经充分搅拌后成胶泥状，这种以膨润土为充填料的止水材料与前面加入橡胶粉的止水材料相比，具有黏性强，附着力高，不流动、不需立模，施工更为简便等突出优点。该工程修补处理后，已经汛期运行考验，效果较为理想。

4 在新建工程中的应用探讨

前面阐述了弹性聚氨酯密封止水材料的基本性能和江西省水利科学研究所在渡槽、输水涵管伸缩缝在止水失效后或形成断裂缝修复补强中的应用。由此联想到这种材料完全可能用作新建的水工建筑物（诸如各种输水涵管、渡槽的伸缩缝、闸底板或下游护坦、面板堆石坝、浆砌块石重力坝的防渗面板）的分缝止水上。下面着重就弹性聚氨酯密封止水材料用作输水管道和渡槽槽身的伸缩缝在设计和施工中的方法谈点意见和设想。

水工建筑物中的输水管道和渡槽的设计与施工，按照规范规定，一般应在 10～15m 设置一条伸缩缝。在复杂地基条件下，分缝长度更短，以适应温度和不均匀沉陷引起的变形，而伸缩缝按传统的做法，多采用内埋紫铜片或柏油麻丝杉板。由于内埋紫铜片不仅成本高，而且极难施工，不易达到要求。柏油麻丝杉板，易腐烂，与混凝土的黏接性能差，和混凝土易脱开，往往建好运行时间不长，甚至刚刚过水就产生渗漏，止水失效后，必须重新设置伸缩缝，需要对接头凿槽，易损伤混凝土的接口，且影响工程运行，而弹性聚氨酯密封止水材料具有上述优良的性能，且施工简单。寿命到期需要更换也很方便（输水管道的直径需在 80cm 以上），所以是一种较为理想的伸缩缝止水材料。

（1）伸缩缝的结构形式和尺寸（输水涵管和渡槽）伸缩缝采用如图 4 所示的结构形式和尺寸较为适宜。

$$h_1=\left(\frac{1}{4}\sim\frac{1}{5}\right)h(\text{一般 } 3<h_1<10\text{cm})$$

$$d_1=(1.5\sim2)d$$

$$d_2=(0.25\sim0.5)d$$

$$d=1.5\sim2\text{cm}(\text{管接隔间隙})$$

式中：h 为管壁厚度。

防渗面板，护坦的分缝止水可参考上述结构形式和尺寸。

（2）伸缩缝的施工。混凝土管接口端在浇筑时，应浇筑成如图 4 所示的台阶形，以便在嵌填弹性聚氨酯密封止水材料时，先嵌进一衬条以堵住材料的向外挤出，控制弹性聚氨

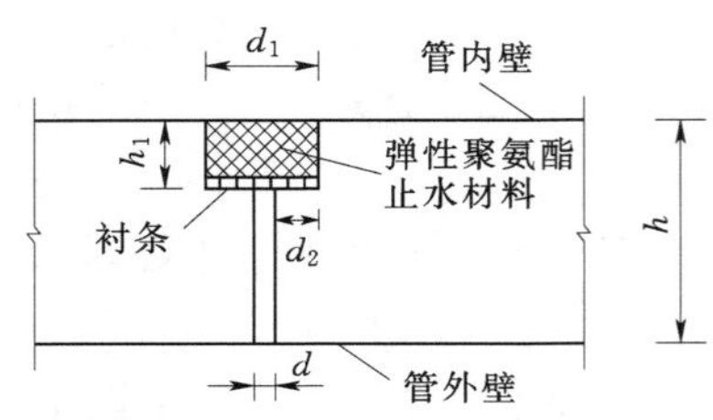

图 4　伸缩缝的结构形式和尺寸

酯密封止水材料的嵌填深度，使管接两头有足够和均匀的黏接面，同时避免因三面黏接，造成纵向伸缩约束，引起应力集中而降低使用的耐久性。

（3）弹性聚氨酯密封止水用于伸缩缝的施工方法如前所述，在施工过程中有两点需要特别注意：

1）对黏接基面的清理：必须清理至混凝土的新鲜面，将表面脏物刷洗干净，并力争黏面干燥后嵌填弹性聚氨酯密封止水材料，因为对于同一种嵌填材料，其黏接强度主要取决于黏接基面的上述条件，如果黏接基面不新鲜、不干净、潮湿，其黏接强度将明显降低，弹性聚氨酯密封止水材料嵌填后应在空气中养护，若水中养护黏接强度也将明显降低。在黏接基面无法达到干燥时，可在黏接基面涂刷环氧基液，以提高黏接强度。

2）弹性聚氨酯密封止水材料嵌填时，一定要让其高出管内壁一定的尺寸，然后立模加压，以消除弹性聚氨酯嵌填料体中的孔隙和气泡等，且形成对黏接基面的压力，使其达到足够的黏接强度。

5　结语

（1）预制的渡槽、涵管等的接缝由于不能埋设止水铜片，选用弹性聚氨酯密封止水材料嵌缝尤为合适。

（2）弹性聚氨酯密封止水材料不仅性能优越，而且费用与传统的几种处理方法相比也是最低的，经粗略统计，处理每米伸缩缝的材料成本为：弹性聚氨酯密封止水材料（断面5cm×3cm）80 元左右，环氧贴橡皮（宽 20cm）120 元左右，U 形紫铜片（宽 500mm、厚 1.2～1.5mm）约 110 元，前者比后者节省费用 40%～30%，且施工费用也比后二者低。

（3）弹性聚氨酯密封止水材料在江西省输水涵管和渡槽的伸缩缝止水补强处理，虽然取得了较好效果，但因应用时间不长，大规模地采用和对新建工程的采用尚无先例。因此，对其长期使用效果尚待观测。

（4）就目前而言弹性聚氨酯密封止水材料虽然有其他材料不可比拟的优点，但由于它是一种易水解的物质，虽然聚醚型（江西省水利科学研究所选配的弹性聚氨酯密封止水材料属聚醚型）比较氨酯型耐水性更好。但即使是纯聚醚型也会缓慢水解，只是水解的速度快慢而已。这就决定了这种材料只具有一定时间的寿命，而并非一种一劳永逸的材料。因此采用现用的配方处于常温水下的耐久性和水解问题，尚待经过实践的检验，并在实际应用中不断改进配方，以延长使用的寿命。

由于对弹性聚氨酯密封止水材料的研究和应用时间不长，且不够深入，水平也有限，错误在所难免，本文只起抛砖引玉之用，欢迎广大同行讨论研究，并对我们的错误和不足提出批评和指正。

溪霞水库坝下涵管顶进技术及事故处理

张开明

江西省水利科学研究所

1 工程概况

溪霞水库位于新建县溪霞乡境内，是一座以灌溉为主的中型水利工程。水库建于1958年，大坝为土坝，最大坝高21m，最大库容4630万m^3。坝下埋有南北两座输水涵管。原南涵管为钢筋混凝土圆管，管线沿山脚布置，呈折线形，进出口位于大坝桩号0+50和0+97.5处。内径0.6m，进口底高程34.00m。由于自然老化，年久失修，涵管断裂漏水严重，直接威胁大坝安全。因此，在大坝除险加固中，决定将该涵管堵塞，采用顶管法重建新涵。

新涵轴线位于0+105处。管身为钢筋混凝土圆管，内径1m，壁厚0.1m，管节长2m，由厂家采用离心生产法预制。管底高程34m，底坡为平底。涵管总长108.7m，顶管长度82.19m，其中坝外顶进60.15m（从坝下游往上游顶），坝内顶进22.04m（从坝上游往下游顶），明挖长度26.51m。管身穿越土体为大坝填土，由于缺乏经验，施工过程中，先后出现了管线偏位，土洞坍方、后座破裂等险情。后经多方设法，采用冲抓钻换土，管外纠偏，改变顶进方向等措施，险情才予解除。工程于1987年12月6日动工，1988年5月25日结束，前后历时5个多月。用于挖洞顶进的实际时间为118d，共76个台班，其他时间用于管周灌浆，基坑开挖，后座施工及险情的处理。

2 顶管设计

2.1 管线布置

以往，顶管施工法的一般工程都是将管轴线选在土质比较好的原生土基上，以防坍方和流土。溪霞水库新建的南涵管要满足此要求颇为困难，要在原生土基上顶进，管线只有布置在原涵管右侧山头上。而右侧山头为风化花岗岩，山体较厚，管线长，挖洞困难。因此，管线只有布置在原涵管左侧0+105处，此处为老河床漫滩处。管线穿越的土体为人工填土，管基为亚黏土和亚砂土。

2.2 管端接缝

目前，无压管接缝主要为平口接缝，即管端面为平面，在接缝处铺贴2～3层油毡，或在接缝处嵌塞沥青浸透的麻绳及石棉水泥，并在管内壁设置凹槽，铺贴钢丝网水泥砂浆

本文发表于1990年。

或环氧橡皮。溪霞水库采用三毡四油接缝，管口没有设置凹槽，顶管完毕后，再在接缝尚有空隙的地方，用沥青麻布或棉花水泥塞缝。这种方法，施工简单，使用方便。油毡一方面可以补偿管端面的不平整，使顶力较均匀分布在管节端面上，另一方面能使管道具有柔性，以适应地基变形。据了解，广东台山县等地大多数顶管工程都是采用这种接缝，效果很好。溪霞水库从一年多的运行实践来看，接缝也没有异常情况。

2.3 荷载计算

用顶管法施工的放水涵管。所受荷载主要有：①垂直土压力；②水平土压力；③管内水重；④管内水压力；⑤管外水压力；⑥管身自重；⑦回填灌浆压力；⑧顶推力。

上述荷载中，①、⑧二种荷载与一般涵管有所不同，下面予以简单介绍，其他荷载可参考一般涵管设计计算进行。

2.3.1 垂直土压力

垂直土压力的计算，目前我国多按土拱理论进行，溪霞水库是采用消力拱法，消力拱示意图见图1，计算公式为：

$$G=nrh_cD_1 \tag{1}$$

$$B=D_1[1+\tan^2(45°-\varphi/2)]$$

其中

$$h_c=\frac{B}{2f}$$

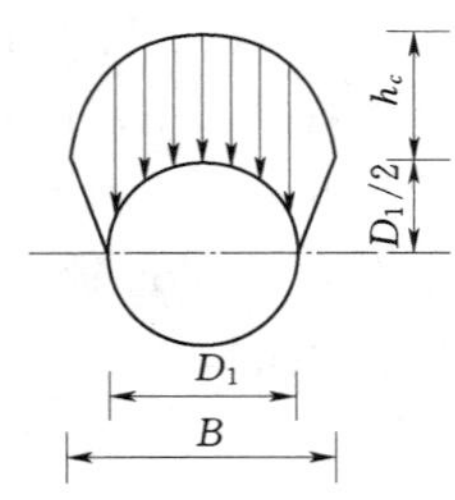

图1 消力拱示意图

式中：G 为管顶垂直土压力，t/m；r 为土壤湿容重，t/m^3；n 为超载系数，采用1.4；h_c 为消力拱高度，m；B 为消力拱跨度，m；D_1 为涵管外径，m；φ 为土壤内摩擦角；f 为土壤坚固系数。

计算成果为，土拱高度1.53m，管顶荷载4.86t/m，与施工中出现的塌方情况较符合。

2.3.2 顶力计算

顶力计算，目前还没有一套完整或接近实际的计算方法。由于土质及施工方法不同，计算时考虑的因素也不尽相同。溪霞水库采用了以下几个公式进行计算。

$$P=f_1\pi D_1L \tag{2}$$

$$P=f[2(P_v+P_y)D_1L+G_n] \tag{3}$$

$$P=ql+K[\pi r_2(ab-R_1^2)+q_h]l\tan\varphi \tag{4}$$

$$P=FG \tag{5}$$

式中：P 为顶力，t；f_1 为管道外壁单位面积摩擦力，t/m^2；黏土为1.5t/m^2；D_1 为管外径，m；L 为顶进总长，m；f 为摩擦系数；P_v 为垂直土压力，t/m^2；P_y 为水平土压力，t/m^2；G_n 为管身总重，t；q 为切削环周边单位长度切土力，t/m，黏土为5～7t/m，砂土为7～10t/m；φ 为土壤内摩擦角；l 为切削环周边长度，m；q_h 为单位管重量，t/m；F 为经验系数，一般为2～5。

以上公式中，式（2）是按摩擦桩考虑的，计算值变化幅度较大。式（3）是按沟埋管计算，主要是考虑涵管在顶进过程中碰到坍方，管顶土拱范围内的土全部压在管上的情况，因此，计算一般偏于安全。式（4）适用于管头加装切削环的情况。式（5）是经验公式。多式计算成果及最大实测值见表 1。顶力过程线见图 2、图 3。

表 1　　计　算　结　果　　t

顶进方式＼计算公式	式（2）	式（3）	式（4）	式（5）	实测值	备注
坝外顶进	339.97	451.85	423.24	103.94～259.8	203.5	管长为 60.15m
坝内顶进	124.57	165.56	169.43	38.18～95.2	61.26	管长为 22.04m

图 2　坝外顶进顶力曲线

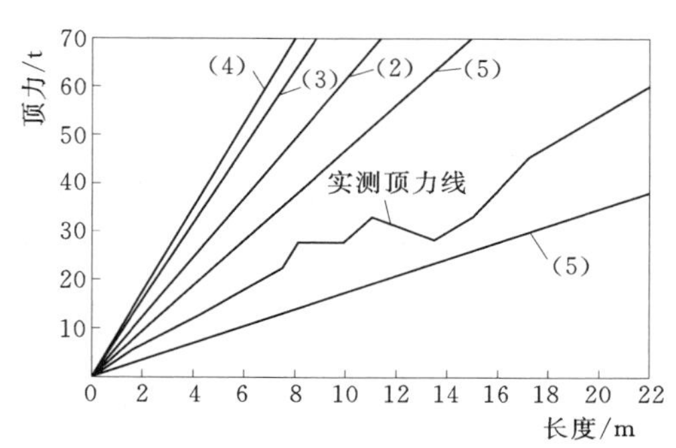

图 3　坝内顶进顶力曲线

注：图 2、图 3 中（5）有两根顶力线，表示按式（5）计算的顶力大小范围。

由表 1 和图 2、图 3 知道，上述各式计算结果各不相同，与实测顶力也相差甚多，因此，计算公式应据土质和具体情况进行选用，其计算值也仅供选择顶力设备时参考。溪霞水库选用了 2 台 200t 的油压千斤顶即可满足施工要求。

2.4　轴向强度和管端局部应力

顶管是用千斤顶将管道逐节顶入土中的，随着顶管长度的增加，顶力将越来越大，因此，管道要有一定的轴向强度。同时，管道在顶进过程中，管端与千斤顶接触处及管道接口处有时难以均匀受压会产生较大的局部应力。因此，管道要能承受由上述原因产生的局部应力。

轴向强度可通过计算允许轴向力来确定。允许轴向力必须大于顶力，否则，应加厚管壁或改变承压方式。允许轴向力可由式（6）确定。

$$F_r = \sigma_m A \tag{6}$$

式中：F_r 为允许轴向力，t；σ_m 为混凝土允许压应力，t/m^2；A 为管道有效承压面积，m^2。

承压面积与承压方式有关。承压方式一般有图 4 所示几种形式。其中以图 4（a）承压面积最大，其次顺序为图 4（b）、（c）、（d）、（e）。溪霞水库承压方式是图 4（b）所示形式，混凝土标号 30$^{\#}$，经计算允许轴向力为 604.45t，大于最大顶力，因而轴向强度满足要求。

对管端局部应力的处理，常用方法是采用管端加筋或使用应力导入管。溪霞水库采用的是管端加筋法，即在每节管端加 2 圈Φ8@5 的环向筋。但在施工时，管端混凝土还有压碎的现象。究其原因，可能是环向筋的效果不够理想。据介绍，局部加压下，以架立筋为好，全面加压下，以螺旋筋为好。

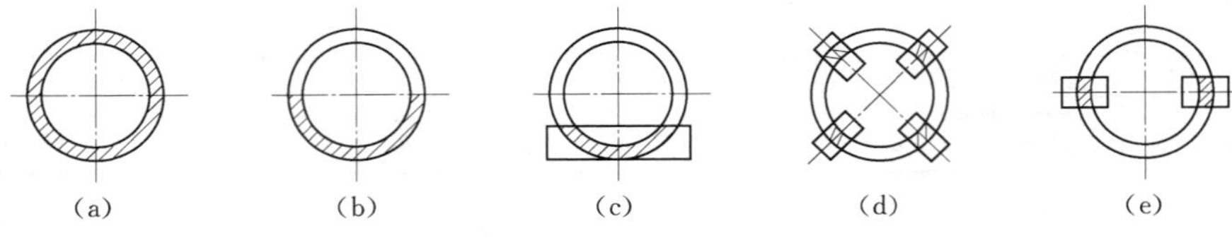

图 4　承压方式

2.5　工作坑与后座设计

2.5.1　工作坑设计

工作坑是安装顶进设备，摆放测量仪器的场所，也是涵管就位，出土弃土的交通要道口。因此，是顶管施工中的一个重要枢纽。其尺寸可参考表 2 采用。

溪霞水库顶管工作坑尺寸为 $l=6.5$m，$B=4$m，即满足了施工要求。

2.5.2　后座设计

后座是承受顶力的，要有足够的刚度和强度，保证不裂缝、不位移。后座形式应视墙后土质情况而定。一般情况下，当土坑体为天然山坡原状土时，后座墙可采用钢木混合结构，当土坑体比较单薄或为填土时，后座墙可采用钢筋混凝土、混凝土、浆砌块石等结构。溪霞水库后座采用了两种形式：在坝外顶进时，土坑体为坝下公路，后座墙采用横竖三排枕木及四层 10mm 厚钢板，并在土坑体上堆压 100m³ 的石块，以增加坑力。坝内顶进时，土坑体为上游含水量较大的契形体填土，后座墙采用浆砌块石和钢板枕木结构，并将基坑开挖土及原部分护坡石块堆压在契形体上。溪霞水库顶管后座示意图见图 5。

表 2　　工作坑尺寸表

项别	计　算　式	说　　明
长度	$l=l_1+l_2+l_3+l_4+l_5$	l_1 为尾管压在导轨上的最小长度；l_2 为出土工作长度；l_3 为每节管长；l_4 为千斤顶长度；l_5 为后背长度；D_1 为管外直径
宽度	$B=D_1+(2.4\sim3.2)$	

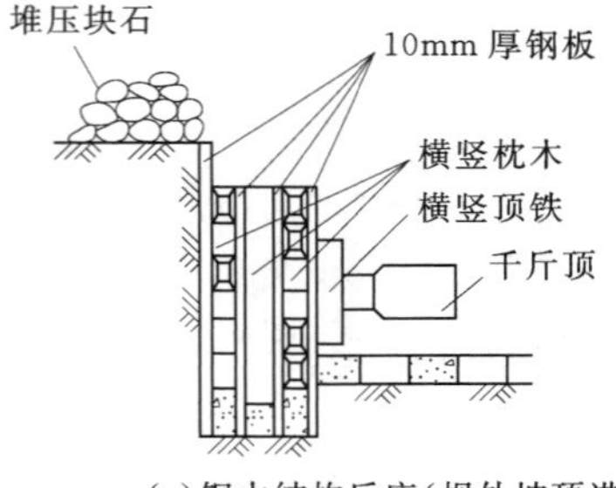

(a)钢木结构后座(坝外坡顶进)

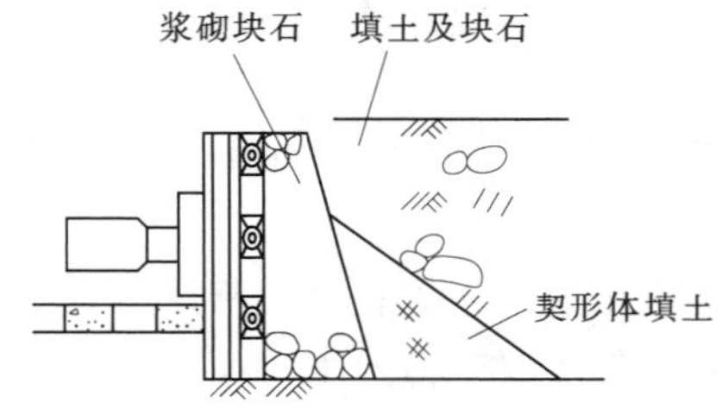

(b)浆砌块石及钢木结构后座(坝内坡顶进)

图 5　溪霞水库顶管后座示意图

2.5.2.1 后座尺寸的确定

后座的平面尺寸（宽×高）可按计算顶力和墙后土壤允许承载力确定，计算公式为：

$$A=\frac{T}{K[\sigma]} \tag{7}$$

式中：A 为后座面积，m^2；T 为最大计算顶力，t；K 为安全系数，一般取 1.5；$[\sigma]$ 为土壤允许承载力，t/m^2。

后座面积确定后，尚需进一步计算后座的宽度及高度，其宽度及高度分别用式（8）和式（9）计算。

$$B=2b+c \tag{8}$$

$$H=\frac{A}{B} \tag{9}$$

式中：B 为后座宽度，m；c 为顶点间距，根据涵管外径确定，m；b 为顶点至侧缘之距，可按顶点间距一半考虑；H 为后座墙高度，m。

2.5.2.2 土抗力计算

后座土抗力一般可按被动土压力公式计算。据经验介绍，采用式（10）计算，其结果与实际情况较接近。

$$R=2B\left(r\frac{H^2}{2}K_p+hHK_p\right) \tag{10}$$

式中：R 为反力，t；B 为后背宽度，m；H 为后背高度，m；r 为土的密度，t/m^3；h 为后背墙上面土的高度，m；K_p 为被动土压力系数，$K_p=\tan^2(45°+\varphi/2)$。

如溪霞水库在坝外顶进时，设计要求在土坑体上面堆石 $100m^3$ 石块。但施工时没有实施，当涵管顶至 51.15m 时，顶力为 154t，后座即出现裂缝（土坑体上不堆压石块计算土抗力为 119.1t）。后来在土坑体上堆压了 $70m^3$ 石块，重新开顶后，顶力达到 243t，后座也没有问题。因而，该公式计算结果是比较符合实际的。

3 顶管施工

溪霞水库顶管是采用人工挖土顶进的，原设计是采用坝外单向顶进，由于后座施工时没有按设计要求进行，致使管道在顶进 51.15m 时，后座出现裂缝，在顶进 60.15m 时，尾管爆裂，不得不放弃坝外顶进，改为坝内顶进。顶进示意图见图 6。

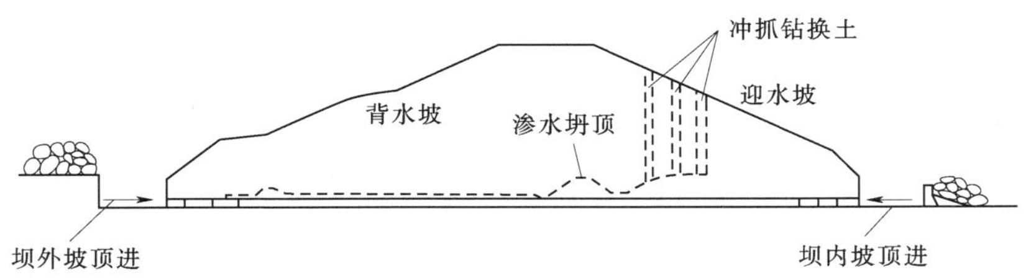

图 6 溪霞水库顶进示意图

3.1 坝外顶进施工

顶进开始时，土质较干燥，进展较顺利，最高进尺为 2.1m/班。在顶至 7～8 节管时，管基出现局部下沉，采用管底下垫石块枕木处理。过了 12 节管后，土质越来越差，挖洞常有渗水，特别是在坝轴线附近，土体挖洞常会坍方、管基下沉、管线方向难以控制。在顶至 51.15m 时，由于后座没有按设计要求施工，后背土坑体已超过弹性范围，后背两角各出现一条长 3m、宽 1cm 的裂缝。事后虽对后座进行了加固处理，在土坑体上堆压了石块，并对钢板枕木进行重新整平安装，但由于后座的变位已对前面几节管线造成了一定影响。于是，在继续顶进时，第 5 节和第 6 节管口出现 7cm 宽错位咬死，顶管被迫停顶纠偏。由于坝轴线附近土质较差，含水量较大，加之停顶时间较长，土拱作用消失，管内涌入大量流土，造成管内纠偏困难。后采用强行顶进方法穿过流土 2m 多，第 5 节、第 6 节管错位增大，第 11～12 节、第 12～13 节、第 13～14 节、第 16～17 节、第 17～18 节管端接口处出现环形压碎，首节管端也出现严重偏位现象，下沉 16cm，偏左 32cm。因此，继续顶进已不现实，需进行彻底的纠偏处理。于是，在多个方案比较下，采用冲抓钻换土，进行管外纠偏。具体方法是，在第 5～6 节管两侧和首节管口用冲抓钻换入黏性土，并在第 5～6 节管口各打开 20cm，让人钻出管外进行纠偏，纠偏方法和管内纠偏基本相同，主要是利用木棒、小千斤顶配合主千斤顶进行。基本调合后，试图继续顶进，顶力达到 243t，为平时顶力的 1.5 倍，导轨上的第 31 节管与顶圈结合处出现粉碎性爆裂，因此，坝外顶进不得不终止。

3.2 坝内顶进施工

坝内顶进是在坝外顶进被迫终止，在上游采用的顶进方法。由于上游土体含水量较大，因而要特别注意工作坑和后座的稳定及挖洞时的坍方。为此，施工时，工作坑采用变坡开挖，并用毛竹石块支护。后座采用浆砌块石与钢板枕木结构，并在契形土坑体上堆放块石。对坍方问题采用冲抓钻换土处理，即在管轴线方向上，采用冲抓钻造心墙的办法对管轴线位置的土体抓出，换入新的黏性土。考虑到冲抓钻换上工作量较大，因此，换土工作只进行了 10m 左右，共 5 节管，其他填土不太高的管节就没有采取这一措施。这一方法，从施工效果来看，基本上可以防止塌方流土。因此，在坝内顶进中，工作进展较顺利。只是在最后一节管接管时，由于坝外顶进时的前几节管有一定偏差，接管情况不够理想，水平方向相差 27cm，高程方向相差 6cm。后采取破碎坝外顶进首节管的办法，才使接管成功。

3.3 管周灌浆

管周灌浆具有回填、固结、防渗三方面作用。因此，是顶管中一道重要工序，应认真对待。

3.3.1 灌装孔布置

灌浆孔采用对称布置，即在一节管的中部断面上按 45°、225°布置两孔，在另一节管的中部断面上按 135°、315°布置两孔。同一方向的两孔相隔 4m。对特殊地段和特殊位置的灌浆孔进行加密。如地基软弱段、塌方较严重段及防渗帷幕处，都另外加孔。防渗帷幕是考虑顶管施工设置截水环较困难而设置的，共有两道，第一道距渐变段首端 16m，第二道为 26m，每道四孔，按 45°、135°、225°、315°布置。灌浆孔在顶管完成后，采用风钻钻进，回填孔以打穿管壁为止，帷幕孔伸入土体 60～100cm 以上。

3.3.2 灌浆工艺及参数

浆液选择纯水泥浆，水灰比为1∶1～1∶3（重量比）。灌浆压力分为三段，涵管两端段拟用0.5kg/cm^2，坝坡中段拟用1.3kg/cm^2，坝顶段拟用2.0kg/cm^2，帷幕灌浆拟用2.2kg/cm^2。灌注方法遵循分序加密原则，采用三道孔序进行灌注。在施灌前，先做压水试验，了解吸水情况，选用适用稠度的浆液进行灌注。一序孔采用逆水流顶灌法，从下游灌起，二序孔采用循环式灌注，三序孔也是循环式灌注。各序孔的终孔标准为，在设计压力下，最小吸浆量为0.3L/min，维持30min以上。灌浆完毕后，待浆液基本凝结（一般为7d），再对灌浆效果进行检查。选择有代表性的地段，打孔检查，发现有的地方灌浆效果较好，管外壁有3～10cm厚的水泥浆层；有的地方不够好，管外壁没有明显的水泥浆层。因而，再对薄弱地段进行了补灌。灌浆共消耗水泥48.06t。

4 结语

（1）溪霞水库在坝体填土上采用顶管技术修建坝下涵管，虽然在顶进中遇到许多困难，施工期相应延长，但顶管法还是优于破坝法。经计算，顶管法较破坝法减少造价30%，节约土石方80%。因此，利用顶管技术在坝体填土上修建涵管还是大有潜力的。

（2）溪霞水库在顶管过程中，出现了许多困难，作为经验教训，今后应认真做好以下几项工作：

1）认真注意后座工作是否保持在弹性阶段范围内，尤其要注意土坑体的可靠性。如土坑体不太可靠，应做好加固工作，加强观测，并采用刚度大强度高的后座结构，以免出现事故，影响顶进工作。

2）认真注意管段的偏差是否在允许范围内。一般情况下，管线偏位超过1cm就要进行纠偏；否则，偏差累积，处理起来就较困难。

3）顶进中不要停顶，停顶时间越长，会使顶力显著增加。一般情况下，启动力为正常工作顶力的1.1～1.2倍，停顶时间越长，启动力越大。另外，停顶时间过长，还会造成土洞坍方、流土。

（3）溪霞水库在顶管实践中，摸索了一些解决问题的实际办法。这些办法，虽然不很完善，但对今后顶管可能会有所借鉴，主要有：

1）当管线偏位较大，管内纠偏困难的情况下，采用管外纠偏法，不失为一种有力的措施，但管外纠偏是让人钻出管外在土洞中纠偏，因此，要特别注意安全。

2）冲抓钻换土对改善地基性能，防止土洞坍方流土是一种好办法，但其工作量大，成本高，施工时需进行方案比较选用。另外，改善地基性能还可通过提前放空水库、降低浸润线，灌黏土水泥浆等措施来达到。

3）当管线较长，采用对顶法能减少顶力。特别是当一端单向顶进失败时，对顶法是一种好的补救措施。但对顶必须特别注意顶进方向、及时纠偏，使涵管偏斜度尽可能小。否则，最后接管就比较困难，也影响管内流态。

参考文献：

[1] 江西省水利科学研究所．坝下涵管顶管技术的总结与改进［M］．1989，11.

[2] 广东省水利电力局．利用顶管法重建土坝放水洞涵管［M］．北京：水利出版社，1980.
[3] 郝杰，钟维达，高延敏．顶管技术在渠下涵中的应用［J］．安徽水利科技，1982，3.
[4] 宜春地区水利电力局工管科，万载县水利电力局．万载县东溪水库顶管施工总结材料［R］．1985，9.
[5] 江西省水利科学研究所．溪霞水库南涵管改建工程管身及出口建筑物设计报告［R］．1987，9.

云山水库坝基砂卵石层高压喷灌板桩防渗墙的试验研究和工程应用

何思源，张文捷

江西省水利科学研究所

摘　要： 本文扼要介绍高压喷射灌注板桩防渗墙在永修县云山水库砂卵石坝基的现场试验和工程应用效果，同时还阐述了高压喷射灌浆技术的基本原理、主要设备、施工方法，高喷灌浆固结体的形成、形状、抗压强度、渗透性能等。

关键词： 高压喷射灌浆技术、抗压强度、渗透性能

1　概述

云山水库位于永修县滩溪乡境内，在潦河支流龙安河上，集雨面积为 172km²，总库容 5290m³（相应校核洪水位 55.63m）。1958 年动工，1964 年基本建成。主坝为黏土心墙坝、最大坝高 18m，坝顶长度 510m。坝基为砂卵石层，厚度约 0.5～5.0m，其下为千枚状绢云母板岩。由于坝基黏土截水槽清基质量差，蓄水后，土坝下游离坝脚不远处出现漏水，曾多处发生碗口大的泡泉，影响土坝安全。

1981 年由江西省水利规划设计院进行加固设计，坝基砂卵石层采用静压灌浆方法处理，1983 年由江西水电建设公司施工。因静压灌浆质量较难达到预期效果，经研究于 1985 年改为高压喷灌板桩防渗墙处理，首次在我省引进全套技术设备，并于同年 9 月进行高压喷射灌浆现场试验，10 月坝基高压喷灌板桩防渗墙正式施工，1986 年元月沿坝长约 460 余 m 的板注墙，全部结束，完成造孔总进尺 6792m，成墙总面积为 2009m²。从现场试验开挖观测、试样室内物理力学性质检验，以及工程质检注（压）水试验等成果表明，高喷灌浆效果良好。浆液使砂卵石胶结成类似混凝土结构，抗压强度为 122～199kg/cm²，渗透系数由 6.6×10^{-1}～1.0×10^{-2}cm/s，降为 4.7×10^{-5}～1.65×10^{-7}cm/s（由 ω 换算），达到了预期目的。

高压喷射灌浆法是近 20 多年来土壤加固和地基处理的一项新技术。20 世纪 60 年代由日本提出，随后意大利、西德、苏联等国相继进行研究。1972 年我国开始试验，现已广泛用于工业与民用建筑、市政、交通、铁路、冶金、煤炭和水利等建设工程。它适用于黏土、黄土、淤泥、砂土和砂卵石层等地基的处理。在提高土的强度方面，可作成旋喷桩，解决软土地基的承载力和沉陷问题；在防渗方面，可作成防渗墙或防渗帷幕、解决堤

本文发表于 1986 年。

坝渗漏或砂卵石地基漏水和管涌问题。尤其对处理已有建筑物地基强度、稳定或渗漏问题，显示出不损坏原有建筑物和维护正常运转的极大优点。

高压喷射灌浆法，施工技术简单、进度快、材料广泛、无公害、造价低，可按工程需要制成各种形状。如旋喷成大直径圆柱桩、扩底（大头）桩、竹节桩、或定喷和摆喷成平接型板桩墙、折线型板桩墙等，是一种具有广泛应用和发展前途的新的地基处理方法。

2 基本原理

高压喷射灌浆法，是在静压灌浆法的基础上，利用高压喷射水力采煤的基本原理发展起来的，但它不同于静压灌浆。静压灌浆是在一般较低的压力作用下，使浆液沿着土或岩石的裂隙、裂缝和空隙进入其内，填充堵塞，并凝胶成整体。而高压喷射灌浆法，是利用高压喷射流破坏土的结构，使浆液与土粒搅混，重新排列，形成新的胶结体，不受可灌性的限制。它解决了细粒土或砂卵石土层静压灌浆不易灌入或灌浆效果不理想的难题。高压喷射灌浆方法基本原理如下。

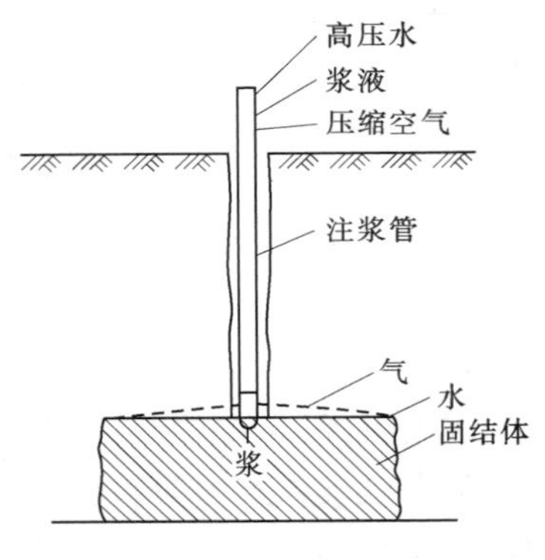

图 1 高压喷射作用示意图

2.1 高压喷射流破坏地基土的结构

高压喷射灌浆法是将具有特殊构造的喷头，安装在注浆管的底部，置入钻机成孔的地基土层设计深度，利用 250～350kg/cm^2 的高压产生的喷射流，强力地冲击、切削破坏土的结构，使其形成空穴或沟槽，而灌入浆液与破坏的土粒搅拌混合，重新排列（图 1），凝成固结体。

2.2 充填挤压作用

浆液与土粒混合物，沿着高压射流在土中形成的空穴或沟槽进行充填，经过凝结固化，在地基中形成具有较高强度和防渗性能良好的固结体。而在冲击破坏土体结构的四周，由于高压射流能量有所衰减，不能直接破坏土体，但仍有一定挤压力，使土体产生压密作用，减少土的空隙。在黏土和细粒土地基中，充填和压密作用是加固地基的主要功能（见图 2）。

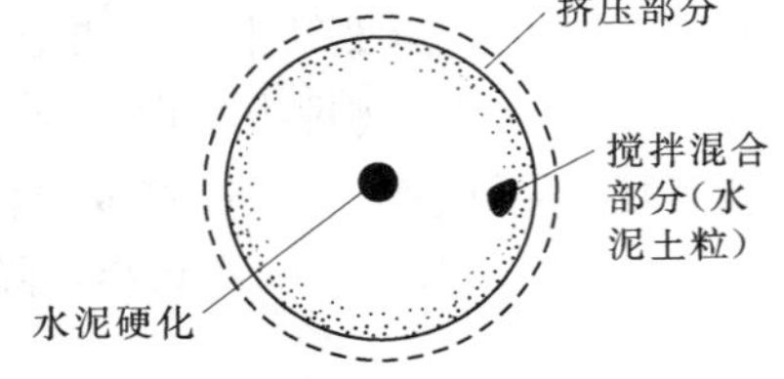

图 2 黏土中高压喷灌固结体示意图

2.3 渗压作用

在砂土地基中，由于高压喷射流的作用，破坏土的结构，大部分土粒与浆液搅拌混合，较大粒径的土粒挤向外侧边缘，浆液在喷射动力作用下，渗入土粒之间的空隙，形成过渡层，使固结体与四周土体紧密结合，称为渗压作用。对于砂卵石地基中的卵石，体积大、颗粒重，相互之间嵌固紧密，高压射流不能冲破嵌固结构，使卵石移动重新排列，只能将卵石之间细颗粒冲走，形成空隙，使浆液充满卵石周围，凝成固结体。因此，对砂类土特别是砂卵石地基加固，渗压充填作用是主要的（见图 3）。

3　高压喷射灌浆法的分类

高压喷射灌浆法，根据喷射介质和喷射施工技术，可分为以下几种方法。

3.1　按高压喷射介质分类

3.1.1　单管喷射灌浆法

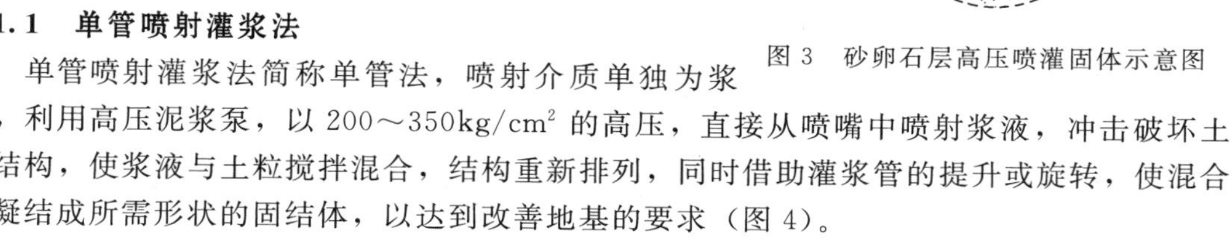

图 3　砂卵石层高压喷灌固体示意图

单管喷射灌浆法简称单管法，喷射介质单独为浆液，利用高压泥浆泵，以 200～350kg/cm^2 的高压，直接从喷嘴中喷射浆液，冲击破坏土的结构，使浆液与土粒搅拌混合，结构重新排列，同时借助灌浆管的提升或旋转，使混合物凝结成所需形状的固结体，以达到改善地基的要求（图 4）。

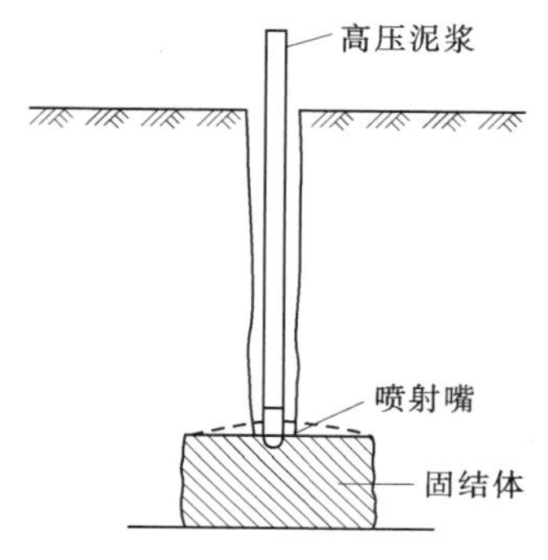

图 4　单管法喷射灌浆示意图

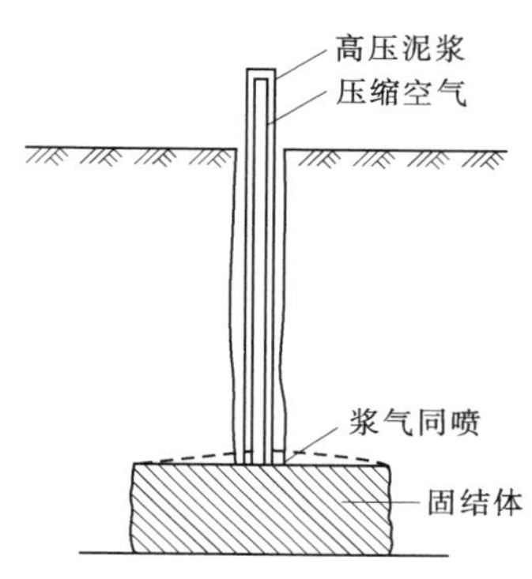

图 5　双管法喷射灌浆示意图

3.1.2　双管喷射灌浆法（双管法）

喷射介质为浆液和空气二种，采用二根灌注管，在一根较大的灌注管内插入一根小的灌注管，一根灌注浆液，一根注入空气，应用高压泥浆泵灌注浆液，空气压缩机压入空气，在喷射出口处同轴喷射，使空气在浆液周围形成气幕，增强冲击能量，提高加固效果（见图 5）。

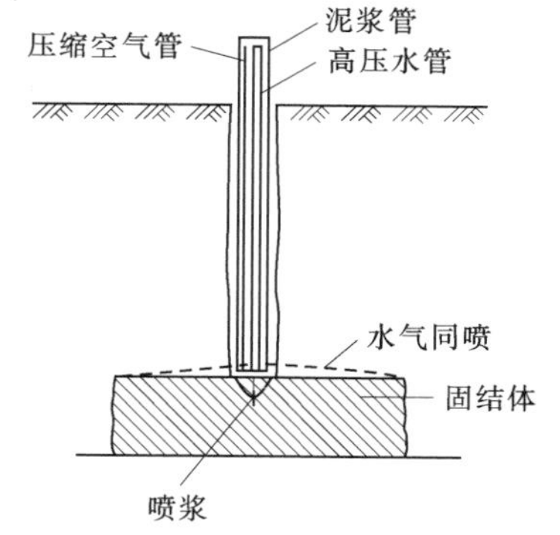

图 6　三管喷射灌浆法示意图

3.1.3　三管喷射灌浆法（三管法）

喷射介质为水、空气和浆液三种。在一根大的灌注管内，放入三根小的灌注管，其中两根管分别灌注高压水和压缩空气，在出口处使水、气同轴喷射破坏土体，另一根注射浆液（见图 6）。这种方法，比单管法和双管法为好，不仅加固范围增大，而且不需高压泥浆泵，可用高压水泵和普通泥浆泵代用，对机具磨损小。

3.2　按喷射转动方式分类

3.2.1　旋喷法

灌浆管置入钻机成孔的地基设计深度后，进行高压喷射灌浆作业，喷射时，一面提升喷射灌浆管，一面旋转灌浆管，使浆液与被扰动土体搅拌混合，凝结成圆柱形固结体，成为旋喷桩。桩的直径可达 0.7～1.6m，适用于提高地基承载力和减少地基压缩性，作为桩基础。

3.2.2 定喷或摆喷法

喷射灌浆时，只提升灌注管，但管不旋转，或左右往返摆动一小角度，浆液与破坏土体搅拌成混合物，则在定向沟槽内凝成板桩墙，可作防渗墙或地基防渗帷幕。板墙宽度为1.5～2.8m，墙的厚度为0.15～0.6m。

4 高压喷射灌浆设备

高压喷射灌浆主要设备，由造孔、制浆和喷射灌浆等机具组成，设备简单，而且是一般施工常用的机械，只要适当添制一些喷嘴、喷头和孔口装置，即可进行高压喷射灌浆施工。

4.1 造孔机具

高压喷射灌浆法和静压灌浆法相似，先用钻机钻孔，钻孔机具常用ZJ100型或SH—30型轻便钻机。钻孔后，将喷射注浆管插入孔内，进行喷灌成桩。在砂土或砂卵石地基内，钻进时灌入黏土泥浆护壁，防止孔壁塌陷。这次施工采用ZJ—100型钻机造孔。

4.2 制浆机及泥浆泵

浆液制备采用WJG80—1型混合搅浆机，利用搅拌与旋转作用，使制浆材料与水混合拌和，过筛后用泥浆泵压入灌浆管喷射。这种制浆设备可制水泥浆，黏土浆或水泥—黏土混合浆。泥浆泵为HB—80/10型。

4.3 高压泵

高压泵是高压喷射灌浆法的关键设备之一，通过它产生200～350kg/cm^{2}的压力，使介质水或浆液从喷嘴猛烈喷射，冲击破坏土体结构。为获得较大射流量，可将两台同类型泵并联使用。喷射泥浆介质的高压泥浆泵，我国采用石油勘探作油田固井的SNC—H300型水泥浆灌浆泵，这种设备投资高，而且易损耗。现在一般在三管喷射灌浆法中均采用以水为介质喷射的高压泵，用于喷射破坏土体结构。常用高压水泵为3W—6B、3ZB等系列三柱塞泵，本次试验采用3ZB型高压泵。

4.4 空气压缩机

为了扩大高压载能破坏土体的范围，提高高压喷射灌浆效果，对于双管和三管高压喷射灌浆法，常用可动式YV—3/8型空气压缩机，以1.8～3m^3/min风量和7～8kg/cm^{2}压力，将空气作同轴喷射，保护水射束能量不过早扩散，使浆液沿喷射方向掺混，升扬置换，改变土层颗粒级配和排列。

4.5 喷头及灌浆管

喷头是高压灌浆机具的特殊装置，安装在灌浆管的下端，喷头装有直径一般为1.6～3.0mm喷嘴，喷嘴用合金钢或45号钢制成，使高压介质通过喷嘴变成高速喷射流喷入地基土层内，破坏土的结构，因此，它是高压喷射灌浆最重要的部件之一。灌注管是注入浆液、空气、水介质的钢管，使介质注入地基土层内。单管法仅有一根浆液灌注管；双管法有两根灌注管，一根为压缩空气注入管，另一根为浆液灌注管；三管法为三根灌注管，比双管法多增一根高压水灌注管。

4.6 其他设备

（1）导流装置：导流器安装在灌注管的顶部，是双管法和三管法使浆液、空气、水等介质，由泥浆泵、空气压缩机和高压水泵分别送至各个灌注管的连接装置，设有推力、向心球轴承和橡胶密封圈。

（2）高压胶管：高压胶管用钢丝缠绕和浸胶制成，可承受 300～600kg/cm^2 高压，用于输送高压泥浆或高压水至灌注管内的中间管路。

（3）气压和水压胶管：用 3～8 层帆布缠绕浸胶制成，可承受 8～20kg/cm^2 的压力，用作输送压缩空气和低压泥浆至灌注管的中间管路。

5 灌浆材料

高压喷射灌浆材料，以水和水泥制成水泥浆液时，用在提高地基承载力和变形模量，减少地基沉陷的地基处理工程中；作为防渗板桩墙，常用水泥浆液、水泥—黏土浆液，提高固结体的防渗性能和强度，减少地基渗漏和渗透变形。本工程采用纯水泥浆喷灌。

浆液应具有以下特性：

（1）良好的可喷性：浆液稠度过大，往往导致主管路或喷嘴堵塞，可喷性较差，而且容易磨损泵体和机具。

（2）良好的力学性能：浆液和土体凝成固结体后，无论作为桩基础或防渗板桩墙，均应具有一定的力学强度，抵抗一定外力作用。

（3）稳定性：浆液与土体凝成固结体的质量与浆液稳定性有关，浆液稳定性好，它的析水率小，则固结体的强度和抗渗性良好。

（4）结石率高：浆液与土体搅拌混合凝成固结体，结石率要高，应具有良好的黏接性能，使固结体以及与四周土体紧密结合，其耐久性亦好，不易受湿度的变化而破坏。

6 施工方法

根据工程处理目的和要求，高压喷射灌浆的施工程序，一般按下列次序进行（见图 7）。

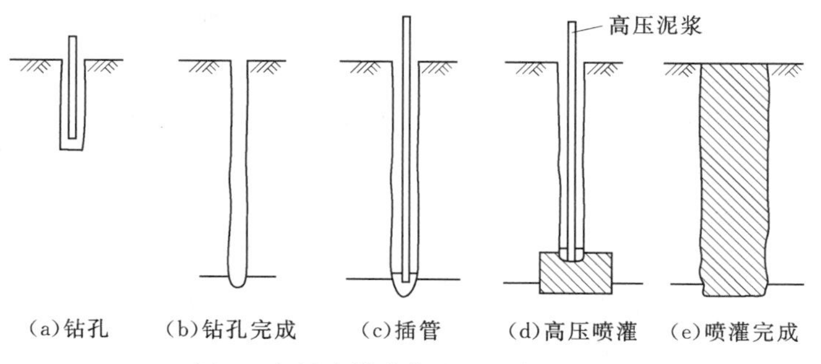

图 7 高压喷射灌浆施工程序示意图

（1）施工场地平整及施工布置：为了便于施工和机架移动，施工场地应进行平整。高压泵，空压机、泥浆泵、搅浆机、制浆材料堆放场地和风、水、浆管路，动力线架设等，应进行施工总布置，以便正常施工，防止事故发生，尽可能减少各种设备随钻、灌孔作业前进而搬移次数太多，影响进度。

(2) 钻孔放样：钻孔位置按设计图纸在现场放样，用小木桩埋设定位，最好分序放样，第一次放第一序钻孔。第二、第三次分别放第二、第三序钻孔位置，以免机架移动时破坏标致。放样时，应先用测量仪器准确定出孔位轴线，二排以上钻孔，尽可能采用梅花形布孔。

(3) 钻机就位及钻孔：钻机安设在设计孔位上，钻头要对准孔位，钻杆保证铅直，钻架应校正水平，使钻孔不致偏斜。钻进后，经常注意钻杆偏斜情况，及时纠正。在砂土及砂卵石层地基钻孔时，要灌注黏土浆液护壁，防止孔壁塌落堵孔，造成埋钻或灌注管不能放置设计深度。

(4) 浆液制备：按设计规定制浆材料配比或浆液比重要求，进行制浆，搅拌时间要充分，防止浆液沉淀。水泥浆应随制随灌注，灌浆作业结束后或中途停止灌浆时间很长，应将多余浆液放出，用清水冲洗，以防水泥浆凝结硬化，制浆机不能运转。

(5) 插管：钻孔完成后，喷射装置就位，校正机架的水平和喷射灌注管的垂直度，开动高压泵，在地面用中等压力试喷，检查喷灌系统是否畅通，再将灌浆管插入钻孔地基设计深度。

(6) 喷射灌浆：开动高压泵、空压机、泥浆泵进行喷射作业，根据工程要求所需固结体的形状，一面喷灌、一面提升，一面旋转、摆动或定向。提升速度和旋转速度根据地质条件和设计规定确定。喷灌到达设计桩长后，停止喷射，继续灌注浆液，进行静压。灌浆管提出孔口后，在孔内间歇采用人工灌浆，避免固结体顶部形成凹形。

(7) 移动喷射机架及冲洗管路：完成一个孔的喷灌作业后，将喷射机架移至新的孔位，开始新的钻孔喷灌施工。若完成一个孔的喷灌作业后，新的孔位尚在钻孔，不能连续进行喷灌时，喷灌系统及管路，应及时冲洗干净，直至冲出清水为止，不得残留浆液。

7 高压喷灌现场试验

高压喷射灌浆法在砂土及砂卵石层地基中的试验和工程应用，目前国内进行了一些工作，积累了一定的资料，取得了许多经验。但高喷各项工艺参数，对云山水库砂卵石层地基结构的适应性，技术可行性，必须通过现场试验探求，找出适合本工程的技术参数，为水库坝基处理采用高压喷灌加固设计和施工参考。因此根据坝基地质勘探资料查明，沿坝轴线河床段砂卵石层比较厚，卵石粒径较大，卵石含量多，采用高压喷射灌浆处理，难度较大的特点。所以选在老河床 0＋410～0＋417 段下游离原坝轴线约 90m 处砂卵石层地基，进行高压喷射灌浆现场试验。

7.1 现场试验地质概况

为了便于施工，在卵石层表面填筑厚度 1.0m 的人工素填土，填土顶面与坝趾地面相平，其下为厚度 4.0～4.3m 的砂卵石层，再下为千枚状板岩。砂卵石呈胶结状态，浸水易崩塌，粒径大于 10cm 的约占 30%～40%，个别最大粒径达 50cm，卵石含量多，渗透系数为 6.6×10^{-1}～1.0×10^{-2}cm/s，钻机造孔钻进慢，泥浆固壁要求较高。

7.2 灌浆孔布置及主要技术参数

为探求高压旋喷、高压摆喷灌浆对云山水库坝基砂卵石层加固效果，以及提升速度、旋转速度、摆喷角度、喷灌压力、固壁泥浆稠度、浆液比重、孔距等各种参数的最佳适应性，在现场布置了两孔高压旋喷桩和六孔高压摆喷板桩构成围井试验。另在大坝左岸（桩号 0+010 附近）坝体黏土心墙内，布置两孔高压旋喷桩和一孔摆喷板桩墙试验，以便测定高压喷灌对黏土的加固效果，提供高压灌浆板桩墙伸入黏土心墙成墙状态的参数。以上试验布置如图 8 所示，有关各孔试验技术参数见表 1。

表 1　各喷灌孔主要技术参数表

土层名称	灌浆孔号	喷灌形式	旋转角度/(°)	旋转速度/(r·min^{-1})	提升速度/(cm·min^{-1})	高喷水压力/(kg·cm^{-2})	水量/(L·min^{-1})	气压/(kg·cm^{-2})	气量/(L·min^{-1})	浆压/(kg·cm^{-2})	浆量/(L·min^{-1})	浆液比重
砂卵石	1	摆喷	30	10	7	360	75	7.5～8	90～96	1.5	80	1.6
	2	摆喷	35	10	5.5	360	75	7.5～8	90～96	1.5～2.0	80	1.6
	3	摆喷	25	10	5.5	330	75	7.5～8	80	1～2	80～90	1.6
	4	旋喷	360	10	10	380	75	7.5～8	84～96	1～2	80	1.6
	5	摆喷	30	10	5.5	360	75	7.5～8	84～96	1.5～2.0	80	1.6
	6	摆喷	30	10	6.0	360	75	7.5～8	90～96	1.5	80	1.6
	7	摆喷		10	9.0	330	75	7.5～8	85	1～2	80	1.6
	8	旋喷		10	10.0	330	75	7.5～8	80	1～2	80	1.6
黏土	1	摆喷	30	10	5.5	360	75	7.5～8.0	84～90	2～3	80	1.55～1.65
	2	旋喷	360	10	5.5	360	75	7.5～8.0	80～90	4～5	80	1.62～1.70
	3	二重喷	360	10	5.5	360	75	7.5～8.0	80～90	4～5	80	1.62～1.65

注　1. 高压水喷嘴直径 1.9mm，气喷嘴直径 9mm，浆液为纯水泥浆。
2. 3#、7#孔，基岩部位 0.5m 为旋喷，以上至地面为摆喷。

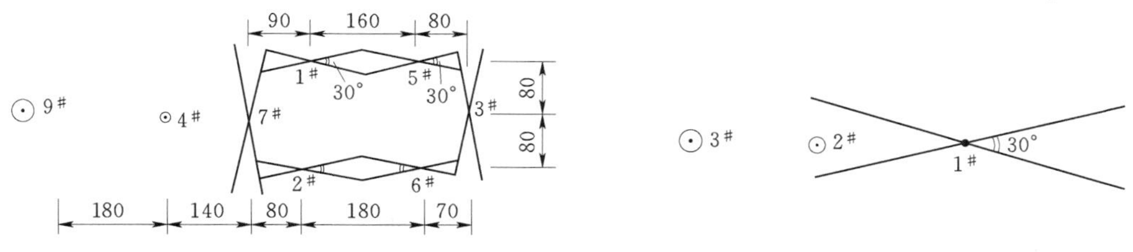

(a)土坝下游砂卵石地基高喷试验布置示意图

(b)坝体黏土层高喷试验布置图

图 8　试验布置图

说明：
1. 图（a）中 1#、2#、5#、6# 为摆喷孔，3#、7# 孔，基岩内为旋喷、砂卵石层内为摆喷，4#、9# 为旋喷孔。
2. 图（b）中 1# 为摆喷孔，2# 为旋喷孔，3# 为两次旋喷孔。

7.3　试验成果

7.3.1　现场试验开挖

为了检查高压喷射灌浆在砂卵石和黏土层中凝成板桩防渗墙和旋喷桩的形状、连接情况、扩散范围和加固质量，对现场高喷试验进行开挖。

（1）左岸土坝黏土层高压旋喷桩和摆喷板桩开挖后，可以看出其断面形状规则，水泥浆与黏土搅拌混合均匀，形成圆柱桩和板桩，均呈灰色，与水泥浆色彩相同，桩体凝结密实，无夹层或松散体，但有极少细粒黏土块夹在其中，外围无明显的浆液渗透区，但有一薄层水泥成分相对少，黏土成分相对多的凝固层，彼此组成不同的界面（见图 9）。这可

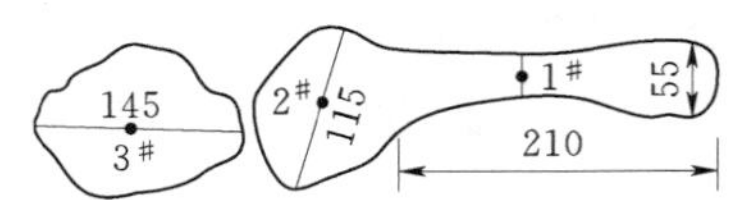

图 9 左岸土坝黏土心墙高压旋喷桩和摆喷板桩试验开挖素描示意图

能由于动量随时流过程的远近差异，因而在断面中心与外围边缘的组成有所差别。桩外与黏土接触面较粗糙，无光滑感觉。坝体黏土层 2# 旋喷桩最大直径为 1.15m，3# 旋喷桩采用两次高压水喷孔灌浆，最大桩径为 1.45m，比一次喷灌成桩直径增加约 30cm 左右。摆喷板桩墙长度为 2.1m，灌浆管中心处墙厚为 13cm，摆喷末端厚度为 55cm。在黏土层的成桩与有关资料介绍高喷情况相同。

（2）土坝下游砂卵石地基高喷板桩围井和旋喷桩开挖情况：将上部人工填土（层厚 1.0m）及围井中砂卵石挖除，发现砂卵石粒径和含量比原地勘资料增大，局部并夹有漂石。灌浆管直径范围内，基本上是水泥浆固结体，其外均为水泥浆与砂卵石形成的凝固体，卵石之间空隙充填水泥与砂粒，其胶结成的混合物，与混凝土相似。围井板墙和旋喷桩周表面，卵石凸凹镶嵌，形状不规则，个别粒径为 50cm 左右的漂石，被固结体包裸 2/3。摆喷板桩墙 3# 孔系先喷灌，成墙厚度和长度不受其他板桩墙影响，其长度为 2.7m，最大厚度 55cm，最小厚度 37cm。其他摆喷 1#、5#、2#、6# 孔板桩墙为对接连接型式，墙的最大厚度为 63cm，最小厚度为 39cm。4#、9# 孔为旋喷桩，其直径为 1.5～1.6m。围井摆喷板桩墙相互连接较好，墙与基岩接合也较紧密。围井内砂卵石挖出后，抽干井内集水，在井内外水位差 3m 左右时，四周板桩墙无集中漏水，仅有两处墙的连接处，有细小渗透滴水点。据了解是由于喷灌时中途停电断灌造成。桩体强度很高，用小手锤敲击，声音响亮，没有损坏，围井及旋喷桩开挖平面图如图 10 所示。

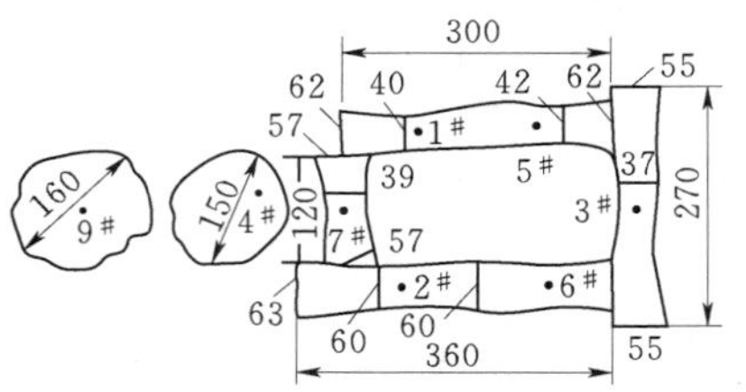

图 10 砂卵石层地基高喷板桩围井及旋喷桩试验开挖示意图

7.3.2 现场压水试验

为检查高压喷射灌浆固结体的防渗效果，在现场进行了注水和压水试验。为检查高压喷射板桩墙与基岩接触处的防渗性能，在围井内的基岩面上铺填厚度 30cm 的砂卵石层，其上现浇一块厚度 40cm 的混凝土盖板，并预埋一根 2# 测压管伸出至地面高程，作注水及压水试验用。对于检查围井四周板桩墙的防渗情况，是将第一块混凝土板作为底板，其上再铺一层厚 60cm 的砂卵石层，再现浇一块厚度 70cm 的混凝土板作盖板，通过盖板在砂卵石层预埋另一根 1# 测管，作为注水及压水试验用（见图 11）。旋喷桩体防渗检查，在旋喷的中心用钻机钻一孔，深度为 2m，尚留有 1m 多未钻，作为孔底。在孔内埋一根测压管。孔口上端用水泥砂浆浇注封密，管口再用橡胶管连接，作压水试验用。注水试验因测压管长度较短，再在其上接长，高出地面太高，很难操作，所以只有 2m 水头作用。压水试验压力为 1.5～8.0kg/cm²，因设备限制，未能进行破坏压水试验。从压水试验资料（表 2）看出，单位吸水率 ω 值在 0.015～0.0006L/(min·m·m)，均小于 0.03L/

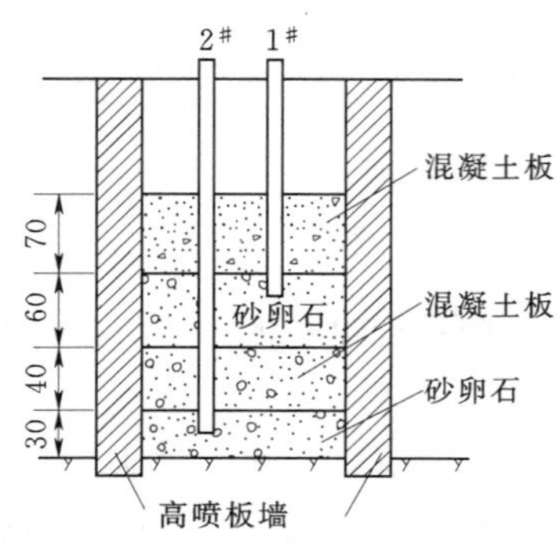

图 11 高压喷灌板桩围井现场压水试验示意图

(min·m·m)，换算渗透系数 K 值为 $3.8\times10^{-5}\sim1.57\times10^{-6}$cm/s，对于中小型水利工程，按有关规范介绍，$\omega$ 小于 0.05L/(min·m·m)，可认为基本不透水。由此可见，在砂卵石地基中采用高压喷灌防渗墙效果良好，可以满足工程要求。

表 2　　　　现场高压喷灌试验围井及旋喷桩压水试验成果表

试验日期/(年.月.日)	孔号	全压力/(kg·cm^{-2})	单位吸水率 ω 值/[L·(min·m·m)$^{-1}$]	渗透系数 K/(cm·s^{-1})	备　注
1985.12.22	1#	1.5	0.00083	2.17×10^{-6}	因管路漏水，取三个读数，不渗水
1985.12.22	2#	2.0	0.0031	8.1×10^{-6}	
1986.1.18		1.7	0.0038	9.9×10^{-6}	
1986.1.18		4.2	0.0021	5.6×10^{-6}	
1986.1.19		6.2	0.0027	7.06×10^{-6}	
1986.1.19		8.2	0.0028	7.3×10^{-6}	
1985.12.22		1.5	0.0028	7.32×10^{-6}	
1985.12.23		2.0	0	0	
1985.12.23		2.5	0.0006	1.57×10^{-6}	
1986.1.18		1.7	0.018	4.71×10^{-5}	
1986.1.18		2.2	0.0145	3.79×10^{-5}	
1985.12.23	9#	2.5	0.000063	1.65×10^{-7}	

注　1. 1# 孔检查高压喷灌板桩墙形成围井的墙体防渗性。

2. 2# 孔检查围井板桩墙与基岩连接处防渗性。9# 系旋喷桩检查钻孔。

3. $\omega=\dfrac{Q}{Hl}$，$K=0.525\omega\lg\dfrac{al}{r}=1.57\omega$（按巴布什金公式换算）。

7.3.3　试样室内物理力学试验成果

从左岸土坝黏土心墙高压喷灌试桩和老河床段下游坝脚砂卵石层高喷试验围井墙取出试样，在室内进行了渗透、抗压强度等主要项目试验，其成果列于表 3。

表 3　　　高压喷射灌浆现场试验的旋喷桩、摆喷板桩墙室内物理力学试验成果表

地层名称	试验编号	比重	干容重/(g·cm^{-3})	饱和容重/(g·cm^{-3})	空隙率/%	孔隙比	吸水率/%	饱水率/%	抗压强度/(kg·cm^{-2})	弹性模量/(kg·cm^{-2})	渗透系数/(cm·s^{-1})	破坏坡降	备注
黏土	1	2.71	1.01	1.63	62.6	1.69	59.9	60.97	103	4.07×10^4	9.73×10^{-8}	1266.67	左岸黏土心墙旋喷体取出试样
	2	2.73	0.89	1.55	67.52	2.09	73.4	75.46	66	3.22×10^4	5.53×10^{-8}		
	3	2.74	0.94	1.57	66.0	1.93	68.0	69.0	105	3.77×10^4	3.62×10^{-8}		
	4	2.74	0.90	1.55	67.20	2.04	72.1	71.3	73	3.98×10^4	4.86×10^{-8}		
砂卵石	摆$_1$								122		4.11×10^{-6}	2666.67	下游坝脚处围井墙体取出试样
	摆$_2$								199		1.49×10^{-8}		
	摆$_3$								92		1.84×10^{-6}		
	摆$_4$										2.76×10^{-7}		
	摆$_5$										5.75×10^{-7}		
	摆$_6$										1.487×10^{-8}		
	旋$_1$								163				下游坝脚处旋喷体取出试样
	旋$_2$								170				
	旋$_3$								119				

从表3看出，加固效果良好。

8 云山水库坝基砂卵石层高压喷灌板桩防渗墙的应用

8.1 坝基地质概况

根据地勘资料查明，在原坝轴线位置，沿坝长桩号0－024.4～0＋013坝基为基岩，0＋013～0＋020左右为钢筋混凝土涵管位置，0＋021～0＋138为砂卵石层，0＋138～0＋167.1为基岩，0＋167.1～0＋482.4为砂卵石层。0＋482.4～0＋537.4为基岩。在0＋167.1～0＋482.4范围内，其中0＋407.1～0＋469.5区段为老河床段，该段砂卵石层厚度约2.5～4.0m，卵石含量高粒径较大，其余滩地段卵石层厚度约为1.5～2.5m，粒径较小。

8.2 高压喷射灌浆孔的布置

土坝为黏土心墙坝，心墙底部设有不完整梯形黏土截水齿槽（右岸台地未设），其底宽约0.8～2m左右，齿槽中心线离坝轴线偏上游约5.0m，高压喷灌防渗墙中心线布置在坝轴线偏上游1.0m，位于齿槽中心线与原坝轴线之间。高压喷灌板桩墙主要设置在砂卵石层地段内，从上述地质资料看出，本工程土坝坝基砂卵石层地段为桩号0＋021～0＋138及0＋167.1～0＋482.4两段。为使砂卵石层处理后的防渗墙与相邻基岩段坝体相互连接紧密，加固长度伸入基岩段适当范围。因此高压喷灌防渗墙布置于0＋021～0＋138及0＋167.1～0＋485.5两个区段。各孔之间板桩墙连接分别采用对接和折线连接，墙的上端伸入黏土心墙内2.5m，下端嵌入基面以下0.3m。灌浆孔的孔距和排距，参考国内采用高压喷灌板桩墙处理砂卵石地基防渗工程资料，以及本工程现场试验初步资料，滩地段喷灌孔设置单排，老河床段0＋407.9～0＋468.7，考虑地基复杂，喷灌孔布置成双排，第二排孔的轴线离第一排孔轴线偏上游0.5m，各排孔距均为1.6m，第一、第二排孔成梅花形。单排板墙连接为折线形，摆喷角为25°；双排板墙为平接形，摆喷角为30°。喷孔布置如图12所示。

8.3 工艺参数

为使高压喷灌板桩墙与基岩结合紧密，减少接触渗流，扩大该处板桩防渗墙的厚度，在基岩面以下30cm至基岩面以上50cm，先用高压旋喷灌浆，随后将喷管插回到基岩面以下30cm处进行摆喷并上提，通过砂卵石层伸入黏土心墙内2.5m。喷灌方式，采用三管法，以摆喷为主，摆喷角度滩地段与布孔轴线成25°，河床段摆喷中心角成30°。喷灌按二序孔施工，第一序孔与第二序孔相隔时间为2～5d以上。高喷水压力，一般为350kg/cm^2，河床段卵石含量多和粒径较大，采用水压大于350kg/cm^2，输水量为75L/min以上，气压为8kg/cm^2，气量为80m^3/h，浆压为1.5～2.0kg/cm^2，浆量为80～85L/min。提升速度，在基岩和砂卵石层内采用5～5.5cm/min，黏土心墙内为7.0～7.5cm/min。旋喷转速为10r/min，摆喷速度5次/min。水喷嘴直径为1.9mm，气喷嘴直径为9mm，浆液为水泥浆，比重为1.6，水泥标号为425及525两种。

8.4 加固效果检查

检查高压摆喷板桩防渗墙处理效果，在滩地段和河床段分别选用围井方法和钻孔方法进行抽样检查。滩地段在桩号0＋048.2～0＋049.8的板墙侧，加喷两孔摆喷板桩组成四

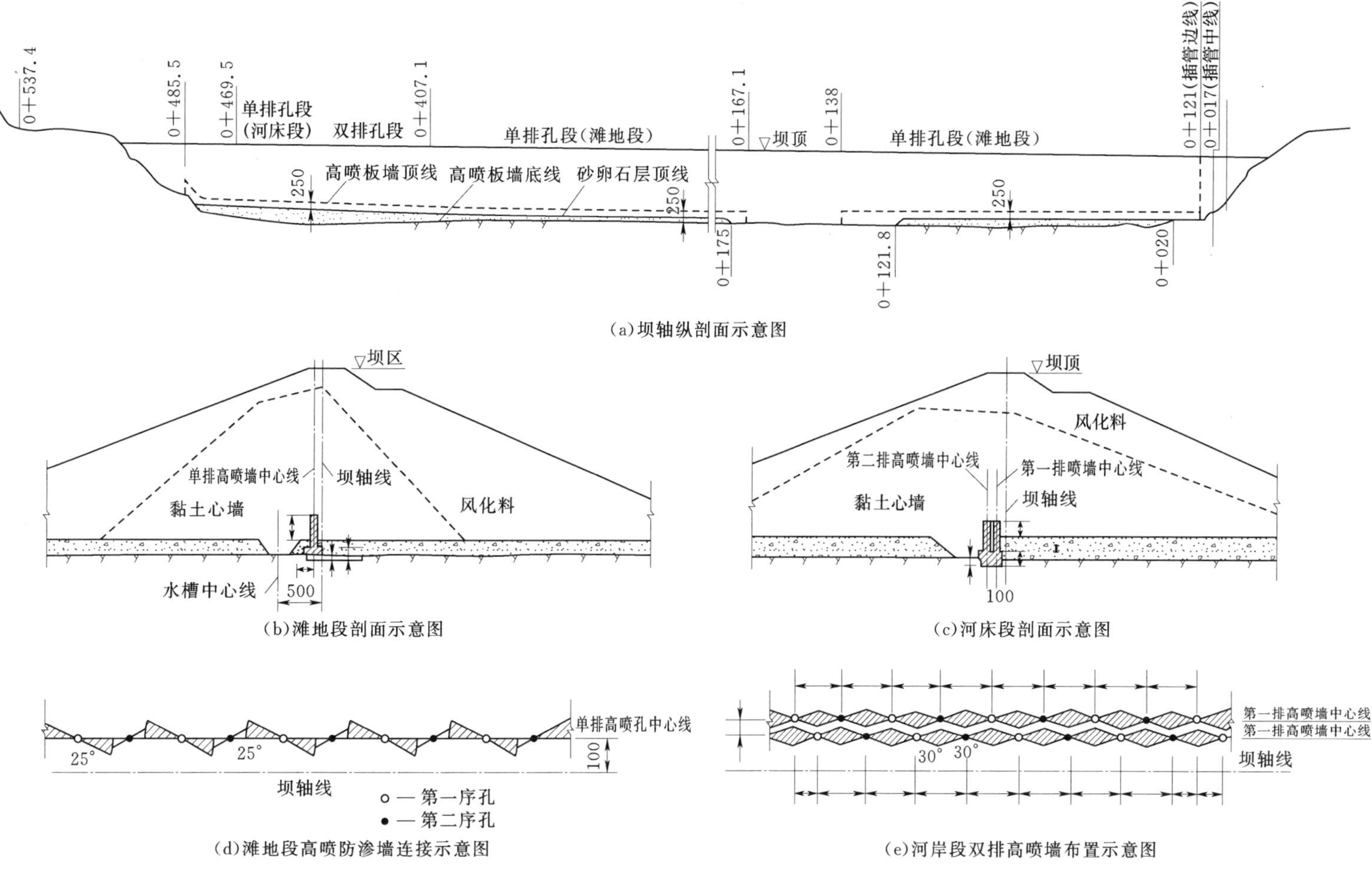

图12 坝基高压喷灌布置示意图

边形围井，在围井中间板墙顶部加一孔旋喷桩盖板。检查时用钻机从坝顶将盖板钻孔并用套管固壁埋设测压钢管，由砂卵石层伸至坝顶，套管与测压管之间深度处，用粗砂细砾石填充，边填边拔套管，在盖板深度位置用黏土球填充、盖板以上套管与测压管之间用水泥球填充至坝顶，全部套管拔出。经 3～4d 后，进行注水及压水试验，测定单位吸水率 ω 值。河床段在桩号 0+430.7、0+437.9 和 0+466.7 处的双排梅花形高喷孔之间，进行钻孔取样和钻孔注水试验，质检孔布置见图 13。原计划质检钻 1#、2#、3#孔，取出高喷固结体进行室内抗压强度、渗透等项目试验，并在钻孔内进行注（压）水试验。由于仅有 2#钻孔，在 4m 多厚的高压喷灌体内有 3m 多范围取得固结体，其余 1#、3#孔均未取得柱状固结体，故又补钻 4#、5#、6#质检钻孔、但亦未取得理想固结体。分析原因，有两种意见：一种认为，静压灌浆质检均用钻孔方法，高压喷射灌注双排板桩墙，在排距 0.5m、单排墙厚 0.35～0.6m 所组成的双排板桩墙理应连成整体，同时底部基岩以上 50cm 厚的砂卵石层为旋喷灌浆，其直径可达 1m 多，因此应用钻孔方法取样，理应取出高喷柱状固结体，现已钻 6 个孔，仅有一个孔在 3/4 的范围内取得高喷固结体，其余 5 个孔，未取出固结体，表明施工质量存在问题，另一种意见认为，高喷灌浆板桩墙的厚度仅 0.3～0.6m，双排喷灌孔为梅花形交叉布置，未必能组成叠加式单排厚墙，而钻孔从坝顶钻至坝内高喷体深度，有 20 余 m，尤其钻进卵石层，钻孔较难保证不发生偏斜，虽然基岩以上 50cm 厚砂卵石层为旋喷灌浆，但层厚很薄，用湿钻法钻进，经钻头旋磨和水冲，也难以获得柱状固结体。在现场试验测定旋喷体的透水性时，曾在旋喷桩体内，用钻机钻入 2m 孔深，埋管进行压水试验。但在 2m 长的钻孔内，仅取得断续 20～30cm 固结体，其余均为卵石及水泥碎块由钢丝钻头带出，细粒料由水冲出孔口。所以高压喷射灌注板桩墙质检工作，国内现已一般不采用钻孔取样，采用围井方法进行抽（压）水试验，作为一种质检方法。为进一步了解坝基砂卵石层高压喷射灌浆成墙厚度、连接可靠程度、高喷体施工质量，拟在枯水期进行竖井开挖检查。

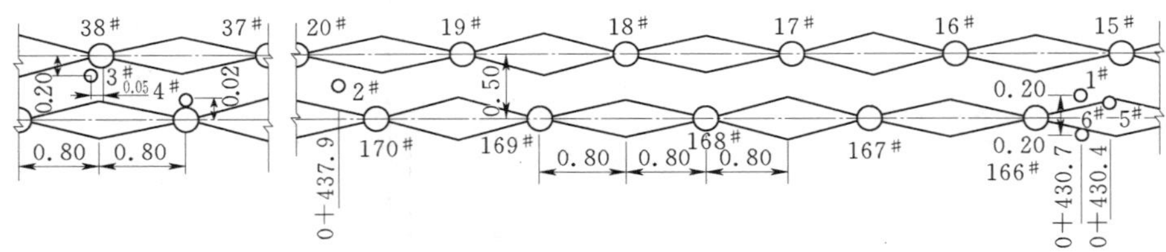

图 13　河床段“高压摆喷”防渗墙检查钻孔布置示意图（单位：m）

各孔注水试验测得单位吸水率 ω 值，列于表 4。从表中列出 ω 值看出，透水性均比较小［ω 小于 0.05～0.1L/(min·m·m)］基本达到防渗要求。

8.5　材料消耗及工程造价

云山水库坝基高压喷射灌注板桩防渗墙，总面积为 2009m²。钻机造孔总进尺为 6791.8m，水泥用量 937.7t，总造价 32.23 万元。每平方米单价为 110 元（穿过坝体的造孔费用在外），水泥用量为 0.467t/m²。参考国内采用高压喷灌防渗工程的材料消耗，水泥最高用量 0.45t/m²，而本工程略大于该值。其原因由于本工程地层较为复杂，提升速度较慢，所以材料消耗略高。同时材料价格调高，相应单价较高。但与 YKC 防渗墙相比，造价和进度均显得非常优越。

9 结语

云山水库主坝砂卵石地基通过高压喷射灌注板桩防渗墙现场试验和工程的应用，初步得出以下看法：

（1）现场试验和工程抽样检测表明，云山水库砂卵石地基的渗漏，采用高压喷射灌注板桩防渗墙处理，技术上是可行的，取得良好效果，达到预期要求，是一种处理砂卵石地基渗漏的新方法。

（2）云山水库砂卵石地基卵石含量较多、粒径也较大，其复杂条件是目前国内采用高压喷射灌浆处理地基工程中少见的实例。本次实践表明，对于卵石粒径较大的地层，只要改进工艺技术参数，适当增大喷射压力，减缓提升速度，采用高压喷射灌浆法还是适用的。其成墙机理不同于一般土层冲击破坏土的结构，使颗粒重新排列，而是浆液通过较大粒径卵石骨架空隙，进行充填包裹，凝成胶结体。因此认为，对于较大粒径砂卵石地基，采用高压喷射灌浆处理，必须首先进行现场试验，得出有关技术参数，以供工程参考。

（3）现场试验开挖看出，砂卵石层内高压喷射灌浆成墙厚度为35～60cm，长度可达1.8～2.6m，旋喷成桩直径为1.4～1.5m；黏土层内成墙厚度为13～50cm，长度为2.4m，旋喷成桩直径为1.15m，比砂卵石内加固范围略小。板桩墙连接紧密、连续，胶结体坚硬，防渗性好。

表4　高压喷射灌注板桩墙质检注（压）水试验成果表

时间/(年-月-日)	桩号	孔号	试验段深度/m	水压力/($kg\cdot cm^{-2}$)或水柱高度/m	单位吸水率 ω 值/[$L\cdot(min\cdot m\cdot m)^{-1}$]	说明
1985-12-24	0+048.2～0+049.8		坝顶以下18.5～19.47	2.0（$kg\cdot cm^{-2}$）	0.00013	左岸滩地围井压水试验
	0+048.2～0+049.8		坝顶以下18.5～19.47	2.5（$kg\cdot cm^{-2}$）	0	左岸滩地围井压水试验
1986-04-15	0+466.3～0+467.15	3#	坝顶以下17.80～21.35	7.82（m）	0.042	河床段注水试验
1986-04-17	0+437.9	2#	坝顶以下17.99～20.66	6.3（m）	0.027	河床段注水试验
1986-04-17	0+437.9	2#	坝顶以下20.66～22.16	6.3（m）	0.0206	河床段注水试验
1986-04-18	0+437.9	2#	坝顶以下22.16～23.56	6.3（m）	0.0975	河床段注水试验
1986-04-20	0+430.4～0+431.0	1#	顶坝以下17.59～23.69	7.4（m）	0.0149	河床段注水试验
1986-04-22	0+466.3～0+467.15	4#	坝顶以下17.40～19.71	8.7（m）	0.0085	河床段注水试验

续表

时间/(年-月-日)	桩号	孔号	试验段深度/m	水压力/(kg·cm^{-2})或水柱高度/m	单位吸水率 ω 值/[L·(min·m·m)$^{-1}$]	说明
1986-04-23	0+466.3～0+467.15	4#	坝顶以下19.71～23.26	4.7（m）	0.0252	因套管接头未旋紧，漏水
1986-04-25	0+466.3～0+467.15	5#	坝顶以下18.00～23.70	6.25（m）	0.0393	河床段注水试验
1986-04-28	0+466.3～0+467.15	6#	坝顶以下20.20～22.05	7.55（m）	0.0055	河床段注水试验

（4）高喷灌浆试验成墙及成桩取样室内检测成果，黏土层中固结体的抗压强度66～105kg/cm^2，渗透系数3.62～9.73×10^{-8}cm/s，破坏坡降1266.7，弹性模量3.22～4.07×10^{4}kg/cm^2；砂卵石层中固结体抗压强度120～199kg/cm^2，渗透系数4.11×10^{-6}～1.49×10^{-8}cm/s；破坏坡降2666.67。现场压水试验 ω=0.018～0.00083L/(min·m·m)。坝基砂卵石层经高压喷射灌浆后，现场钻孔注水试验测得 ω 值一般均小于0.03L/(min·m·m)，个别段最大为0.0975L/(min·m·m)，最小为0.00013L/(min·m·m)。从强电、诱水性能等来看，有效地改善了砂卵石层地基的性质。

（5）开挖后，砂卵石层内高压喷射灌浆板桩墙及旋喷秒的表面，明显看出，水泥浆与砂卵石胶凝成坚硬固结体，类似混凝土结构。固结体表面水泥浆嵌固着卵石，凹凸不平，与周边未处理砂卵石层紧密结合，周边摩擦力较钢管灌注桩，有所增大。

（6）高压喷射灌浆方法用于砂卵石层地基防渗处理，技术简单，操作方便，施工进度快，比静压灌浆可靠，不受可灌性影响、可以随工程需要控制喷灌范围和成墙型式，单位造价比YKC防渗墙低很多。

（7）高压喷射灌浆法用于软土地基可提高土的强度，可作成直径为0.8～1.4m圆柱桩，桩的型式可控制为底部大的扩大桩、竹节桩和上大下小的锥体桩等，单桩承载力较普通灌注桩高。

10　今后有待研究的主要问题

高压喷射灌浆方法是一项新的技术，无论在理论上、工艺设计和施工等方面，有待进一步继续研究和总结，逐步完善和提高，但当前应着重探讨以下主要问题：

（1）浆液浪费问题。目前高压喷射灌浆施工，孔口冒浆量很大，虽然采取某些措施，通过滤筛回收，重新利用一部分，但总的浪费量还有20%左右；因此，必须研制较好孔口装置减少冒浆量，降低单位造价。

（2）浆液材料问题。高压喷射灌注板桩防渗墙，一般采用水泥浆液材料，水泥用量较多。为减少水泥用量，对于地基要求防渗方面，在保持固结体有足够强度条件下，研究水泥—黏土混合浆液成墙的物理力学性质，以及在提高地基强度方面，研究水泥—粉煤灰混合浆配比成桩的物理力学性质，使这项新技术更广泛地为工程建设服务，有重要意义。

（3）黏土浆防渗墙的应用研究。我省广大堤防地基泡泉问题，比较突出，过去采用压浸台方法处理，工程量大，亦未消除隐患。因此，研究黏土浆或以黏土为主掺合少量水泥混合浆液，采用高压喷射灌注成墙方法处理，具有实用价值。

（4）提高单管喷射灌浆效果的研究。为适应城市狭小施工场地和已有建筑物加固开挖的限制，单管喷射灌浆法比较方便，但以水泥浆为介质的单管高喷灌浆方法，由于浆液有一定稠度，喷射范围，必然较以水为冲击破坏的三管法为小，所以对研究提高单管高压喷射灌浆效果，解决城市房屋密集区，三管法设备多的缺点，相当迫切。

（5）坝基高喷成墙的质检技术研究。坝顶以下深层地基土壤高喷灌浆成墙的施工质量检查，采用围井压（注）水试验方法，难以了解地下成墙厚度、连接可靠程度、工程质量等，开挖竖井检查，既费时，造价又高，因此，应着手研究新的地下成墙质检技术。

参考文献：

［1］ 铁道部科研所，解放军00069部队，等．关于单管旋喷法的试验和运用［R］．1979，10.

［2］ 电力工业部科技情报研究所．软基处理的旋转喷射法［R］．1980，6.

［3］ 王吉望，熊大玉，等．高压喷射桩改良软土地基［J］．建筑结构学报．1981，6.

［4］ 查振衡．高压喷射灌浆法的应用．水利工程管理技术［J］．1984（2）．

［5］ 大冶水库坝基砂砾层高压定喷灌浆防渗．水利工程管理技术［J］．1984（2）．

［6］ 王宣玉，等．高压定向喷射灌浆构筑防渗板墙技术的研究［J］．岩土工程学报，1984（6）．

［7］ 铁道部旋喷注浆科研协作组．旋喷注浆加固地基技术［M］．北京：中国铁道出版社，1984，8.

［8］ 山东水利科学研究所．高压喷射灌浆防渗技术的研究［R］．1985，7.

四、工 程 管 理

SHUIGONGANQUANYUFANGZAIJIANZAI

峡江水利枢纽水工钢闸门的焊缝超声波探伤

胡碧辉，戴国强

江西省水利科学研究院

摘　要：峡江水利枢纽水工钢闸门主要通过焊接而成，焊缝质量决定了水工钢结构产品的质量，本文介绍峡江水利枢纽水工钢闸门的焊缝超声波探伤。

关键词：峡江水利枢纽；钢闸门；焊缝；超声波探伤

江西省峡江水利枢纽工程是鄱阳湖生态经济区建设的重点水利工程之一，也是目前江西省投资最大的水利工程，其中的钢结构主要包括钢闸门，启闭机和拦污栅。峡江水利枢纽水工钢闸门的种类有人字闸门，平面闸门及弧形闸门，用作进水闸、泄水闸、船闸、临时挡水闸门及检修闸门，闸门厚度最高达 56mm。这些闸门主要是通过焊接组合而成的，所以焊缝质量的检测和评价对整个结构的质量控制起得极其重要的作用。金属结构的焊缝探伤主要有射线探伤、磁粉探伤和超声波探伤。由于射线探伤危险系数较高，磁粉探伤只能探测表面有缺陷的焊缝，故峡江水利枢纽钢闸门焊缝内部质量主要通过超声检测来完成，探伤比例为一类焊缝 100%，二类焊缝 50%。本文就峡江水利枢纽的水工钢闸门的焊缝超声波探伤作一些探讨。

1　焊缝超声波探伤

1.1　焊缝等级的评定

评定一条焊缝焊接是否合格，首先需要对这条焊缝的等级进行评价。在闸门产品的设计与制作中，水利水电行业标准将焊缝按其“重要性”进行了分类，这种分类主要是依据焊缝的受力条件。在《水工金属结构焊接通用技术条件》（SL 36—2006）中规定：在动载荷或静载荷下承受拉力，按等强度设计的对接焊缝、组合焊缝或角焊缝属于一类焊缝；在动载或静载荷下承受压力，按等强度设计的对接焊缝、组合焊缝或角焊缝属于二类焊缝；除上述一类、二类焊缝以外的焊缝为三类焊缝。检测依据行业规范《水利工程钢闸门制造安装及验收规范》（DL/T 5018—2004）来评定焊缝等级。

1.2　T 形焊缝

T 形焊缝是通过腹板和翼缘板组合而成，分为组合焊缝和角焊缝，峡江水利枢纽工程水工钢结构中 T 形焊缝很多，很多被列为重要焊缝。由于 T 形焊缝工艺更复杂，所以 T 形焊缝中存在缺陷的概率更高。常见的缺陷有气孔、夹渣、未焊透、未熔合、层状撕裂等

本文发表于 2013 年。

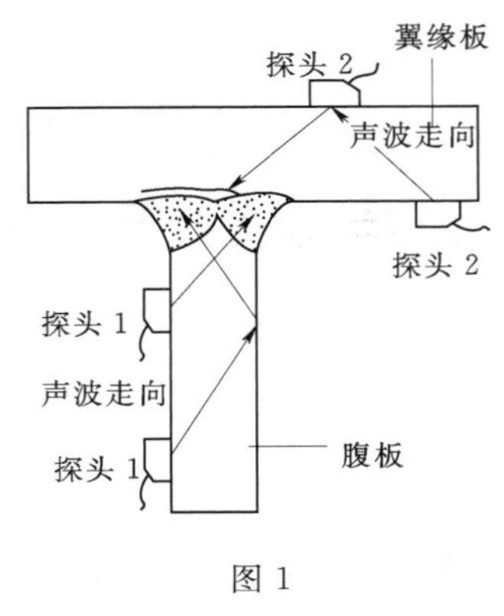

图 1

多种，所以较之平板对接焊缝一般采用两种方式进行测量，如图 1。

方式一：采用斜探头在梁腹板上通过一、二次波进行探测，探测未焊透、未熔合及其他缺陷。这种探伤时，频率选 2.5～5.0MHz，探头 K 值可根据梁腹板厚度参照表 1。

方式二：采用斜探头在翼缘板外侧或内侧进行探测。探头在外侧时用一次波探测，外侧一次波通常灵敏度更高，纵向、横向的缺陷都可以探，并且找出缺陷比较方便。缺点在于焊缝位置不明确，需要在翼缘板外侧画出腹板中心线及焊缝位置。

表 1　腹板厚度与探头 K 值

腹板厚度/mm	≤25	25～50	≥50
探头 K 值	2.5	2.5、2.0	1.0、1.5

1.3　探伤结果评定

依据《钢焊缝手工超声波探伤方法和探伤结果分级标准》(GB/ T 11345—2013)，超声检验有 A 、B、C 三个等级，检验工作的难度系数按 A、B、C 顺序逐级增高。按照 DL/T 5018—2004规范要求，将检验等级选做 B 级。原则上采用一种角度斜探头在焊缝的单面双侧进行检验；按《水利工程钢闸门制造安装及验收规范》(GB/T 11345—2013) 规范规定，一类焊缝 BⅠ级合格，二类焊缝 BⅡ级合格。

2　非人为结果影响因数

2.1　耦合剂材料的影响

检测时，为提高检测效果，需在探头与工件表面之间施加一层透声介质，即耦合剂。一般选机油、水和化学糨糊，但这三者中化学糨糊的声耦效果最好，所以我们在检测过程中推荐使用化学糨糊作为耦合剂，但检测完后需及时处理，否则会锈蚀钢结构工件表面，对钢结构产生破坏。

2.2　工作温度的影响

由于超声波在以化学糨糊为介质的传播声速变化较大，所以在实际检测操作中，如果室温度与现场温度相差不大时，可直接用室温下实测 K 值的探头进行检测，不需要修改；而当室温明显低于或则高于工作温度时，就必须模拟工作温度对探头性能进行测试，即在工作温度下完成 DAC 曲线的制作。

2.3　焊缝余高的影响

在检测过程中，焊缝余高产生的干扰波在焊缝探伤中是比较常见的干扰，特别是在双面自动焊焊缝。当声波入射到焊缝下表面余高上时会产生波形转换，转换成反射横波和反射纵波，沿余高表面，声波的入射角不同，所产生的反射波的传播方向也不同。反射波传到上表面，又在上表面余高上反射，部分沿原路径返回被探头接受，会对操作者的判断产生干扰。这种干扰不管你选的探头斜角度多大，都不能避免，这就需要检测人员认真观察焊缝外形，更换探头角度，必要时打磨焊缝，以免造成误判。

3 结语

峡江水利枢纽工程中水工钢闸门的质量对整个工程的质量起着至关重要的作用，不仅不要误判，也不允许遗漏，所以研究钢结构的焊缝探伤非常之必要。故焊缝探伤需要探伤人员经验丰富，对很多干扰因素能够精准识别，因为误判后要求返工，不仅会耽误工程进度，还有可能会产生不可计量的经济损失，因此焊缝探伤需要结合多种因素才能下定结论。在评判前采用不同探头及多种探伤方式进行探伤，做到不误判，不遗漏，不放过。

在峡江水利枢纽水工钢闸门的制造质量检测过程中，发现的缺陷大多数为未焊透，或未熔合，同时也发现焊接方式不同，存在缺陷的比例也不相同。该枢纽水工钢闸门面板大部分是通过埋弧自动焊焊接，而主梁腹板对接或翼缘板对接则是通过手工气保焊焊接，我们发现埋弧自动焊焊接而成的构件存在缺陷的比例明显更低，这是因为人工操作更容易产生失误，所以建议制造厂家尽量通过埋弧自动焊的方式进行焊接。

基于最大熵原理的贝叶斯方法在水库渗漏预测上的应用

喻蔚然

江西省水利科学研究院

摘　要： 最大熵原理是Jaynes于1957年提出的，在处理不确定问题上简捷、快速。大坝的渗流监测存在各种不确定性因素，通过运用最大熵原理，建立库水位与渗漏量的贝叶斯模型，同时引入匹配系数，实现预测模型的优化反分析，从而可以预测未来水位条件下对渗漏量的影响趋势。通过对港河水库渗漏量预测模型的建立，将实测观测数据与预测值进行比较，结果表明误差小于2%，运用最大熵原理的贝叶斯反分析得到的反演参数能够反映大坝实际渗流状况，结果是合理可行的。

关键词： 最大熵；贝叶斯；反分析；渗漏预测

1　前言

大坝监测设施的埋设有一个时间过程，同一个监测项目（如渗流、位移等）的不同测点的观测基准日期可能相差数月或者1年以上，而单个测点的实测资料分析仅能反映局部的监测变量的空间和时间分布规律，而难以将所有测点的数据有机联系起来，综合分析同一基准条件下的大坝的安全性态。针对基准状态不一致的情况，工程中大多采用凭借经验进行简单修正的方法，但这种方法随意性较大，而且并不能真实反映大坝安全性态。目前，较为流行和先进的方法是基于最大熵原理的贝叶斯反分析方法，即在充分考虑各种不确定性因素的基础上，利用实测资料，进行坝体和坝基材料参数反演，尽可能消除不确定性因素影响，实现同一监测项目各测点的一致性修正。

2　最大熵原理的贝叶斯反分析方法

2.1　最大熵原理[1]、[2]

熵是一个函数，用以描述概率系统状态的不确定性。对于随机变量X，其概率分布为：

$$P\{X = x_i\} = p_i,\text{且}\sum_{i}^{n} p_i = 1(i = 1,2,\cdots,n) \tag{1}$$

则熵可以定义为：

本文发表于2012年。

$$H(p_1,p_2,\cdots,p_n)=-k\sum_{i}^{n}(p_i\ln p_i) \tag{2}$$

1957 年，Jaynes 提出了最大熵原理，即在给定试验条件下存在一个使熵取极大值的分布，这种分布是系统最常见的分布（或称为优势概率分布）。采用拉格朗日乘子法，使熵 H 达到最大，建立如下拉格朗日函数：

$$L=H(X)+(\lambda_0+1)\left[\int_R f(x)\mathrm{d}x-1\right]+\sum_{i=1}^{n}\lambda_i[x^i f(x)\mathrm{d}x-M_i] \tag{3}$$

令 $\partial L/\partial f(x)=0$，则可求出密度函数：

$$f(x)=\exp\left(\lambda_0+\sum_{i=1}^{n}\lambda_i x^i\right) \tag{4}$$

式（4）即为最大熵密度函数的解析形式。

2.2 最大熵原理的贝叶斯反分析方法

此方法是最大熵原理在专家系统中的规约推理，即：有了新的知识后，只要用它修改原有的结论就可以得到新的结论，而不必把新、旧知识再推理一遍[3]。

贝叶斯方法的原理是：把参数 x 看作具有某种验前概率密度 $p(x)$ 的随机变量，通过观测与该函数有关联的其他变量，设法从输入输出数据中提取关于 x 的信息。可以归结为由参数 x 的验后概率密度函数 $p(x\mid u)$ 的计算来推断这个参数。根据最大熵原理，其模型为：

$$\int p(x)\ln[p(x\mid u)]\mathrm{d}x=\max \tag{5}$$

在对大坝坝体和坝基的参数反分析中，贝叶斯法是用于量测误差分布和待辨识参数分布已知情形下的参数估计，其前提条件有：

$$\begin{cases}u^*=u(x)+\varepsilon_u\\ \varepsilon_u\sim N[0,\mathrm{cov}(\varepsilon_u)]\\ x\sim N[\mu_x,\mathrm{cov}(x)]\\ \mathrm{cov}(x,\varepsilon_u)=0\end{cases} \tag{6}$$

式中：u^*，u 分别为观测数据列阵及计算列阵；ε_u 为量测误差列阵；cov（·）为协方差列阵；μ_x 为待求参数 x 的均值列阵。式（6）中 cov(·)、；μ_x 均为已知矩阵。

因此，根据式（5）以及上述已知的前提条件，可推出贝叶斯反分析法的准则函数，即

$$J=(u^*-u)^{\mathrm{T}}\mathrm{cov}^{-1}(\varepsilon_u)(u^*-u)+(x-\mu_x)^{\mathrm{T}}\mathrm{cov}^{-1}(x-\mu_x) \tag{7}$$

为了使贝叶斯法的主观信息和客观信息之间有一个合理的匹配，引入一个比例系数 β，并假设

$$x=\mu_x+\varepsilon_x,\ \varepsilon_x\sim N[0,\mathrm{cov}(x)/\beta] \tag{8}$$

于是，贝叶斯反分析法的准则函数相应调整为：

$$J=(u^*-u)^{\mathrm{T}}\mathrm{cov}^{-1}(\varepsilon_u)(u^*-u)+\beta\cdot(x-\mu_x)^{\mathrm{T}}\mathrm{cov}^{-1}(x-\mu_x) \tag{9}$$

可以看出，通过比例系数 β 的调节，较好地解除了传统的贝叶斯法在应用方面的制

约，尤其是观测信息和先验信息不匹配的缺陷，从而可由观测数据的质量和数量推算模型的最佳阶数[1]。

2.3 优化计算

最大熵原理的贝叶斯反分析模型建立后，随机参数的辨识问题就转化成了最优化问题。优化方法主要有直接法和解析法两大类，两类方法各有所长，各有所限。对于大坝及坝基材料进行反演时，工作量巨大，直接法有一定的限制，而解析法收敛速度快，收敛性较好。其中DFP法是一种变尺度法，目前被公认求解极值问题最有效的方法之一，具体技术方法不再赘述。

3 工程实例

以江西省崇仁县港河水库大坝为例，说明如何利用最大熵原理的贝叶斯反分析方法实现坝体和坝基渗漏量的预测。

3.1 工程概况

港河水库位于崇仁县河上镇港里村，距崇仁县城20km，坐落在抚河水系崇仁河左港支流元家桥水上。坝址以上控制流域面积37.5km²，水库总库容为2748万m³，是一座以灌溉为主，兼顾防洪、发电、养殖等综合利用的中型水利枢纽。水库枢纽工程主要建筑物有：均质主坝、副坝、溢洪道、泄洪洞、灌溉发电洞、导流放空洞、坝后电站等。港河水库2010年完成了除险加固，完善了大坝安全监测系统，其中包括增设了量水堰。量水堰设置在排水棱体脚排水沟内，排水沟底部为混凝土结构，两侧岸坡排水沟将坝面以及岸坡雨水直接排出坝体；因此，通过坝基和岸坡的渗水无法进入量水堰内，量水堰仅反映坝体渗漏情况。

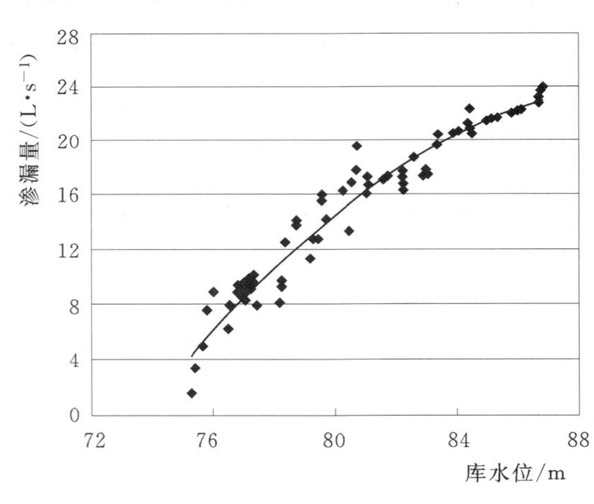

图1 库水位—渗漏量关系曲线

3.2 渗漏量预测模型

经过一年的运行和观测，积累了一些渗漏量的观测数据。为全面了解工程渗流现状，并预测未来大坝渗漏量的变化，建立了渗漏量预测模型。

3.2.1 数据统计

通过渗漏量的观测数据进行整理，渗漏量和库水位的关系曲线，见图1。

对库水位、渗漏量进行数理统计分析，见表1。

表1 观测数据数理统计结果

变量	均值	标准差	分布类型
库水位/m	84.27	2.1	正态
渗漏量/(L·s^{-1})	15.76	3.0	正态

3.2.2 模型及参数

从图 1 可以看出，渗漏量和库水位呈非线性关系，大致符合双曲型模型变化规律。双曲型模型可表示为：

$$Q=\frac{H}{a+bH} \tag{10}$$

式中：Q 为渗漏量；H 为库水位；a 和 b 为模型参数。

令 $y=1/Q$，$x=1/H$，则式（10）变换为：

$$y=b+ax \tag{11}$$

根据最小二乘法原理，a、b 分别由样本序列确定，公式为：

$$a=\frac{\sum x_i y_i-(\sum x_i \sum y_i/n)}{\sum x_i^2-(\sum x_i)^2/n} \tag{12}$$

$$b=\frac{\sum y_i}{n}-\frac{a}{n}\sum x_i \tag{13}$$

港河水库库水位—渗漏量的样本序列从 2010 年 5 月 16 日开始，至 2011 年 9 月 22 日止，观测频次约为 4 次/月，部分月份稍多，样本总数为 82 个。计算过程如表 2。

表 2　双曲型模型计算过程

序号	日期 /(年-月-日)	库水位 /m	归零的库水位 /m	渗漏量 /(L·s^{-1})	(4) × (5)	(4) × (4)
(1)	(2)	(3)	(4)	(5)	(6)	(7)
1	2010-05-16	79.64	8.44	16.1	135.88	71.23
2	2010-05-23	81.80	10.60	17.8	188.68	112.36
3	2010-05-30	82.30	11.10	17.7	196.36	123.21
⋮						
80	2011-09-2	79.7	8.5	14.2	120.70	72.25
81	2011-09-12	79.5	8.3	12.7	105.41	68.89
82	2011-09-22	78.2	7.0	11.7	81.90	49.0
Σ			812.81	138.3	14581.2	9013.8
	$a=1.58$，$b=0.39$					

则双曲型渗漏量预测模型为：

$$Q=\frac{H-71.20}{1.58+0.39(H-71.20)} \tag{14}$$

3.2.3 优化计算

采用 DFP 法，计算步骤为[4]：

(1) 给定初始点 $x^{(1)}$，允许误差 $\varepsilon>0$。

(2) 置 $H_1=I_n$（单位矩阵），计算出在处的梯度 $g_1=\nabla f(x^{(1)})$，置 $k=1$。

(3) 令 $d^{(k)}=-H_k g_k$。

(4) 从 $x^{(k)}$，出发，沿 $d^{(k)}$ 方向搜索，求出步长 λ_k，使其满足：

$$f(x^{(k)}+\lambda_k d^{(k)})=\min_{\lambda\geqslant 0} f(x^{(k)}+\lambda d^{(k)})\text{，并令 } x^{(k+1)}=x^{(k)}+\lambda_k d^{(k)}$$

(5) 检验是否满足收敛准则，若收敛则停止迭代，得到点$\bar{x}=x^{(k+1)}$；否则进行步骤(6)。

(6) 若 $k=n$，则令 $x^{(1)}=x^{(k+1)}$，返回步骤 (2)；否则进行步骤 (7)。

(7) 令 $g_{k+1}=\nabla f(x^{(k+1)}), p^{(k)}=x^{(k+1)}-x^{(k)}, q^{(k)}=g_{k+1}-g_k$，利用

$$H_{k+1}=H_k+\frac{p^{(k)}p^{(k)^{\mathrm{T}}}}{p^{(k)^{\mathrm{T}}}q^{(k)}}-\frac{H_kq^{(k)}q^{(k)^{\mathrm{T}}}H_k}{q^{(k)^{\mathrm{T}}}H_kq^{(k)}} \tag{15}$$

求出 H_{k+1}，置 $k=k+1$，返回步骤 (3)。

3.2.4 最大熵原理的贝叶斯反分析

结合双曲型渗漏量预测模型，根据前述的最大熵原理的贝叶斯反分析方法，实现对影响渗漏量的库水位进行反分析，见表 3。

表 3　反分析结果及匹配系数

日期/(年-月-日)	①库水位/m	②反分析库水位/m	匹配系数 β	①与②的差值/m
2010-05-16	79.64	79.73	100	−0.09
2010-7-03	86.31	86.42	100	−0.11
2010-10-15	81.29	81.06	100	0.23
2010-12-21	77.35	77.42	100	−0.07
2011-03-09	79.08	79.04	100	0.04
2011-06-13	85.87	86.01	100	−0.14

注　数据较多，此处仅截取部分具有代表性时间的库水位的反分析结果。

从表中可以看出，运用最大熵原理的贝叶斯反分析得到的库水位值与实测值较为接近；因此，可以认为计算结果可靠性较好。

3.2.5 预测渗漏量

将上述反分析得到的库水位代入双曲型渗漏量模型中即可预测大坝渗漏量，选取上述时间与库水位相对应的渗漏量进行反分析，其结果见表 4。

表 4　渗漏量反分析与预测结果

日期/(年-月-日)	①实测渗漏量/(L·s^{-1})	②预测渗漏量/(L·s^{-1})	②与①的差值/(L·s^{-1})	②相对①的误差/%
2010-05-16	16.1	16.51	0.41	2.5
2010-07-03	22.5	22.86	0.36	1.6
2010-10-15	17.1	16.98	−0.12	−0.7
2010-12-21	10.3	10.17	−0.13	−1.2
2011-03-09	11.2	11.35	0.15	1.3
2011-06-13	21.9	22.11	0.21	1.0

可见，预测渗漏量与实测渗漏量的误差不足 3%，且绝大多数低于 2%，基本上运用反分析方法预测的大坝渗漏量可代表实测渗漏量。

3.3 合理性分析

影响大坝渗漏量的因素有：降雨量、库水位（或上下游水头差）、时效以及温度等。通常，温度的影响较小，可不考虑。

大坝渗漏受降雨的影响较大，实际运行过程中也显示，降雨明显导致量水堰过流量增大。降雨的影响体现在两方面：①即时影响，即降雨入渗后经坝体渗漏流出坝外；但据管理人员介绍，渗漏量的观测一般选择没有降雨的时间，因此可以认为，实测渗漏量受降雨的即时影响有限；②时效影响，即降雨入渗后坝体土有个非饱和—饱和—排水的过程，在降雨结束后一段时间内，会有部分储存在坝体内的雨水逐步渗出，形成坝体渗漏。

降雨的时效影响取决于两个因素：坝体土质（指渗透系数的大小）和排水设施。一般土质致密，渗透系数偏小，排水设施堵塞严重时，时效影响越大；反之越小。港河水库大坝于 2010 年完成除险加固，设混凝土心墙，上、下游坝壳土相对较差，透水性较好，坝后排水设施运行良好，因而降雨的时效影响持续时间不长。

故对于港河水库大坝，影响坝体渗漏量大小的因素最主要的是库水位；当然，不可否认的是，降雨对坝体渗漏的即时影响大于库水位的影响。在正常运行工况下，运用上述分析方法预测大坝渗漏量是合理的。

3.4 预测结果分析评价

经过对 16 个月的观测数据进行整理、分析、建模、预测，渗漏量随着库水位的变化而产生相应的改变，大坝渗漏量与库水位表现出较好的相关性，总体而言，渗漏量符合大坝渗流正常性态，说明大坝进行除险加固取得了明显的效果。同时，通过运用最大熵原理的贝叶斯反分析方法建立了渗漏量的预测模型，预测结果与实测结果比较吻合，能够反映现实状况，该模型对于今后水库大坝安全运行将起到指导性的作用。

4 结论

最大熵原理目前已经广泛应用于工程技术领域，它能用比较简捷的方法推导已有的结果，在解决矩阵问题、偏微分方程方面具有明显的优势，在处理不确定问题上更具优越性。大坝的渗流监测存在各种不确定性因素，通过运用最大熵原理，建立库水位与渗漏量的贝叶斯模型，同时引入匹配系数，实现预测模型的优化反分析，从而可以预测未来水位条件下对渗漏量的影响趋势。这一过程解决了传统的贝叶斯法实测数据与先验信息不相匹配的问题，实现了两者的有机结合和对应。通过对港河水库渗漏量预测模型的建立，将实测观测数据与预测值进行比较，结果表明运用最大熵原理的贝叶斯反分析得到的反演参数能够反映大坝实际渗流状况，结果是合理可行的。

参考文献：

[1] 杨杰．大坝安全监控中若干不确定性问题的分析方法与应用研究［D］．南京：河海大学，2006.

[2] 马文丽，苏怀智，游艇．基于最大熵理论的大坝安全监控指标［J］．水电能源科学，2011，29（6）：77－79.

[3] 吴乃龙，袁素云．最大熵方法［M］．长沙：湖南科学技术出版社，1991.

[4] 陈宝林．最优化理论与算法［M］．北京：清华大学出版社，2005.

[5] 梁国钱，郑敏生，孙伯永，徐家海．土石坝渗流观测资料分析模型及方法［J］．水利学报，2003，（02）：83－87.

资溪县山洪灾害特征与防治对策研究

王小笑，傅　群

江西省水利科学研究院

摘　要： 本文介绍了江西省资溪县暴雨山洪的特性及存在的问题，通过研究山洪灾害的历史及现状，对资溪县的山洪灾害危险区进行划分，并针对山洪灾害提出防治对策。

关键词： 山洪灾害；防治对策；预警系统

1　概况

资溪县位于江西省东部闽赣交界的边缘山区，大体呈东南高、西北低的趋势。全县总面积为 $1251km^2$，其中山地面积占总面积的 83.1%。境内地跨信江、抚河两大流域。东部以泸溪为主流，属信江水系，境内控制流域面积 $837km^2$；西部以许坊河为主流，属抚河水系，境内控制流域面积 $414km^2$。境内多年平均降水量 1929.9mm。受季风和地形影响，降水多集中在每年的 4～7 月，降水量占全年总降水量的 67.1%。

2　暴雨山洪特性

资溪县地处武夷山脉西麓，属山区，地形复杂。山洪为典型暴雨洪水，洪水出现的季节特性与暴雨出现的季节特性相同，大洪水主要由锋面、气旋雨及台风雨形成。由于小流域山体陡峭，坡度较大，暴雨来时，极易形成量大而集中的地表径流，使溪水在短时间内暴涨，形成山洪，从而造成山体滑坡、崩塌、泥石流等灾害。资溪县小流域的山洪具有以下特点：

（1）突发性。由于暴雨强度大，山丘区坡陡谷深、高程起伏大，河道比降大，因此，汇流迅速，洪水涨势猛，极易突发成灾，从降雨到山洪形成一般只需几个小时，最短的甚至仅 1h，往往防不胜防，预测预报难度大。

（2）区域性。由于暴雨的区域性加上地形、地貌、地质特点，一般中上游发生山体滑坡、崩塌、泥石流等灾害的概率明显大于下游河谷地区。

（3）季节性。由于降水时空分布不均，年际和年内变化较大，资溪县的降水主要集中在 4～7 月，多以暴雨形式出现，暴雨在年内具有明显的季节性，因此山洪及山洪灾害的发生也就相应地具有明显的季节性。另外，7—9 月，受台风影响，也可能产生局部的强降雨引发山洪灾害。

本文发表于 2011 年。

（4）破坏性。山洪来势凶猛，致使其危害性、破坏性极强。山洪成灾很快，顷刻之间，就可造成房屋倒塌、耕地被毁、公路中断甚至溪河改道，严重危及人民生命财产安全。其破坏性强的突出表现是造成人员伤亡和基础设施损坏严重，恢复难度很大，有的甚至具有毁灭性。

3 历史山洪灾害概况

资溪县境内的溪涧均属于源头水流，集雨面积不大，汇流时间较短。洪峰与暴雨同频率遭遇，过程的持续时间较短，洪水来势凶猛，退却迅速，若遇山洪暴发，伴随岗崩、滑坡、泥石流，容易瞬间酿成灾害。历史上资溪县多次发生山洪灾害，特别是 2010 年 6 月 19 日，遭遇罕见特大强降雨，引发特大山洪灾害，造成全县交通、供电、供水、通讯全部瘫痪，引发山体滑坡，道路冲毁，县城及多个乡镇被淹，全县 12 个乡（镇、场）受灾，受灾人口 10.65 万人，基础设施损坏严重。

4 山洪灾害防治现状

资溪县山洪灾害防御尽管采取了一系列措施，取得了一定效果。但总体上看，资溪县对山洪灾害特别是小流域洪水还缺乏一个完整、系统、科学的防治体系：

（1）山洪灾害防治难度大。对小流域容易形成的洪水灾害防治缺乏系统的规划和研究，对小流域洪水灾害缺乏科学的预测和积极主动的预防、抢险措施。

（2）资金投入不足，工程建设标准低，基础设施薄弱，防灾能力低。

（3）山洪灾害预警预报系统不够完善。由于站网密度不足，预报和信息传递手段落后，气象、水文预报预见期短，时效性和预报精度都不能满足灾害防治要求。

（4）山洪灾害防治体系和管理体制不够完善。

5 山洪灾害危险区的划定

危险区是指受山洪灾害威胁的区域，一旦发生山洪、泥石流、滑坡，将直接造成区内人员伤亡以及房屋、设施的破坏。危险区一般处于河谷、沟口、河滩、陡坡下、低洼处和不稳定的山体下。

通过对资溪县历史山洪灾害发生区域的调查，并结合本地区的气候和地形地质条件、人员分布等，资溪县山洪灾害危险区主要分布范围（见图 1）：处于山洪灾害危险区的集镇、村落和人口居住集中的区域，重点是滑坡点和泥石流危险区；处于河沟两侧、河口顶冲和地势低洼的集镇、村落；处于工程建设项目附近的区域及开矿区周围的群众。

6 山洪灾害防治对策

6.1 加密水雨情站网监测密度

资溪县现有自动雨量站 5 个，自动水位雨量站 2 个，结合已有站点和气象部门站点情况，通过适当加密水雨情站网监测的站点数量，使监测网密度达到 20～50km^2/站，可满足全县监测预警的需要。

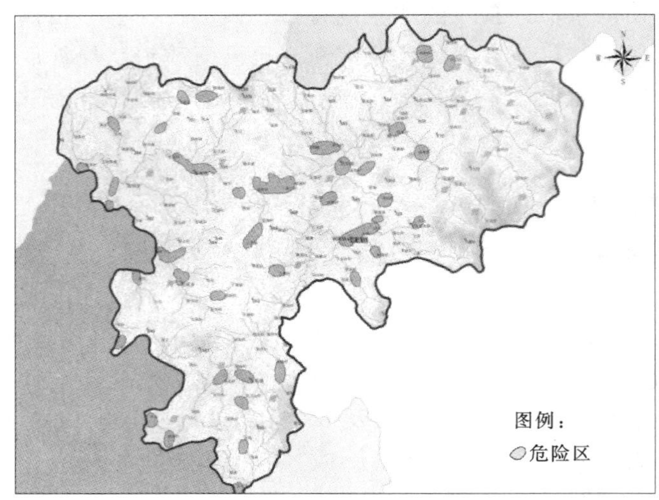

图 1　资溪县危险区调查分布图

6.2　保障监测预警信息传输

为确保县防汛指挥中心准确、实时接收到监测站水雨情信息，可采用两条途径接收监测站发来的数据，一是通过抚州市水情分中心与县防办的防汛专网，将市水文分中心接收到的数据转发到县防汛指挥中心；二是通过托管服务器将数据通过无线方式转发到县防汛指挥中心。监测站信息传输流程如图 2 所示。

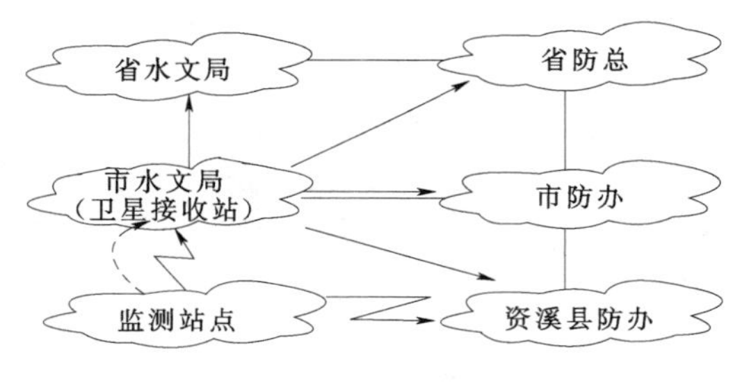

图 2　资溪县监测站信息传输流程图

通过上述两条路径接收监测站数据，可互为备份，使网络安全性得到保障。

6.3　加紧山洪灾害预警体系的建立

6.3.1　预警信息服务系统

建立预警信息服务系统有利于县防汛指挥中心全面掌握山洪灾害信息，并对信息进行加工、处理、交换、发布等综合业务处理。系统通过提供实时水雨情、工情、社会经济等信息服务，实现山洪灾害预警信息分析、发布、响应，是一套集山洪灾害监测、预警和应急响应于一体的山洪灾害防御指挥决策体系（见图 3）。

6.3.2　预警信息通信方式

根据资溪县山洪灾害特点，可用于预警信息传输的通信方式有电视、广播、互联网、电话、传真、移动通信、短信、报警器、锣鼓号等。各地可根据当地经济状况、现有通信资源条件以及各种通信方式的适用性，并考虑山洪灾害预警信息传输的时效性和紧急程度，选用适宜的通信方式组建山洪灾害预警信息传输通信网。对山洪灾害危险区可重点考虑采用无线广播系统实施预警，辅以铜锣、警报器等设备实施报警。

无线广播发送站：选用手机和电话两种不同通信方式，确保信息传输到无线发送站。

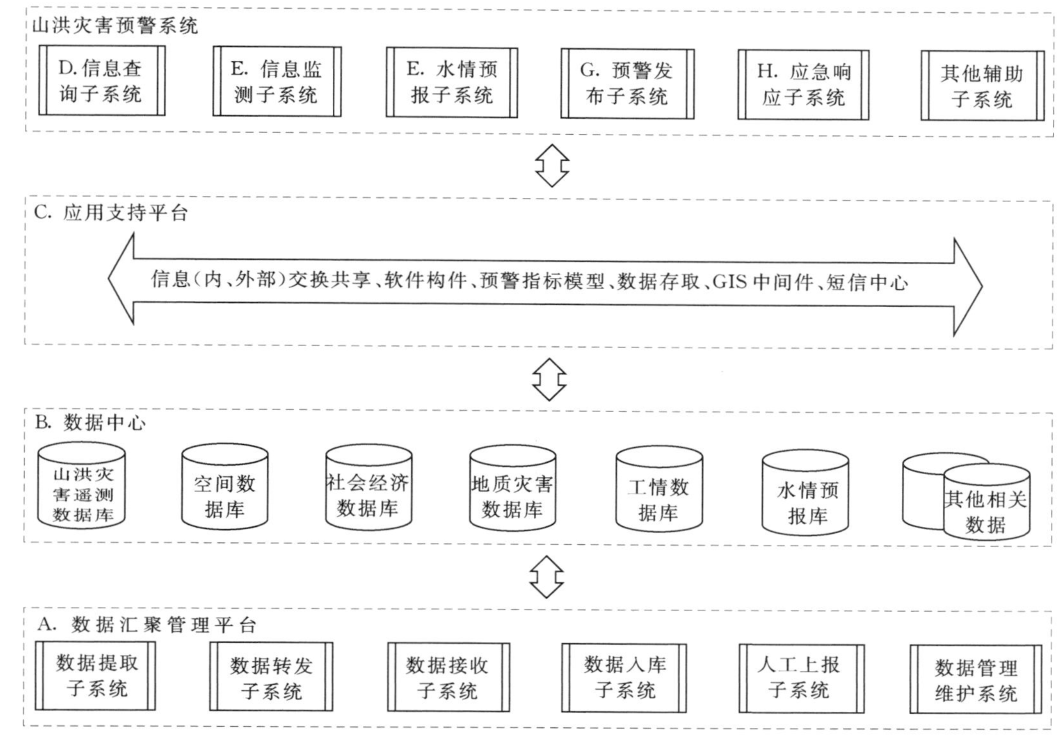

图 3　资溪县山洪灾害预警系统结构示意图

无线发送站接收到预警信息，通过发射机及时发送到接收机上，同时通过扩音机、喇叭，把预警信息传到附近区域，完成信息预警功能。

无线广播接收站：具备实时接收发送站发送的信息和人工语音输入预警信号功能。将接收到的或人工语音预警信息，通过扩音机、喇叭传到组、户。一旦广播系统遭到破坏，采用手摇报警器、铜锣等人工的方式进行预警信息传送。

6.4　落实防御预案的编制

资溪县行政区划设 5 镇、2 乡、5 场，全县总人口数为 12 万人。防御预案是防御山洪灾害实施指挥决策、调度和抢险救灾的依据，是基层组织和人民群众防灾、救灾各项工作的行动指南。资溪县防御预案编制需结合山洪灾害普查结果，以县级、乡（镇）级进行单独编制；村级预案以简要、易记、易实施的图表形式编制。

6.5　加强山洪灾害防御宣传、培训、演练

印刷《山洪灾害防治知识宣传手册》，用通俗易懂的语言，图文并茂，宣传山洪灾害防治知识。印刷《山洪灾害防治明白卡》，内容包括防治对象名称、各级负责人、避险地点、避险路线、联系电话等。制作山洪灾害防治常识光碟、录音带，内容包括山洪灾害的成因、危害、特点、防治组织机构、预警信号、避险注意事项、预警监测设施的保护等。设置警示牌、宣传牌，在划定的危险区、安全区以及转移线路、设立固定式混凝土警示牌；在较大居民点设立宣传牌，公布防治山洪灾害工作的组织机构、防御示意图和宣传山

洪防治知识。组织责任人员和监测技术人员进行相应的培训，每年在危险区组织防治山洪灾害演练，使危险区群众熟悉预警信号、转移路线、临时安置点等，使山洪灾害防治工作家喻户晓。

7 结语

资溪县山洪灾害具有季节性、突发性、区域性、破坏性等特点，只有通过提高山洪灾害易发区水雨情信息采集、传输和洪水预警预报能力，加强应急、预警响应体系建设，完善山洪灾害防御预案，强化群测群防体系，大力宣传防御知识等措施，才能有效地提升山洪灾害防御水平和防汛减灾服务水平，最大限度地减轻灾害带来的损失，确保当地人民群众的生命财产安全，促进区域经济社会的可持续性发展。

参考文献：

[1] 富海文，陈彬，吴晓翔．临安市山洪灾害防御现状分析及对策研究 [J]. 水利发展研究，2008，(12)：55-57.

[2] 王锁云，张金祥．岐山县山洪灾害状况与防御对策初探 [J]. 陕西水利，2010，(2)：125-128.

[3] 张茂省，薛富平，王晓勇．陕西省山洪灾害特征及防治对策 [J]. 西北地质，2004，(3)：96-101.

江西省水库大坝注册登记管理的实践及其思考

傅琼华[1,2]，平其俊[3]

1. 江西省水利科学研究院；2. 江西省大坝安全管理中心；3. 江西省水利厅

摘　要： 水库大坝注册登记工作是对水库安全监管的重要措施，制定明确的注册实施方案和工作要求，规定注册登记的范围、职责分工、工作程序、时间要求和工作方法，强调各级水行政主管部门对各水库大坝注册登记信息审核，保障了江西省水库大坝注册登记工作顺利进行，达到了水库大坝安全监管工作科学规范化管理的目的。

关键词： 水库大坝；注册登记；信息核实

水库大坝注册登记的目的是掌握水库大坝的安全状况，是加强水库大坝安全管理和监督的基础工作，通过注册登记，水行政主管部门不仅可以准确掌握水库大坝的数量、安全状况和基本情况，还可以为水库运行调度提供准确、合理的基础资料。同时，也是促进水库安全管理与监督法制化、规范化、科学化和信息化的一项重要工作，是加强水利基础设施管理的重要措施。随着近年病险水库除险加固工作的顺利开展，水库大坝的安全类别发生了较大变化。开展水库注册登记工作，对及时掌握水库大坝安全现状，加强大坝安全监管，十分必要和迫切。为此，江西省水利厅于 2006 年 6 月开始，对全省行政区域范围内已建成的小（2）型及以上水库（电站）大坝进行了新一轮的注册登记工作。

江西省大坝安全管理中心作为全省水库大坝注册登记机构，采用制定明确的水库大坝注册登记工作方案、规定注册条件、加强注册信息审核、注重登记审批等工作方法，通过注册登记申报、审核、登记发证、公布结果等工作程序，截至 2007 年 12 月 30 日，已建成的小（2）型及以上水库（电站）大坝，全省上报申请注册水库共有 9799 座。经审核，具备注册登记条件已注册的水库共有 9716 座，另有 83 座水库因不具备注册登记条件，未予注册。

1　注册登记工作思路

江西省是水库大省，水库数量众多，分布面广，如何实事求是、扎实地做好注册登记工作，尽量使登记工作做到数据准确、少漏登、不错登、不重复，是本次注册登记工作的重点。

1.1　编制工作方案，明确目标任务

根据水利部《水库大坝注册登记办法》和《江西省水库大坝注册登记办法》等有关规

本文发表于 2009 年。

定，结合江西省实际，制定了《2006年水库大坝注册登记工作方案》，明确了目标任务、职责分工、申报范围、申报要求、工作程序、水库注册编码、注册条件和工作措施，并制定了详细的水库注册登记表及填报说明，为全省水库大坝注册登记工作起到了指导性作用。

1.2 加强领导，落实责任

明确各级水行政主管部门要加强对大坝注册登记工作的领导，落实主管领导和具体工作部门及专门工作人员工作责任，按照全省大坝注册登记工作的总体要求，制订严密的工作计划。对本行政区域内水库基本情况进行一次全面的核查，加强与水库管理单位，特别是加强与本行政区域内中央、省属企业，省直有关部门，私营业主水库管理单位的联系与沟通，督促和指导水库管理单位做好大坝注册登记申报工作，确保工作不留“盲区”，不遗漏一座水库。

1.3 强化培训力度，规范注册登记

为确保注册登记工作的规范化和科学化，提高注册登记工作质量和效率，以及确保申报资料的完整性和准确性，举办了水库大坝注册登记培训研讨班，对全省11个设区市水利（水务）局负责水库大坝注册登记的部门负责人和具体负责该项工作的技术人员进行了大坝注册登记相关业务培训，规范了注册申报及注册登记表填写标准。

1.4 分级负责，逐坝审核，确保注册登记资料准确

注册登记工作采取分级负责、属地管理的方式进行。江西省大坝安全管理中心为全省水库大坝注册登记工作部门，具体负责全省水库大坝注册登记汇总和指导工作，负责全省大中型水库大坝注册登记工作，并向省级水行政主管部门报告水库大坝注册登记工作情况。

各设区市水行政主管部门负责注册登记本行政区域内小（1）型水库，县（市、区）水行政主管部门负责注册登记本行政区域内小（2）型水库，并对注册登记结果进行汇编、建档，逐级上报大坝中心核定备案。

1.5 认真做好“四无”电站水库大坝注册登记工作

对于暂不具备注册登记条件已经完工并投入运行的“四无”电站，要求其大坝注册登记由业主单位按照大坝现状进行申报。强调各级水行政主管部门要按照水利基本建设程序有关规定督促业主单位并限期完善项目审批和验收工作，必要时可以组织专家对电站大坝的安全状态进行认定。各级大坝注册登记机构依据有关项目审批、竣工验收或大坝安全状态认证等材料并经审核后办理注册登记。

2 注册登记工作方法

水库大坝注册登记按申报、审核、登记发证和公布结果等工作步骤进行。

2.1 明确申报范围

全省行政区域已建成运行的小（2）型及以上水库（电站）大坝（以下统称为“水库大坝”）必须申报注册登记。所指大坝包括永久性挡水建筑物以及与其配合运用的泄洪、输水等建筑物。

2.2 注册信息资料要求

水库大坝注册登记信息资料以2006年6月30日为期限。①已实施除险加固的水库，以其竣工验收报告资料填报；②正在实施除险加固但未经竣工验收的水库，以批复的初步设计报告填报，但在竣工验收后一个月内应向有关大坝注册机构申报办理注册变更，申请换证；③已进行安全鉴定的水库，以安全鉴定资料填报；④正常运用的水库，以水库工程现状数据填报。

2.3 注册登记主要内容

注册登记表是全面准确反映水库基本信息及安全状况的主要依据，因此，分别对大中型、小型水库制定了统一的水库大坝注册登记表，表格内容主要包括：水库基本情况、水文特征、三大建筑物特征值、调度运用、工程效益、安全监测、管理机构、大坝安全鉴定、加固扩建、大坝安全状况及各级管理单位审查意见。注册表一式四份，经各级主管部门签章后分别存于省大坝中心、市级及县级水行政主管部门、申报单位归档保存。

注册登记证是水库对外展示和长期保存的凭证，统一由江西省水利厅颁发。分正本、副本二式，正本备有水库大坝注册复审、变更、注销登记表，水库管理单位存档保留；副本上墙展示。

2.4 资料准确性审核

各市（县）水行政主管部门受理申报单位注册登记申请后，及时对申报单位提供的材料进行核实。对不合格的申报材料，告知申报单位进行补充和完善；经审查合格的，提出并签署审查意见。

大中型水库的注册信息资料由省大坝中心技术人员进行逐座审核；对于小型水库，以2003年“江西省病险水库除险加固规划”普查统计数9268座为基准数进行审核，对增加的12座小（1）型及减少的5座水库，逐座进行了大坝建设、竣工时间、建设审批文件等资料核实；对增加的570座小（2）型及减少的170座小（2）型，提出审核原则要求，由各设区市级水行政主管部门进一步核实确认。

各县（区）水行政主管部门完成本区域内水库大坝注册登记申报材料审核后，按照注册登记权限，将本行政区域内水库大坝申报材料汇总并上报至上一级水行政主管部门审查。各设区市水行政主管部门完成本行政区域内大中型水库注册登记申报材料的审核、小型水库注册信息汇总，并逐级审查签章报省大坝中心汇总。

2.5 注册审批条件

（1）大中型水库大坝注册登记由省级水行政主管部门审批，审批条件为符合下列条件之一：

1）具有批准的设计文件和竣工验收文件。

2）具有“三查三定”结论。

3）具有审批权限的水行政主管部门的批准文件。

4）水库大坝安全鉴定结论。

经省水行政主管部门审定，第一批符合注册登记条件的大中型水库为249座。

（2）小型水库注册由各设区市按以下原则进行审定，经各设区（市）审批准予注册的小型水库，上报至省大坝管理中心审查备案。

小型水库注册原则：

1）2003年规划普查（9268座）已认定的小型水库，经各设区市审定后，可准予注册登记。

2）对新增的小（1）型水库（122座），按以下原则进行注册登记：①1989年12月底前建设的55座水库，主要为扩建或原为电站或其他部门管理，且大都已运行20年以上，考虑水库建设年限已久，经各设区市按有关规定审定后，可准予注册登记。②1990—1999年间建设的12座，因建设较早，按基建程序增补相关批文有一定难度，应进行全面大坝安全鉴定，并经各设区市审定后，可准予注册登记。③2000年1月后建设的55座水库，应按建设程序补齐相关批文及竣工验收手续，并进行全面的大坝安全鉴定，经各设区市水行政主管部门验收或大坝安全鉴定并审定后，准予注册登记。

3）对新增的小（2）型水库（570座），按分级管理原则，由各设区（市）参照以上小（1）型水库注册登记原则，进行审核注册登记。

2.6 及时公布注册信息，接受社会监督

水库大坝注册登记信息核实审定后，由省大坝安全管理中心颁发了统一印制的注册登记证，注册登记工作经省水利厅验收后，以省水利厅文件形式，公布全省第一批注册登记大中型水库名单，共249座水库，其中大型25座，中型224座；14座中型水库因缺乏竣工验收批复等资料，不具备注册条件，暂不准予注册。

按注册登记分级负责制，小（1）型水库的注册登记发证工作由各设区市水行政主管部门负责审批公布，小（2）型水库的注册登记发证工作由各县（市、区）水行政主管部门负责审批公布。经各设区（市）审定批复：小（1）型水库符合注册登记条件的水库共1426座，26座不具备注册条件的，暂不注册登记；小（2）型水库符合注册登记条件的水库共8047座，43座不具备注册登记条件的，暂不注册登记。

同时明确规定，未注册登记的水库（电站）蓄水属于违章运行，水库（电站）管理单位应根据有关规定，尽快完善相关程序，及时办理水库注册登记。在水库未注册登记之前应空库运行，确保水库下游人民生命财产安全。

3 思考与探讨

这次水库注册登记工作吸取以往的经验，制定了注册实施方案和工作要求，明确规定了注册登记的范围、职责分工、工作程序、时间要求和工作措施，强调各级水行政主管部门对各水库注册登记信息审核，各司其职，各负其责，强化安全监管责任意识，对做好该项工作起到了关键作用。但有的水库管理单位对注册登记工作认识不够，尚有部分已建成运行的水库没有申报注册；有的水行政主管部门对不具备注册条件暂不准予注册的水库监管不足；注册信息更新资料缓慢；注册登记信息化、现代化程度不高等。

3.1 加强注册登记，逐步建立水库运行许可证制度

水法、行政许可法明确规定直接涉及国家安全、公共安全、生态环境保护以及直接关系生命财产安全等特定活动需要按法定条件批准。随着社会主义市场经济体制建立，水库的责任主体发生了变化，水库所有权由过去单一的国有、集体所有转变为国有、集体、私有、股份制等多种形式。水库的安全管理也由过去的政府全面负责，转变为政府监管、水

库所有者负责的形式。而注册登记是政府监管的必备基础，现有法规也有明确规定，所有已建成的水库（电站）必须到水行政主管部门进行注册登记，通过大坝安全鉴定为一类、二类坝的水库方可正常运行；鉴定为三类坝的水库应限制蓄水或空库，不能正常运行。因此，建议各级水行政主管部门应制定有约束力的办法，对尚未申报注册的水库向公众公布其安全信息，并要求其必须空库运行，以确保水库运行安全。特别是对以发电为主的私营电站，要督促各级水行政主管部门及当地政府，加大水库电站监管力度，督促水库电站及时申报注册登记，纳入正常管理，并逐步建立“水库运行许可证”发放制度。

3.2 强化公共安全意识，健全公共安全信息定期发布制度

水库安全涉及公共安全，按《水库大坝安全管理条例》水库大坝在竣工验收后，应每隔6～10年进行一次安全鉴定。已注册登记的水库大坝完成扩建、改建、除险加固的，或经批准升级、降等运行的，或大坝安全鉴定类别发生变化的，应在此后3个月内向原注册登记机构办理变更登记和换证，对不进行变更、换证的水库应有相应的制约措施。因此，水库大坝注册登记是一项长期动态管理的工作，建议水行政主管部门建立水库大坝注册工作长效运行机制，以确保水库大坝注册数据准确无误，及时可靠。同时，应实行“年度大坝安全报告制度”、制定规范的《运行、维护和监测手册》《水库安全管理应急预案》等。同时建立水库安全信息定期发布制度，水行政主管部门会同水库主管部门定期（如每年汛前）对水库安全进行检查评估，及时会同政府向公众发布水库安全信息。

3.3 完善管理信息化建设，提高水库大坝安全信息化管理水平

水库安全动态监管是一项长期的、动态的工作，集中注册登记只是水库大坝系统化管理的开始，应有专管部门和专门人员负责此项工作，常抓不懈，及时掌握水库安全的变化动态。而传统的水库注册登记方式，更新缓慢，准确率低，建议开发相应的注册登记信息管理系统，实现水库注册登记的网络化管理，直接在网上进行查询、统计数据以及办理申报、变更和注销等手续，为水库大坝安全管理与监督提供准确的基础信息。

参考文献：

[1] 孙继昌．中国的水库大坝安全管理［J］．中国水利，2008（20）．

[2] 施伯兴，吕金宝．水库大坝注册登记［J］．中国水利，2008（20）．

[3] 李雷．我国水库大坝安全管理发展趋势和挑战［J］．中国水利，2005（8）．

[4] 施俊跃．关于《水库大坝安全管理条例》修订若干问题的探讨［J］．中国农村水利水电，2006（10）．

基于.Net技术的江西省暴雨洪水计算程序模型开发应用

雷　声[1]，闵青红[2]，王小笑[1]

1. 江西省水利科学研究院；2. 江西省水电工程局

摘　要：根据《江西省暴雨洪水查算手册》，采用面向对象编程方法，利用.Net技术开发出江西省暴雨洪水计算模型，支持二次开发。模型整合了该手册所有的查算表格，并推导出P－Ⅲ型频率曲线、时段单位线的计算方法，应用到Web网站，获得了广泛应用。

关键词：.Net技术；江西；暴雨洪水计算；模型；P－Ⅲ型频率曲线

1　概况

江西省水文局1986年编写了《江西省暴雨洪水查算手册》（以下简称《手册》）。该《手册》在现阶段主要为江西省水文资料短缺的地区，作为中、小型水利水电工程（一般用于集水面积在1000km^{2}以下的山丘区）进行安全复核及新建工程设计洪水计算的依据。《手册》推荐了瞬时单位线法和推理公式法，供工程设计人员结合工程具体情况选用。

江西省水利水电工程数量众多，仅各类水库就有9000多座，占全国1/9，其中许多中、小型水利水电工程处于我省水文资料短缺地区，利用暴雨推求设计洪水成为新建工程设计和已建工程安全复核的重要内容。《手册》得到了各级水利部门的广泛应用。在实际工作中，水利部门很多设计和科研单位都有自己的计算程序，但是这些程序五花八门，有的是Excel计算，有的是早期Fortran语言编写的DOS程序，很多都需要大量的手工输入查表数据，适用面小，效率低，且单位之间程序互不公开。所以当前我省并没有一套标准化高、通用性强、界面可视化程度高、能省去大量查表工作的高效率洪水计算程序。

江西省大坝安全管理中心根据相应规范和《手册》，基于当前流行的.Net2.0技术开发出江西省暴雨推求洪水计算模型。模型利用当前流行的面向对象软件开发方法，采用统一的接口标准开发封装成标准dⅡ类（动态链接文件）。模型把工程作为一个对象并赋予各种属性，如工程名称、设计频率、流域面积等，把计算过程封装成公用函数或类，可以被各种高级编程语言如VC，C＋＋、VB、Delphi等调用，二次开发成单机或Web程序，现已根据该模型做成网页形式供水利工作者使用。

在功能上，模型推导出各类查算表格经验拟合公式（如点面折算系数、降雨径流折算

本文发表于2008年。

关系表等)、P-Ⅲ型频率曲线、时段单位线的计算公式，省去了大量的查表、试算过程。计算一个水库的洪水，用户只需输入《手册》中查得的几个基本参数便可以计算，计算过程由几天时间缩短到几分钟，结果以图、表形式输出，并可导入到 Excel。

2 模型开发

由于《手册》推求暴雨洪水的特点是计算步骤相对固定，其主要工作量为各种表格的查算和繁琐的数据整理。对于《手册》中步骤明确，或已经附有公式的计算过程，由于易于程序化本文不作赘述，如暴雨雨型分配、净雨过程计算、瞬时单位线法中的汇流参数计算、推理公式法概化五点折腰多边形过程线的计算等。以下仅介绍各种图表查算模块的实现。

2.1 曲线的拟合与插值

暴雨洪水计算过程中，部分参数需要通过关系查算图表插值才能得到，且《手册》中没有提供相应计算公式，例如暴雨点面折算系数曲线、不同计算分区的降雨径流（$H+P_a\sim R$）关系表，降雨下渗关系表等；还有部分参数是计算过程中需要插值，例如推理公式法中地面设计洪峰流量和汇流时间的试算。以往这些参数多为人工插值，本文通过引入常用的数学算法，实现参数计算函数化。例如计算点面折算系数时，可以先等密度读出暴雨笼罩面积 F 和点面折算系数 a 关系表，计算折算系数 a 时可利用 B3 样条或拉格朗日算法插值计算，该方法的特点是精度较好，但不能得出经验公式，程序需附带数据库，不利于程序模块化；也可以利用最小二乘法拟合出经验计算公式 a（F），精度通过读数密度和最小二乘法中多项式项数控制，效果较好。

2.2 P-Ⅲ型频率曲线的模比系数 K_p 计算

P-Ⅲ型频率曲线的模比系数 K_p 计算较为复杂，《手册》通过列出常用频率下各种 C_v 对应的 K_p 值，由用户进行插值查算。该方法有一定的局限性：一是表中所列频率个数有限，且仅限于 $C_s=3.5C_v$ 情况，不利于各种频率和 C_v、C_s 下 K_p 的计算；二是需要人工读取，降低了程序效率。

通过下述常规求解方法可以实现任意频率、任意 C_v、C_s 情况下的 K_p 计算，并能保证计算精度。

频率计算中的 P-Ⅲ型频率曲线可以通过 P-Ⅲ型分布的概率密度函数式表示：

$$p(x)=\beta^2/\Gamma[a\cdot(x-a_0)^{a-1}\mathrm{e}^{-\beta(x-a_0)}](a>0,x>a_0) \tag{1}$$

其中 $a=4/C_s^2,\beta=2/[E(\xi)\cdot C_v\cdot C_s],a_0=E(\xi)\cdot(1-2C_v/C_s)$。

经积分可以求得下式：

$$p=P(\xi\geqslant x_p)=\beta^2/\Gamma[a\cdot\int_{x_p}^{\infty}(x-a_0)^{a-1}\mathrm{e}^{-\beta(x-a_0)}\mathrm{d}x] \tag{2}$$

将 ξ 标准化，即令 $\Phi=[\xi-E(\xi)]/\sigma$，则 $x=E(\xi)\cdot(\phi C_v+1)$，这里 x、ϕ 分别是 ξ、Φ 的取值。

式（2）可以进行下述一系列转化：

$$p=P(\xi\geqslant x_p)=P(\Phi\geqslant\phi_p)=a^{a/2}/\Gamma\left[a\cdot\int_{\phi_p}^{\infty}(\Phi+\sqrt{a})^{a-1}\cdot\mathrm{e}^{-\sqrt{a}(\phi+\sqrt{a})}\mathrm{d}\phi\right] \tag{3}$$

令 $N=\sqrt{a}\ (\Phi_p+\sqrt{a})$，则由式（3）可以推得下式：

$$p = 1/\Gamma[a \cdot \int_{N_p}^{\infty} N^{a-1}\mathrm{e}^{-N}\mathrm{d}N] \tag{4}$$

令 $M = \int_{N_p}^{\infty} N^{a-1}\mathrm{e}^{-N}\mathrm{d}N$，又 $\Gamma(a,N) = \int_{0}^{N_p} N^{a-1}\mathrm{e}^{-N}\mathrm{d}N/\Gamma(a)$，则

$$\Gamma(a)\cdot\Gamma(a,N) = \int_{0}^{N_p} N^{a-1}\mathrm{e}^{-N}\mathrm{d}N = \Gamma(a) - M$$

可得 $M=\Gamma(a)\cdot[1-\Gamma(a,N)]$。

代入式（4）得，$p=1/\Gamma(a)\cdot\Gamma(a)[1-\Gamma(a,\sqrt{a}(\Phi_p+\sqrt{a}))]=1-\Gamma[a,\sqrt{a}(\Phi p+\sqrt{a})]$

即 $\Gamma(a,\sqrt{a}(\Phi_p+\sqrt{a}))=1-p$，则

$$\Gamma[4/C_s^2, 2/C_s^2\cdot(\phi_p+2/C_s)]=1-p \tag{5}$$

$\Gamma(a,\ N)$（其中 $a>0$，$N>0$）称为不完全伽马函数。由于 $\Gamma(a,\ N)$ 在 $N\in(0\sim\infty)$ 时，单调递增，故当 a、$\Gamma(a,\ N)$ 的值已确定时可以采用二分法求出 N，进而求得 ϕ_p 值。

$$K_p=\phi_p C_v+1 \tag{6}$$

式中 C_v、C_s、p 均为已知值，根据式（5）、式（6）可编制函数 $K_p(C_v,\ C_s,\ p)$。通过本方法计算结果跟《手册》中 K_p 值表进行比较，结果基本一致，最大误差不超过 1‰，可以满足工程设计要求。

2.3 时段单位线的计算

瞬时单位线推求洪水时，要大量用到各种参数下的时段单位线，《手册》中的方法为先求出流域单位线的调节系数 n，调蓄系数 k，再查算《时段单位线用表》，该方法繁琐且局限性大：一是工作量大，每种工况都对应一组时段单位线，特殊时段的单位线还需要转换；二是所列 n、k 参数个数有限，不能查算任意有意义情况下的单位线，有时还要插值计算；三是人工查表容易引起误差，直接影响最终结果。

实际上当参数 n、k 为已知时，无因次时段单位线 $u(\Delta t,\ t)$ 可以根据瞬时单位线和 S 曲线（瞬时单位线的积分曲积）转换。瞬时单位线是指流域上分布均匀，历时趋于无穷小，强度趋于无穷大，总量为一个单位的地面净雨在流域出口断面形成的地面径流过程线。

瞬时单位线的数学方程：

$$u(0,t)=\frac{1}{k\Gamma(n)}\left(\frac{k}{t}\right)^{n-1}\mathrm{e}^{-\frac{1}{K}} \tag{7}$$

式中：$u(0,\ t)$ 为 t 时刻的瞬时单位线纵高；n 为调节系数；k 为调蓄系数；$\Gamma(n)$ 为 n 的伽马函数。

按 S 曲线的定义有：

$$S(t) = \frac{1}{\Gamma(n)}\int_{0}^{t/k}\left(\frac{t}{k}\right)^{n-1}\mathrm{e}^{-\frac{t}{k}}\mathrm{d}\left(\frac{t}{k}\right) \tag{8}$$

式（8）中，t 为以 $t=0$ 起算的时刻。以不同的 t 代入上式积分，就可得到 S 曲线。将以 $t=0$ 为起点的 $S(t)$ 曲线向后平移一个 Δt 时段，即可得 $S(t-\Delta t)$ 曲线，两条 S 曲线的纵坐标差：

$$u(\Delta t,t)=\frac{1}{\Delta t}[S(t)-S(t-\Delta t)] \tag{9}$$

式中：Δt 为时段长，一般可取 1h、3h、6h、12h、24h。

即为时段为 Δt 的无因次时段单位线，可见时段单位线关键在于 S 曲线的求积。

$S(t)$ 积分的精确求解比较复杂，利用矩形法、梯形法或抛物线法（辛普森法）均可获得近似积分解，其中抛物线法以较短的循环步长可获得较为精确的近似解，效果很好。

经复核模型计算结果跟《时段单位线用表》一致。

2.4 计算参数输入

为了简化操作，减少查表，提高效率，模型中需用户输入的参数被尽可能地减少，中间过程由模型完成。用户要输入参数只包括洪水推求方法、设计频率 P、产汇流分区、流域面积 F、主河道比降 J、河道长度 L、设计历时、计算时段长、暴雨均值 H、变差系数 C_v、偏差系数 C_s 等，其中大部分参数用户只需在下拉菜单中选择。

3 模型验证

模型开发完成后，对《手册》中的 2 个计算实例以及 20 多个水库进行复核，并利用该模型开发出江西省暴雨推求洪水计算程序 Web 网页，放在江西省大坝安全管理中心网站（http：//dam.jxsks.com）上供互联网用户免费测试和使用（见附图），并根据用户的建议对模型进行了修改和优化。网页发布后短短几个月内得到了吉安、九江、鹰潭、新余、上饶、宜春、赣州等市甚至外省市水利工作者的关注和好评，目前已有 376 座水利工程 3100 余次的计算量，反馈的精度和可靠度均可满足工程设计要求（表 1～表 4）。

表 1　　《手册》范例测试计算结果

水库名称	计算方法	流域面积/km^2	频率/%	24h 面暴雨量比较			洪峰流量比较		
				原报告/mm	本程序/mm	相差/%	原报告/($m^3\cdot s^{-1}$)	本程序/($m^3\cdot s^{-1}$)	相差/%
《手册》范例 1	瞬时单位线	161.1	1	258.2	258.01	−0.07	1152.0	1150.3	−0.15
《手册》范例 2	推理公式法	16.3	1	370.4	370.44	0.01	197.0	199.2	1.12
《手册》范例 3	瞬时单位线	36.1	1	272.9	272.77	−0.05	377.0	378.9	0.50
	推理公式法	36.1	1	272.9	272.77	−0.05	339.0	318.9	−5.93

注　《手册》范例 3 表 4－14 计算净雨时 $\sum R_{总}$ 查表出现错误引起洪峰流量误差。

表 2　　乐平县勤俭水库推理公式法测试计算结果

频率/%	24h 面暴雨量比较			洪峰流量比较		
	原报告/mm	本程序/mm	相差/%	原报告/($m^3\cdot s^{-1}$)	本程序/($m^3\cdot s^{-1}$)	相差/%
0.1	495.8	491.91	0.78	203.28	201.6	0.8
0.2	454.9	450.38	0.99	183.48	181.4	1.1

续表

频率/%	24h 面暴雨量比较			洪峰流量比较		
	原报告/mm	本程序/mm	相差/%	原报告/($m^3 \cdot s^{-1}$)	本程序/($m^3 \cdot s^{-1}$)	相差/%
1	357	355.63	0.38	135.46	134.4	0.8
2	314.7	314.1	0.19	114.45	113.9	0.5

注 原报告取值 K_p 偏大造成面暴雨量误差；流域面积 12.4km²。

表 3　吉安市白云山水库瞬时单位线测试计算结果

频率/%	24h 面暴雨量比较			洪峰流量比较		
	原报告/mm	本程序/mm	相差/%	原报告/($m^3 \cdot s^{-1}$)	本程序/($m^3 \cdot s^{-1}$)	相差/%
0.1	549.1	553.89	−0.87	3854.58	3871.4	−0.4
0.2	505.6	507.12	−0.30	3529.22	3525.3	0.1
1	398.1	400.44	−0.59	2650.49	2657	−0.2

注 流域面积 464.7km²。

表 4　铜鼓县大塅水库推理公式法测试计算结果

频率/%	24h 面暴雨量比较			洪峰流量比较		
	原报告/mm	本程序/mm	相差/%	原报告/($m^3 \cdot s^{-1}$)	本程序/($m^3 \cdot s^{-1}$)	相差/%
0.05	446.2	442.6	0.81	6647.87	6539.6	1.6
0.1	411.1	409.86	0.30	5928.57	5913.4	0.3
0.2	379.1	377.11	0.52	5321.85	5282	0.7
1	299.3	299.21	0.03	3860.53	3852.5	0.2
0.1	411.1	409.86	0.30	6116.32	6121.8	−0.1
0.2	379.1	377.11	0.52	5598.09	5578.6	0.3
1	299.3	299.21	0.03	4310.1	4305.4	0.1

注 流域面积 610.5km²。

4　结语

江西省暴雨洪水计算模型由于采用较为主流的面向对象编程方法实现，是一套标准化、通用化、自动化的计算模型，封装后支持各种高级编程语言的二次开发。模型整合了《手册》所有的查算表格，并推导出 P-Ⅲ型频率曲线模比系数、时段单位线的计算方法，可以同时计算一个或多个工程不同参数下的暴雨洪水，省去了大量的查表、试算过程，提高了效率。使用该模型开发的江西省暴雨推求洪水计算网站是我省乃至全国首个公开的暴雨洪水计算网站，界面友好，使用方便，速度快捷，图文并茂，结果可靠，在省内外得到大量应用。

参考文献：

[1]　江西省水文总站．江西省暴雨洪水查算手册［R］．1986，12.
[2]　长江流域规划办公室水文处，水利工程实用水文水利计算［M］．北京：水利出版社，1980.
[3]　张杨波．P-Ⅲ型频率曲线离均系数 ϕ_p 的一种简易求解方法［J］．甘肃水利水电技术，2002，12.
[4]　水利水电科学研究院水资源所，全国雨洪办．时段单位线用表［R］．1983，8.

基于流变统一格式的库岸堆积体斜坡变形监控模型及应用

吴海真[1,2]，吴晓彬[2]
1. 河海大学水利水电工程学院；2. 江西省水利科学研究院

摘　要：库岸堆积体斜坡受库水位变化、时效和降雨的影响显著，研究蠕变对堆积体稳定性的影响可为其准确预测、预报提供科学依据。基于岩土体流变统一格式的推导，结合修正的西原模型，提出多因素作用下的库岸堆积体斜坡变形监控模型，为其时效因子的选取提供了理论依据。实际应用表明，相对于传统的统计模型，该模型可明显提高回归精度，具有一定的实用价值和推广价值。

关键词：库岸堆积体斜坡；流变；变形监控模型

1　引言

在库岸堆积体斜坡的形成过程中，由于应力状态的变化，岩土体将发生不同方式、不同规模和不同程度的变形，并可能在一定条件下发展为失稳破坏[1-2]。库岸堆积体斜坡的运动过程是从变形演化发展到破坏的一个复杂动态力学过程，也是一个变形由量变积累到质变的发生发展过程。工程实践表明，库岸堆积体斜坡的变形一般要经历变形启动→时效变形快速发展→均匀蠕滑→变形趋稳（或加速变形）等几个阶段（如新滩滑坡等）[3]，这与滑床形状、滑带组成、滑体结构、环境因素以及运动历史等密切相关。其中，蠕变对滑坡的稳定性产生了重大影响，因此，研究蠕变对堆积体滑坡稳定性的影响，可为滑坡的准确预测、预报提供科学依据。以蠕变理论为基础的斜坡监控模型最早由日本学者斋藤迪孝于 1965 年提出，近 20 多年来，已有一些学者对松散堆积体的蠕变模型及滑动带土的蠕变特性等进行了试验和研究，但其中大多是基于各种数学处理方法和手段研究蠕变位移加速度阶段的预报，如各种生物增长曲线模型[4]、灰色位移矢量角模型[5]、协同预测模型[6]、时间序列预报模型[7]和分维跟踪预报模型等[8]，这些临滑预报模型因忽略了蠕滑的前期阶段，且时效因子的选择具有较大的主观性，缺乏理论验证，故均未从堆积体的运动特征出发给出其多因素作用下的时变监控模型，一般较难推广应用于实际工程。

本文在前人研究的基础上，推导出岩土体流变的统一格式，并将其作为时效影响因子，提出其多因素作用下的变形监控模型。实际工程应用表明，该模型的精度明显优于传

本文发表于 2007 年。

统监控模型，具有一定的实用价值和推广价值。

2 岩土体流变统一格式的推导

众所周知，松散堆积体既不是弹性体也不是塑性体，而是具有弹性、塑性和黏滞性的黏弹塑性体，即具有流变特性[3]。现有流变模型一般是基于基本流变元件（包括虎克弹性体、牛顿黏壶和圣维南刚性体）串联或并联组合而成。例如，Kelvin 模型由一个弹簧和一个粘壶组成；Burgers 模型是由一个 Maxwell 模型和一个 Kelvin 模型串联而成的组合模型，也称四元件模型等，常用来描述碎石土和粗粒料的流变特性[9-13]，该模型的组成如图 1 所示。

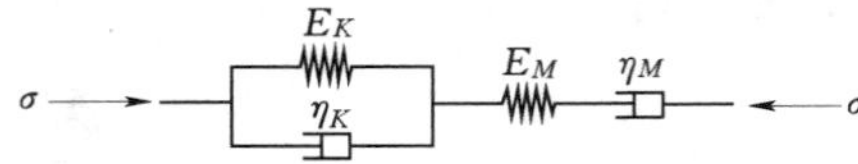

图 1 Burgers 四元件模型示意图

在恒载条件下，$\sigma=\sigma_c$，$\ddot{\sigma}=\dot{\sigma}=0$，由 Kelvin 体和 Maxwell 体的蠕变方程叠加可得蠕变方程为：

$$\varepsilon=\left(\frac{1}{E_M}+\frac{t}{\eta_M}\right)\sigma_c+\frac{1}{E_K}\left(1-\mathrm{e}^{-\frac{E_K}{\eta_K}t}\right)\sigma_c \tag{1}$$

式中：ε 为应变；σ 为应力；E_K 和 η_K 分别为压缩（或拉伸或剪切）模量和黏滞系数；E_M 和 η_M 分别为延时模量和黏滞系数。

传统的流变模型之所以不能描述岩土体加速蠕变阶段，就在于将岩土体视为理想的牛顿流体。实际上，岩土体的流变不仅具有普遍流体的特性，由于其结构的特殊性，还具有非牛顿流体的特性。现引入一种非线性黏滞阻尼器，该黏滞阻尼器所受应力与其蠕变加速度大小成正比，即 $\sigma=\eta_3\ddot{\varepsilon}$。该模型的示意图如图 2 所示。

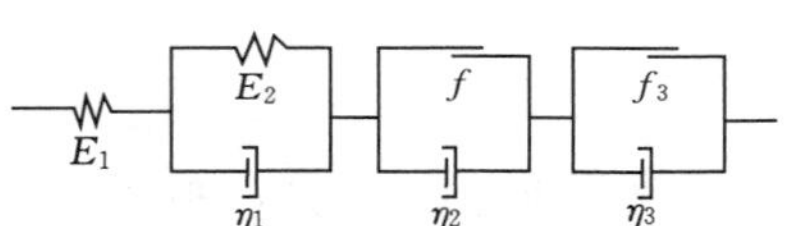

图 2 修正的西原模型示意图

略去推导过程，可得到修正西原模型的蠕变方程为[9]：

$$\varepsilon(t)=\begin{cases}\left(\dfrac{1}{E_1}+\dfrac{1}{E_2}\right)\sigma+\dfrac{\sigma-f}{\eta_2}t-\dfrac{\sigma}{E_1}\mathrm{e}^{\frac{-E_2}{r_1}t} & \sigma<f_3\\ a\mathrm{e}^{-\frac{E_2}{r_1}t}+bt^2+ct+d & \sigma\geqslant f_3\end{cases} \tag{2}$$

其中

$$a=-\frac{\sigma_c}{E_2}$$

$$b=\frac{\sigma_c-f_3}{2r_3}$$

$$c=\frac{\sigma_c-f}{\eta_2}$$

$$d=\left(\frac{1}{E_1}+\frac{1}{E_2}\right)\sigma$$

式中：σ 为应力；ε 为应变；E 为弹性模量；η 为黏性系数；t 为时间。

通常，岩土体黏弹性模型微分形式可表示为：

$$P\sigma=Q\varepsilon \tag{3}$$

其中

$$\begin{cases} P=p_0+p_1\dfrac{\partial}{\partial t}+p_2\dfrac{\partial^2}{\partial t^2}+\cdots+p_n\dfrac{\partial^n}{\partial t^n} \\ Q=q_0+q_1\dfrac{\partial}{\partial t}+q_2\dfrac{\partial^2}{\partial t^2}+\cdots+q_m\dfrac{\partial^m}{\partial t^m} \end{cases} \tag{4}$$

式（4）仅是针对一维状态下的均质各向同性介质而言，三维情况可从一维情况类推得到。类似弹塑性力学中的做法，将一点的应力状态分解为偏张量和球张量，相应的应变分解为偏应变和球应变，若假定体积应变为弹性的，则三维状态下的黏弹性本构模型可表示为：

$$\begin{cases} PS_{ij}=2Qe_{ij} \\ \sigma_{ii}=3K\varepsilon_{ii} \end{cases} \tag{5}$$

式中：K 为体积应变。

对式（5）进行拉普拉斯变换，考虑光滑化假定，得

$$\begin{gathered} \overline{P}\,\overline{S}_{ij}=2\,\overline{Q}\,\overline{e}_{ij} \\ \overline{\sigma}_{ii}=3\overline{K}\,\overline{\varepsilon}_{ii} \\ \overline{P}(s)=\sum_{i=0}^{n}p_i s^i \\ \overline{Q}(s)=\sum_{i=0}^{n}q_i s^i \end{gathered} \tag{6}$$

应用弹性-黏弹性对应原理，得到黏弹性剪切模量的拉普拉斯变换为：

$$\overline{G}(s)=\frac{Q(s)}{P(s)} \tag{7}$$

同理，系统模型蠕变规律的拉普拉斯变换为：

$$\overline{J}(s)=\frac{\overline{P}(s)}{s\,\overline{Q}(s)}=\frac{p_0+p_1s+p_2s^2+\cdots+p_ns^n}{s(q_0+q_1s+q_2s^2+\cdots+q_ms^m)} \tag{8}$$

若 $q_0+q_1s+q_2s^2+\cdots+q_ms^m=0$ 有 m 个相异的实根，则蠕变柔量为：

$$J(t)=A_1+A_2e^{A_3t}+A_4e^{A_5t}+\cdots+A_{2m}e^{A_{2m+1}t}=A_1+\sum_{i=1}^{m}A_{2i}e^{A_{2i+1}t} \tag{9}$$

由此可知，常用流变模型一般采用时间多项式函数和指数函数来描述岩土体的蠕变特性。本文基于上述分析，给出土体蠕变的通用简化数学表达式：

$$\varepsilon(t)=ae^{-Dt}+f(t) \tag{10}$$

式中：$f(t)$ 为蠕变持续时间 t 的多项式表达式（其中含回归系数）；a 和 D 为回归系数。

3　库岸堆积体斜坡变形监控模型的建立

库岸堆积体斜坡主要受库水位变化、温度和时效等因素的影响，其变形监控模型一般采用如下表达式[14-15]：

$$\delta = \sum_{i=1}^{3}[a_{1i}(H_u^i - H_{u0}^i)] + \sum_{i=1}^{m}[a_{2i}(H_{ui} - H_{ui0})] + \sum_{i=1}^{2}\left(b_{1i}\sin\frac{2\pi_{ti}}{365} + b_{2i}\cos\frac{2\pi_{ti}}{365}\right)$$
$$+ \sum_{i=1}^{m} d_i P_i + c_1\theta + c_2\ln\theta + a_0 \tag{11}$$

式中：δ 为堆积体位移；H_u 和 H_{u0} 分别为观测日、始测日所对应的库水深，即库水位与坡前底高程之差；H_{ui} 和 H_{ui0} 分别为观测日、始测日所对应的前期平均水深，这里 $m=4$，分别为前 1d、前 3d、前 7d 和前 15d 的平均库水深；a_{1i} 和 a_{2i} 为水压因子回归系数；b_{1i} 和 b_{2i} 为温度因子回归系数；P_i 为当天和前期平均降雨量因子，这里 $m=5$，分别为当天、前 1d、前 3d、前 7d 和前 15d 的累积降雨量；d_i 为降雨因子回归系数；$\theta=t/100$；t 为监测日至始测日的天数；c_1 和 c_2 为时效因子回归系数；a_0 为常数项。

本文基于库岸堆积体斜坡的蠕滑运动特性，结合上述岩土体流变统一格式的推导和修正的西原模型，提出其多因素作用下的变形监控模型如下：

$$\delta = \sum_{i=1}^{3}[a_{1i}(H_{1u}^i - H_{1u0}^i)] + \sum_{i=1}^{n}[a_{2i}(H_{1u}^i - H_{1u0}^i)] + \sum_{i=1}^{2}\left(b_{1i}\sin\frac{2\pi_{ti}}{365} + b_{2i}\cos\frac{2\pi_{ti}}{365}\right)$$
$$+ \sum_{i=1}^{m} d_i P_i + d_1\theta + d_2\theta^2 + d_3 e^{-D\theta} + a_0 \tag{12}$$

式中：d_1，d_2，d_3，D 为时效因子回归系数。

也可根据需要，时效分量取为指数项与时间的 3 次多项式之和：

$$\delta = \sum_{i=1}^{3}[a_{1i}(H_{1u}^i - H_{1u0}^i)] + \sum_{i=1}^{m}[a_{2i}(H_{1u}^i - H_{1u0}^i)] + \sum_{i=1}^{2}\left(b_{1i}\sin\frac{2\pi_{ti}}{365} + b_{2i}\cos\frac{2\pi_{ti}}{365}\right)$$
$$+ \sum_{i=1}^{m} d_i P_i + d_1\theta + d_2\theta^2 + d_3\theta^3 + d_4 e^{-D\theta} + a_0 \tag{13}$$

式中：d_1，d_2，d_3，d_4，D 为时效因子回归系数。

为叙述方便，把传统的多因素监控模型式（11）称为模型 1，而把本文基于岩土体蠕变统一格式建立的模型式（12）称为模型 2，式（13）称为模型 3。

4 工程应用

江西省某大（2）型水库库岸堆积体位于坝址左岸上游约 70～300m 地段，分布在高程 210.0m 以下至河床（高程 142.00m）一上窄下宽形似“鼻梁”状地段上，边坡产状 N501°E/NW∠25°～30°，底部横向宽度约 220.0m，纵向最大高度 120m，厚一般为 12～15m，最厚近 20m，总体积约 30 万 m^3。经勘探查明，堆积体物质组成以碎（块）石为骨架，内充填黏性土，碎（块）石、砾含量达 40%～50%，母岩成分为石炭系石英砾岩、砂砾岩和石英砂岩等；块径大小不一，最大可达 2m，最小为 2～5mm，一般为 0.2～0.4m，块体具棱角状。堆积体顶部为石炭系岩层（石英砾岩、砂砾岩和含砾砂岩），一般为弱风化～新鲜岩体，岩层反倾；中下部为震旦系变余细砂岩夹板岩，岩层产状 N1021°E/NW∠32°～41°，构造发育，主要由顺坡向层间挤压带（泥化夹层）、断层和卸荷张裂隙 3 组结构面构成稳定的不利组合面。

因该堆积体的稳定性对于大坝安全有重大影响，为监测堆积体的变形性态及运动特征，设有滑 1、滑 3～滑 8 和滑 10 等共 8 个外部变形监测点。规定：x 向水平位移以指向上游为正，反之为负；y 向水平位移以指向河床（垂直水流方向）为正，反之为负。垂直位移（z 向）监测点结合水平位移布置：以下沉为正，上抬为负。图 3 为堆积体位移观测点布置及水平位移空间分布图（包括主滑方位）。

为定量分析该堆积体的变形特性和影响因素，分别基于模型 1、模型 2 和模型 3 对其三向位移测值系列建立了统计模型，典型测点的逐步回归计算结果见表 1。限于篇幅，仅给出滑 1 测点 y 向水平位移的实测、拟合及残差过程线（测值系列为 1996 年 5 月 1 日至 2003 年 12 月 15 日），见图 4。

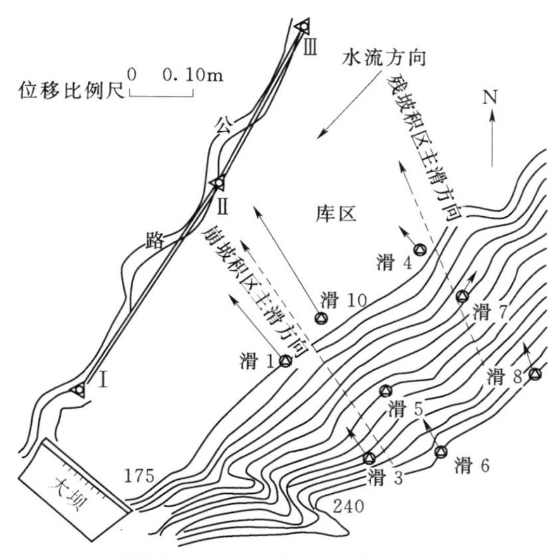

图 3 堆积体水平位移空间分布和主滑方位

由上述图表可以看出，相对于传统的统计模型（模型 1）而言，采用本文建立的统计模型（模型 2 和模型 3）可明显提高回归精度，相应回归残差平方和显著降低，复相关系数均有不同程度的提高。另外，从计算结果可以看出，对于具有明显蠕滑特性的库岸堆积体斜坡而言，一般时效分量中的时间因子采用 2 次或 3 次多项式已有相当的精度，能满足工程精度要求。

表 1　　多因素作用下堆积体斜坡典型测点回归模型计算结果

测点编号	模型名称	残差平方和	复相关系数 R	典型年年变幅时效分量比例/%
滑1（y 向）	模型 1	1720	0.981	77.90
	模型 2	826.2	0.993	82.05
	模型 3	508.1	0.998	83.50
滑 1（z 向）	模型 1	1761	0.892	79.48
	模型 2	766.2	0.957	83.40
	模型 3	433.3	0.994	85.08
滑 10（y 向）	模型 1	2374	0.970	79.60
	模型 2	937.5	0.988	86.51
	模型 3	622.1	0.996	88.13
滑 10（z 向）	模型 1	971.1	0.972	77.06
	模型 2	611.6	0.987	81.34
	模型 3	401.8	0.996	82.85

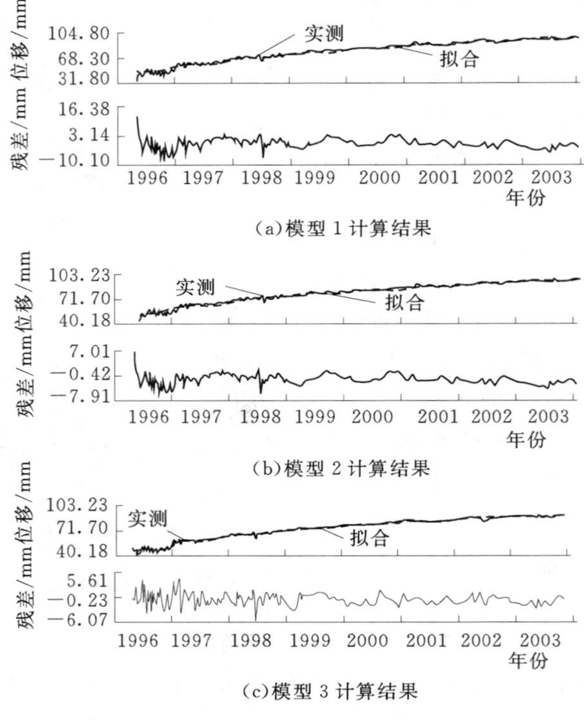

(a)模型 1 计算结果

(b)模型 2 计算结果

(c)模型 3 计算结果

图 4　滑 1 测点 y 向水平位移实测、拟合及残差过程线

5　结语

本文基于岩土体流变统一格式的推导，同时结合修正的西原模型，提出了多因素作用下库岸堆积体斜坡的时变监控模型，为其时效因子的选取提供了理论依据。实际应用表明，相对于传统的统计模型而言，本文建立的统计模型可明显提高回归精度，相应的回归残差平方和显著降低，复相关系数均有不同程度的提高，具有一定的实用和推广价值。

参考文献：

[1] 邓建辉，CHANDH，MARTINCD，等．一例由蓄水诱发的库岸边坡变形［J］．中国地质灾害与防治学报，2002，13（1）：21－24.

[2] 包太，刘新荣，税月．水位下降卸荷诱发库岸边坡快速失稳机理分析［J］．水文地质工程地质，2004，31（5）：7－11.

[3] 田陵君，王兰生，刘世凯．长江三峡工程库岸稳定性［M］．北京：中国科学技术出版社，1992.

[4] 文海家，张永兴，柳源．滑坡预报国内外研究动态及发展趋势［J］．中国地质灾害与防治学报，2004，15（1）：1－4.

[5] 王钊，刘祖德，林鲁生，等．边坡变形的灰色预测模型［J］．岩土力学，2000，21（3）：244－251.

[6] 黄志全，张长存，姜彤，等．滑坡预报的协同－分岔模型及其应用［J］．岩石力学与工程学报，2002，21（4）：498－501.

[7] 许东俊，陈从新，刘小巍，等．岩质边坡滑坡预报研究［J］．岩石力学与工程学报，1999，18（4）：369－372.

[8] 周创兵，陈益峰．基于相空间重构的边坡位移预测［J］．岩土力学，2000，21（3）：205－208.

[9] 孙钧．岩土材料流变及其工程应用［M］．北京：中国建筑工业出版社，1999.

[10] 殷建华．等效时间和岩土材料的弹粘塑性模型［J］．岩石力学与工程学报，1999，18（2）：124－128.

[11] 徐日庆，龚晓南．粘弹性本构模型的识别与变形预报［J］．水利学报，1998（4）：75－80.

[12] 王勇，殷宗泽．面板坝中堆石流变对面板应力变形的影响分析［J］．河海大学学报，2000，28（6）：60－65.

[13] 詹美礼，钱家欢，陈绪禄．软土流变特性试验及流变模型［J］．岩土工程学报，1993，15（3）：54－62.

[14] 吴中如．水工建筑物安全监控理论及其应用［M］．北京：高等教育出版社，2003.

[15] 郑东健，顾冲时，吴中如．边坡变形的多因素时变预测模型［J］．岩石力学与工程学报，2005，24（17）：3180－3184.

枫渡水电站滑坡体变形观测资料分析

王　姣[1]，高桂青[2]

1. 江西省水利科学研究院　大坝安全管理技术研究所；

2. 南昌工程学院　水利工程系

摘　要： 通过对枫渡水电站滑坡体的变形观测资料进行整理分析，得出该滑坡体目前的工作性态，并对其稳定性态进行评价。

关键词： 滑坡体；变形；观测；分析

1　工程概况

枫渡水电站大坝坐落在赣江禾水支流宁冈河上，距永新县城约22km，是一座兼有发电、防洪、养殖等综合效益的中型水库。坝址区是U形河谷，河床平坦，建坝时发现大坝左右坝肩及下游存在滑坡群，其中右岸为一倾倒体、左岸坝肩为一不稳定体，均在大坝施工时按设计已基本挖除；左岸下游450m范围内为一滑坡群，建坝时只做了局部清理。现左岸分布的滑坡体情况如下（以坝轴线为起始桩号，顺河流方向）：

Ⅰ号滑坡体：位于0＋15～0＋115m段，后缘在高程265m以下，见是一弧形滑壁，高程在180.00m左右有明显的山坡隆起现象，体积约$2\times10^5m^3$。

Ⅱ号滑坡体：位于0＋115～0＋215m段，后缘在高程245m以下，见有一长百余米连续环状裂缝，前沿坍塌成55°～60°陡壁现象，体积约$3\times10^5m^3$。

2　左岸滑坡体处理情况

建坝时在挑流鼻坎向下游沿左岸滑坡体坡脚建了一道长117m的混凝土挡墙进行保护（墙顶高程为140.00m），在高程140.00～146.00m之间做浆砌石护坡，并设置了沉陷位移观测点和地下水观测孔进行观测。因1976年暴雨期间，左岸滑坡体在高程265.00m以下，出现了弧形拉裂缝，在高程260.00m以下出现弧状破裂陡壁，后缘最大滑落2.6m；1982年5月9日至6月18日暴雨期间，左岸滑坡体又出现大的局部滑动，滑坡体后缘壁再次下落4.6m，并在高程175.00～223.00m范围内，出现多处小塌滑体。几次下滑，已使其后缘壁下滑最大处达7.2m，前沿隆起，局部塌滑。因此分别在1976年及1984年对左岸滑坡体进行了2次处理，处理情况如下[1]：

1976年高程在173.00m、205.00m和255.00m以上分别沿等高线方向开挖排水沟

本文发表于2007年。

一条；对局部陡坎和已松动的松散体进行少量削坡，坡角定为38°，并高程在173.00m、205.00m、255.00m修建了3条马道，宽度1.5～2.0m，坡面用草皮护坡；对产生的裂缝进行回填；加高桩号0+037.7～0+153的挡土墙至高程147.5m，延长部分挡土墙。

1985年对左岸滑坡体进行减重削坡；设置三排抗滑桩（在主滑段下部，成品字形布置），截面为矩形，长3m，宽2m，长边平行于滑动方向；Ⅰ号、Ⅱ号滑坡体坡面上各设置了一条预制混凝土主排水沟和6条浆砌块石支沟作地表排水用，并在Ⅰ号滑坡体设置了5个平洞，Ⅱ号滑坡体设置了1个平洞；将滑坡体坡脚处的挡墙加高1m；在坡面上铺草皮；在滑坡体中布设了位移观测网（详见图1）。

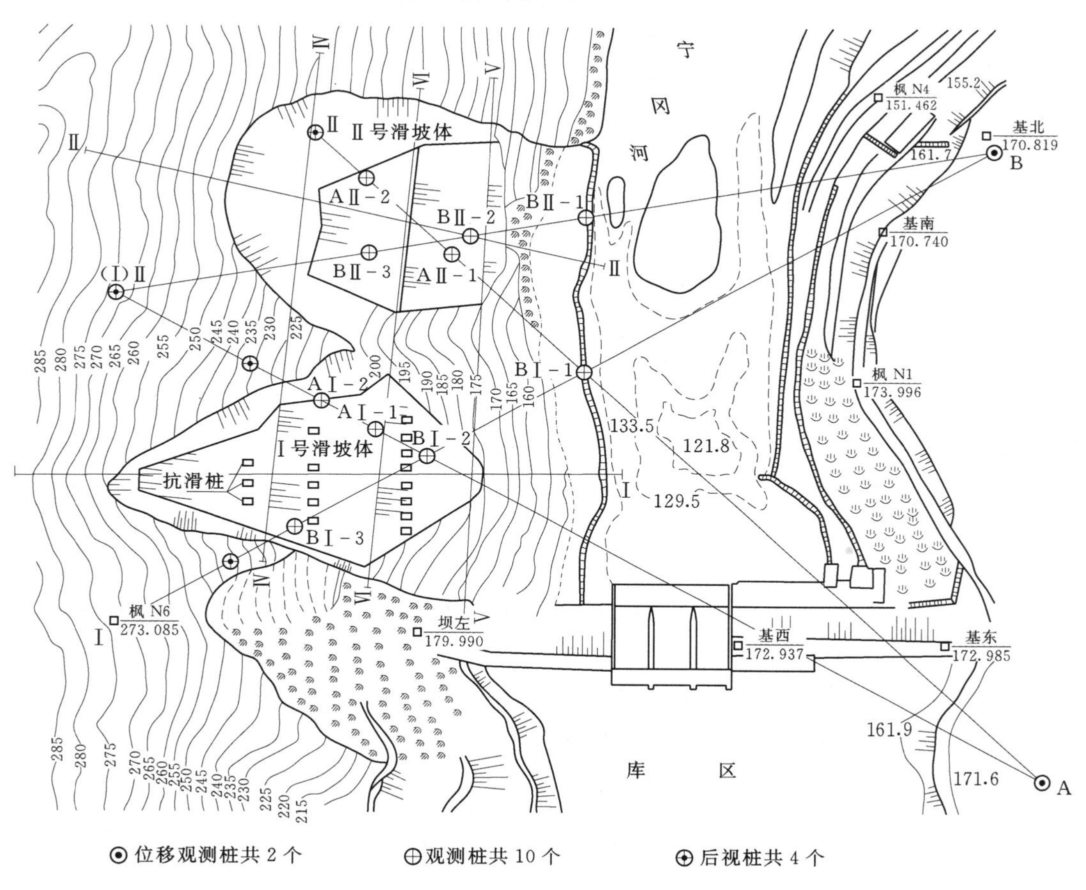

图1 枫渡水电站左岸滑坡体位移观测网布置图

3 滑坡体地质概况

3.1 滑坡体地层岩性

滑坡体区域出露的岩层均属下右生界奥陶系中统韩江-石口组（O_2^{E+S}）轻微变质岩系，岩性为长石石英砂岩、粉砂岩、板岩及第四纪堆积物按新至老顺序排列如下：Q_4 人工堆积，碎块石，分布在河床左岸；Q_4^{noe} 坍滑堆积，碎块石，红色含碎粒黏质组成，分布在Ⅱ、Ⅲ号滑坡体下部；Q_4^{ae} 冲积堆积，砂卵石、蛮石混合组成河漫滩；$Q_4^{u\text{-}de}$ 残积—坡积堆积，棕红色砂质黏土夹碎石，分布在两岸山坡上；Q_3^{ae} 冲积堆积，黄色砂壤土，底部为半胶结卵砾石，主要分布在河床右岸，组成一级阶地；$O_2^{H+S}-4$ 青灰色，灰绿色砂岩夹少量灰绿色板岩；$O_2^{H+S}-3$：灰绿色板岩夹青灰色粉砂岩；$O_2^{H+S}-2$ 青灰色砂岩夹少量灰绿色板岩；$O_2^{H+S}-1$ 深灰色长石石英砂岩。

3.2 滑坡体地质构造

滑坡体区域位于韩江褶断群内复式向斜中更次一级倒转背斜的东南翼轴部附近；且至少受到2次强烈造山运动影响。一次是加里东运动，具体表现在上奥陶流、志留系及下泥盆流地层缺失。中奥陶流地层与上泥盆流地层不整合接触，同时这次运动使奥陶系地层轻微变质。第二次是在泥盆系后的海西运动，具体表现为泥盆系地层受北西～南东的强烈水平挤压而成北东～南西方向之褶皱带，以致造成滑坡体区域内小型褶皱、小断层、破碎带及层间挤压较发育。

4 左岸滑坡体变形分析

4.1 观测设施考证

左岸滑坡体中布设的位移观测网——由4条交叉的视准线组成。在滑坡体内建立了10个观测桩；在滑坡体外的稳定地方设立2个固定点对10个观测桩进行位移观测。滑坡体位移测量仪器使用的是瑞士THE0010B型经纬仪及自制活动觇标，观测采用视准线法[1]。

4.2 左岸滑坡体观测资料分析

1994年左岸滑坡体工程竣工后，从1995年1月开始，每月对滑坡体的位移进行了定期观测，周期为1个月，每月月初观测，从1996—2005年有10年观测资料。工程管理单位每年将其整理成册，进行汇总，因而从总体上看，观测数据连续性好。因而本文主要对左岸滑坡体1995—2005年的变形资料进行分析，并以此评价该滑坡体的稳定性态。

滑坡体位移观测资料中[2]，垂直视准轴向下游为正，向上游为负。各测点的历年垂直位移过程线如图2所示，表1对相应的特征值进行了统计。

表1　滑坡体位移特征值统计表　　mm

测点	滑坡体	最大值	日期/(年-月-日)	最小值	日期/(年-月-日)	最大年变幅	年份	最大年均值	年份	最小年均值	年份	多年平均值
AⅠ-1	Ⅰ	0.98	1996-11-01	−1.6	2003-04-01	1.83	2003	−0.07	1996	−0.50	1999	−0.28
AⅠ-2	Ⅰ	1.17	1996-11-01	−1.81	2004-01-01	2.77	1996	−0.47	1996	−1.13	2003	−0.75
BⅠ-1	Ⅰ	0.92	2000-04-01	−6.33	1997-09-01	6.40	1997	−0.19	1996	−1.65	1997	−0.61
BⅠ-2	Ⅰ	0.37	1999-11-01	−3.05	1999-10-01	3.42	1999	−0.68	1996	−1.25	1999	−0.85
BⅠ-3	Ⅰ	1.00	1997-07-01	−4.08	1996-02-01	4.08	1996	−0.78	1998	−1.52	2003	−1.32
AⅡ-1	Ⅱ	1.90	2001-05-01	−2.17	2002-03-01	3.72	2001	0.59	1998	−1.70	2002	−0.71

续表

测点	滑坡体	最大值	日期 /(年-月-日)	最小值	日期 /(年-月-日)	最大年变幅	年份	最大年均值	年份	最小年均值	年份	多年平均值
AⅡ-2	Ⅱ	1.13	1996-09-01	−2.20	2001-02-01	2.95	2001	−0.12	1998	−1.69	2003	−1.07
BⅡ-1	Ⅱ	1.09	1997-07-01	−1.89	2004-03-03	2.25	1999	−0.05	1996	−1.04	2004	−0.49
BⅡ-2	Ⅱ	1.14	1998-01-01	−2.21	2003-04-01	2.59	1997	−0.36	1996	−1.37	2004	−0.89
BⅡ-3	Ⅱ	1.42	1997-07-01	−2.78	1997-10-01	4.20	1997	−0.19	1998	−1.58	2004	−1.04

注 垂直视准轴向下游为正，向上游为负。

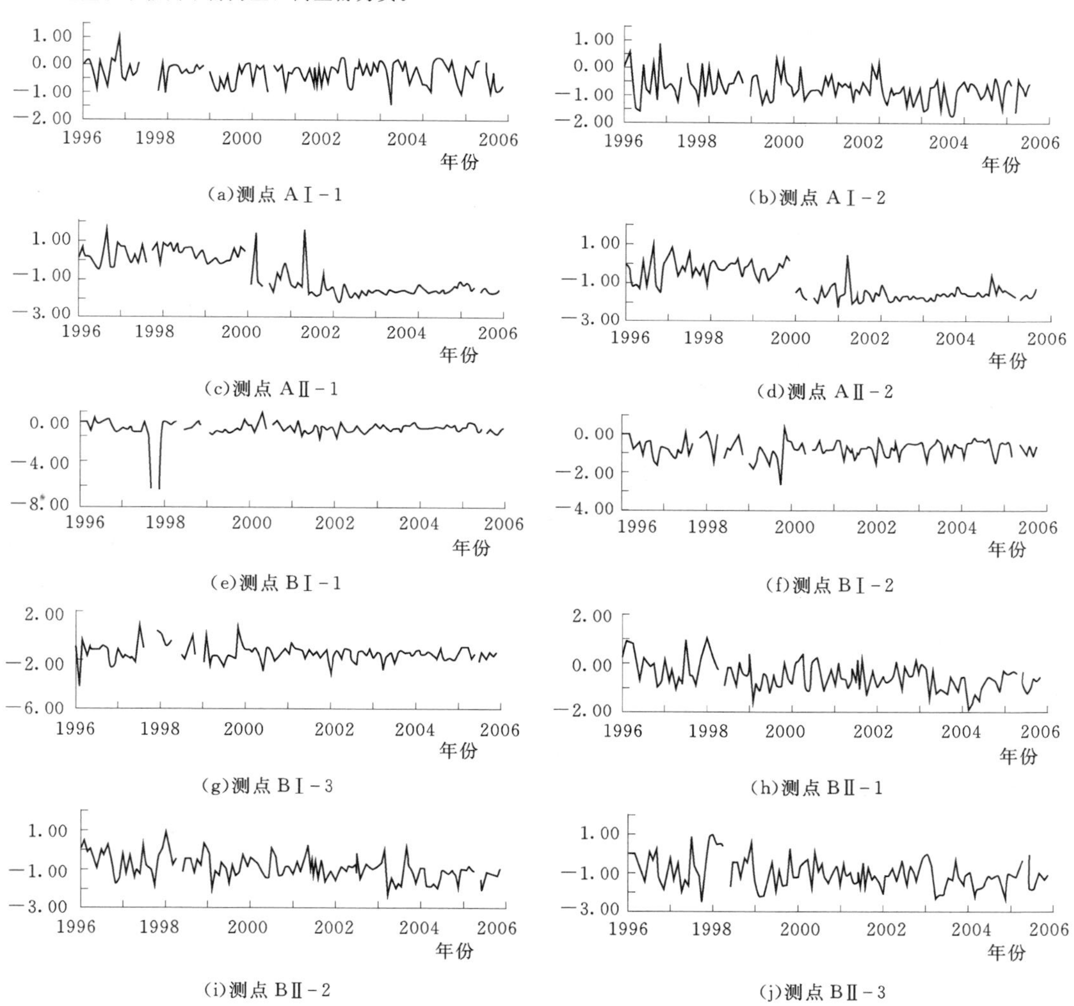

图 2 各测点位移过程线

4.3 左岸滑坡体观测资料分析结论

4.3.1 左岸滑坡体位移极值分布规律

1996—2005 年期间位移最大值，即垂直视准轴向下游最大值为 1.90mm，发生在Ⅱ号滑坡体 AⅡ-1 测点（2001-05-01）；其次为 1.42mm，发生在Ⅱ号滑坡体 BⅡ-3 测点

(1997-07-01)。此外，位移最小值，即垂直视准轴向上游最大值为6.33mm，发生在Ⅰ号滑坡体BⅠ-1测点（1997-09-01）；其次为4.08mm，发生在Ⅰ号滑坡体BⅠ-3测点(1996-02-01)。

4.3.2　左岸滑坡体位移年变幅分布规律

Ⅰ号滑坡体：最大年变幅值为6.40mm，发生在BⅠ-1测点，发生在1997年；其次为4.08mm，出现在BⅠ-3测点，发生在1996年。

Ⅱ号滑坡体：最大年变幅值为4.20mm，出现在BⅡ-3测点，发生在1997年；其次为3.72mm，出现在AⅡ-1测点，发生在2001年。

4.3.3　左岸滑坡体位移年均值分布规律

4.3.3.1　最大年均值

Ⅰ号滑坡体：最大年年均值为－0.07mm，出现在AⅠ-1测点；其次为－0.19mm，出现在BⅠ-1测点，均发生在1996年。

Ⅱ号滑坡体：最大年年均值为0.59mm，出现在AⅡ-1测点，发生在1998年；其次为0.05mm，出现在BⅡ-1测点，发生在1996年。

4.3.3.2　最小年均值

Ⅰ号滑坡体：最小年年均值为－1.65mm，出现在BⅠ-1测点，发生在1997年；其次为－1.52mm，出现在BⅠ-3测点，发生在2003年。

Ⅱ号滑坡体：最小年年均值为－1.70mm，出现在AⅡ-1测点，发生在2002年；其次为－1.69mm，出现在AⅡ-2测点，发生在2003年。

左岸滑坡体各测点的位移变化值都不大，位移最大值为1.90mm，位移最小值为6.33mm，最大年变幅值为6.40mm，可见总体已趋于稳定。由上述统计分析可见，左岸滑坡体位移测值主要受随机因素影响，位移量和年变幅均较小，变形已趋于收敛稳定。从运行情况来看，从1995年至今，枫渡水电站左岸滑坡体没有出现大的滑动现象，说明1985年的处理措施起到了明显的效果，但今后仍需加强观测。

参考文献：

[1]　吉安地区水利水电建筑公司．大坝左岸滑坡体处理竣工报告［R］．1994.

[2]　王德厚．大坝安全检测与监控［M］．北京：中国水利水电出版社，2004.

浅谈水库的降等与报废管理

傅琼华[1]，杨正华[2]

1. 江西省水利科学研究院；2. 南京水利科学研究院

摘　要：结合工程实例，介绍了我国病险水库的现状，水库的除险加固与降等、报废的关系及其管理。

关键词：水库；降等与报废；除险加固；水利管理

1　引言

我国已建成8.4万多座水库工程，这些水库在防洪、灌溉、供水、发电、养殖、旅游及生态等多方面发挥了巨大的效益，为支撑国民经济可持续发展作出了重大贡献。但由于这些水库大多数兴建于20世纪50年代后期到70年代末，受当时财力、物力、技术等因素制约，普遍存在着前期工作粗浅、施工质量较差、防洪标准偏低、工程设施不全等先天不足，加上后期投入不足，工程老化失修严重，不少水库大坝长期带病运行，导致功能萎缩，效益衰减。更严重的是这些病险水库还对下游人民生命财产安全构成了严重威胁，成为防洪工程体系的薄弱环节。党和政府高度重视这一问题，近几年来，国家投资对一大批病险水库进行了除险加固，这一"德政工程"、"民心工程"深受社会认同，特别为广大受益农民所称赞，在收到了显著社会效益的同时，也取得了明显的经济效益。

但还有相当一部分水库，由于水库淤积严重或工程病害复杂，对其进行除险加固技术上已不可行，经济上也不合理。有的水库已部分或完全丧失了按原设计标准运行管理的作用和意义，丧失了原有的工程机能，还对下游构成更大风险。对这部分水库如何科学合理的处置已成为水利管理工作中的一个日益迫切的问题。针对除险加固效益不能抵偿投资代价的工程，对于建设目的、服务对象和运行状况发生重大变化的工程，制定科学的大坝退役政策，定期进行综合评估，实行大坝降等报废制度，已是水利管理的重要内容之一。2003年5月26日，水利部第18号令发布了《水库降等与报废管理办法（试行）》（以下简称《办法》），该办法自2003年7月1日起施行，为规范水库降等与报废工作，加强水库大坝安全管理，确保水库安全运行提供了明确的法规保障。

类似生物和人类生命过程一样，水库大坝工程也有生老病死的自然过程，对此应该得到科学的认同和有效的管理，形成水库大坝从"规划设计—工程建设—运行管理—除险加固—降等报废"全过程的管理体系。如何依法全面地从经济、社会、环境的协调发展目

本文发表于2005年。

标，体现人与自然和谐相处的理念，维护水库大坝工程安全和可持续发展，加强水库大坝的风险管理，对病险水库除险加固项目进行科学、客观的技术经济评价，必要和可行时对水库进行降等与报废处理，是目前水利管理和病险水库除险加固项目建设管理工作中亟待解决的重要课题。

2 水库功能丧失与降等报废

美国最早报废拆坝的实例发生在1912年，到1999年，已经拆除了455座坝。据最新统计，2003年被列入拆除的坝有58座，2000—2002年间共拆除了79座坝，且每年拆坝数量呈上升的趋势。美国近年来虽然拆除了一些坝，但主要是位于支流上的小坝、老坝，服役期在70～100年，坝高一般为10m以下，包括早期建设的用于造纸业、纺织业、灌溉、供水等目标的小型水电站等。这些工程多为经过长期的运行，超过使用寿命，已丧失了原设计的功能，许多工程已不适应现代化大工业生产的需要；或是病险水库，不维修就无法正常运行，而维修费用远高于拆坝费用（一般为拆坝费的3～5倍，甚至高达10倍）。从大坝安全管理的角度出发，考虑工程自身的安全和对下游带来的风险，以及加固费用高等原因，为保障公共安全而拆除的大坝占拆坝总数的2/3左右；另外是坝的作用不大，拆除后可恢复河流的天然状态；还有一些是考虑洄游性鱼类生长的需要与国家自然公园相协调的河道建设而拆。这些拆坝工程，都是在做了详细的分析比较后选择了拆除方案。

我国有一半以上的水库大坝是在20世纪50年代，特别是在大跃进时代建造的，现在都到了经济使用年限的后期，从理论上讲面临的出路：一是投资更新改造，二是降等与报废。由于泥沙淤积、崩岸、地震和工程质量、老化等方面的影响，一般水库大坝寿命只有几十年，最多也不超过百年。据国际大坝委员会的报告，世界范围内水库的泥沙淤积十分严重，每年约有1%的水库淤满报废。这种情况在水土流失情况严重的我国西部地区尤其是西北地区比较突出。据统计，陕西省在册水库中至少有1/3已经不能发挥效益；一些水库运行30多年，从来没有蓄过水，又无经济可行的加固措施；一些水库已被泥沙淤满，完全丧失了水库功能，并且无法采取合理有效的排沙减淤措施恢复库容。这些水库对地方经济发展影响不大，但严重威胁着下游人民生命财产安全，成为水利管理和防洪的一大包袱。一个典型的例子是榆林市韩岔水库，原是一座中型水库，但建成20多年来一直低水位运行，1988年以来更是一直空库运行，仅仅灌溉了下游几百亩土地，而库区已经淤地133hm^2，两岸住有200多户群众，由于水库无泄洪设施，一旦蓄水，不仅使库区133hm^{2}土地无法耕种，还会淹没两岸群众，造成交通中断，从理论上讲，这是一个投资数千万元的水库，但实际上已经没有利用价值了。近几年来，在调查研究、摸清情况的基础上，经过充分论证，按照严格的程序，陕西省有10座水库降等使用，注销了577座水库，其中水毁垮坝164座，报废413座，使全省水库的总数减少到1050座。再如黄土高原各地在沟道中兴建的近3000座小水库，大部淤满失效，被迫改作淤地坝使用。四川省龚嘴水电站库容3.6亿m^3，原设计为蓄水发电，1976年建成，1987年已淤满，改为径流发电。河北省庙宫水库，库容1.83亿m^3，1960年建成，1988年已淤积9650万m^3，占总库容的53%。辽宁省兴修的733座小水库中10年左右已有106座淤满报废。广东省梅州市兴修的3.5万多座山塘和370座小水库，到20世纪80年代后期已分别淤积31.6%和61.5%，其中 505 座山塘库容已全部淤满报废。

江西省至 2002 年底统计现有水库 9268 座，与原水库数 9600 余座差别较大，其中有统计上的误差，也有相当一部分水库降等或报废，如上饶县旭日镇岭背水库，总库容 34.5 万 m^3，坝高 7.2m，灌溉面积 13.33hm^2；以及旭日镇麻车塘水库，总库容 12.9 万 m^3，坝高 6.5m，灌溉面积 10hm^2；2003 年 12 月均因上饶县工业园区建设，征用了水库及下游农田建厂，水库因此报废。新建县流湖乡新毛桥水库，总库容 14.3 万 m^3，坝高 4.0m，灌溉面积 4.56hm^2，因上游新建水库效益完全丧失。兴国县长冈乡九斤坑水库，集雨面积 9.61km^2，为小（2）型水库，因水土流失，严重淤塞水库，已作报废处理。

对失去了水库功能的水库继续按水库管理，既浪费人力财力，又增大下游风险，如按淤地坝管理，则可增加耕地面积，变害为利。对那些有效库容被淤满、防洪标准降低、功能丧失的水库，应采取降等或报废措施，放下管理和防洪的包袱。同时，报废水库也要采取审慎的态度，建大坝取得了综合效益，但也付出相当的代价，包括自然生态环境、河流形态、物种的改变、移民、文物的淹没等，大坝兴建和报废决策的社会化是必然的选择。因此，大坝的建设和报废需要更多的社会、经济、环境等领域的专家，包括相关的公众参与，这是先进国家在这方面已经取得的经验。

3 水库除险加固的经济效益与水库降等报废的关系

大坝建设，对效益与代价的评估是一项社会与自然学科交叉的复杂工作，需对大坝的运行效益与设计目标进行比较，判断原有可替代方案的现实性，并分析大坝对局部与整体环境的影响。

病险水库除险加固，确保水库安全，充分发挥防洪和兴利效益，对促进经济与社会的发展，具有十分重要的意义。水库除险加固目的是增加水库的安全性和进一步挖掘水库自身潜力。除险加固前要进行效益分析，其效益主要有社会效益（防洪保安）和经济效益。防洪效益主要体现在加固后防洪标准的提高，目前常用频率分析法，即通过水库修建（加固）前后发生同频率洪水而引起下游淹没损失的比较，来计算水库的防洪效益。经济效益分析包括估算增加的蓄水量以及由此而增加的防洪、灌溉、供水、发电等效益。同时还要进行费用（包括固定资产投资、年运行费和流动资金）估算和经济效益指标（包括经济内部收益率、效益费用比和经济净现值）分析。

对水库病险严重且除险加固技术上不可行或者经济上不合理的水库，经论证后应降等使用或报废。同时水库降等以后，也为实施除险加固工程奠定了基础。如陕西省延安市孙台水库（中型水库），原总库容 1475 万 m^3，水库淤积严重，放水设施毁坏废弃，泄洪洞设计不合理，进口淤堵频繁，按中型水库加固需要投资 2750 万元，经对水库流域水文、供水等情况分析论证，最后确定按照小（1）型水库加固和运行管理，加固工程投资 750 万元，节省了 2000 万元资金，加固后可以充分利用水资源，发挥工程效益。铜川市福地水库，总库容 1050 万 m^3，存在严重病害，经对淤积现状、水库功能、洪水情况及下游危害等因素认真论证后，将其校核标准由 1000 年一遇定为 500 年一遇，大坝、溢洪道、泄洪洞等建筑物加固标准相应降低，节省投资 1000 多万元。

4 降等与报废的管理

水库大坝老化、安全风险以及报废评价是一项非常复杂的工作，包括技术、经济、社会等诸多方面的因素。美国对那些早期建设的坝的安全及报废问题给予了高度的关注，建立了拆坝许可证制度，并已经编写出有关《导则》，用于科学地指导大坝退役的评价工作，不仅有技术方面的，还考虑了有关环境生态、社会经济方面的因素，同时更加关注坝的拆除过程中的相关技术问题，重视大坝拆除过程中的安全，减少坝拆除后对周围生态环境、社会经济发展的影响，真正立足于可持续发展的目标。拆坝的焦点主要反映在 3 方面：①大坝安全因素；②大坝功能丧失；③考虑洄游性鱼类生长及环境要求。在对已拆除坝的技术经济比较中发现，有些坝的除险加固费用远超过继续运行所发挥的效益或拆除所需的费用。一般开展大坝退役评价要考虑的因素包括大坝安全、美学、鱼类、河流流量减少等，在经济方面要比较大坝继续运行或加固维护的费用。在上述因素中，大坝安全放在首位。其他考虑的方面还有：社区生活对供水或娱乐设施的依赖情况；城市用水与灌溉的水权；生态需水要求；泥沙管理；防洪与水流调节效益。通过对大坝的评价可以得出 3 种结论：①继续运行；②部分功能将退役；③整个工程全部退役。

近 10 年来，我国的已降等或报废的水库大都未履行严格规范的手续，绝大多数是以市、县水利局的文件形式批复的，做得好一点的市、县召开过专家论证会，并形成会议纪要。目前，已出现对以前降等报废水库提出质疑的情况，水利部 2003 年 7 月 1 日起实施的《办法》中第 5 条规定："水库降等与报废必须经过论证、审批等程序后实施。"《办法》规范了水库的降等与报废工作，并分别给出了符合降等或报废的条件。《办法》规定水库降等论证报告内容应当包括水库的原设计及施工简况、运行现状、运用效益、洪水复核、大坝质量评价、降等理由及依据、实施方案。对水库实施报废的管理措施，除应进行降等论证报告中相同内容的工作外，《办法》特别要求论证报告内容则应包括报废理由及依据、风险评估、环境影响及实施方案。《办法》是对我国水库大坝安全管理工作的一种延续和深化，是对水利管理全面与科学化的一个补充，也是完善病险水库及除险加固项目管理的一种科学措施。

如果对水库采取报废的处理方式，有三方面的问题要给予特别地关注与研究。首先是要加强对报废水库的管理，如果管理不当，废弃的水库有可能成为新的致灾因素；其次是要考虑水库周边已建立起来的生态环境因水库报废而发生变化，需要在一个相当长的时间内建立起一个新的平衡，由此对水库周边生态环境带来的影响；最后是因水库报废而产生的对水库上、下游及周边社会经济发展的影响。修建水库大坝是为了兴利除害，而废弃不安全的、丧失功能的水库也同样应达到兴利除害的目的。

水库降等与报废在我国是一个新生事物，还需要有与之配套的、可操作的细则，特别是开展生态环境与社会经济定量的评价指标体系与评估模型的研究，探索出适应于我国国情的、科学的对策与相应的技术措施，以指导水库降等与报废工作的有效实施。

5 结语

江西省目前共建成或基本建成各类水库 9268 座，约占全国水库总数的 1/9，这些水库作为国民经济的重要基础设施之一，对全省国民经济建设发展起着举足轻重的作用。但到 2002 年底还有 3488 座各类病险水库，要对这些不同类型不同病险隐患的水库工程进行全面彻底的除险加固，经测算共需投资 80.79 亿元，其中大型 4.96 亿元，中型 38.05 亿元，小（1）型 28.83 亿元，小（2）型 8.95 亿元。对于我省欠发达的经济状况来说存在较大压力，因此，在除险加固工作中综合实施水库的降等与报废管理具有突出的意义。

目前许多水库大坝（多数为小型水库大坝）的运行期也达到 40～50 年，对于已存在的病险水库问题和将来还会出现新的病险水库问题，今后水库管理工作更多的是面对这些水库大坝的安全管理和除险加固工作，包括部分病险水库的降等与报废工作。因此，结合我国的国情，开展大坝老化、工程寿命、风险管理和降等报废等方面的安全技术研究，科学决策病险水库除险加固与降等报废工作是当前应该考虑的重要课题。

参考文献：

［1］ 李雷，陆云秋．我国水库大坝安全与管理的实践和面临挑战［J］．中国水利，2003，11，A 刊．
［2］ 刘宁．21 世纪中国水坝安全管理、退役与建设的若干问题［J］．中国水利，2004（23）．
［3］ 高杰，王德文．陕西省小型水库安全管理现状与对策［J］．中国水利，2003－2B 刊．

江西省病险水库除险加固规划

傅琼华[1]，王　纯[2]

1. 江西省水利科学研究所；2. 江西省水利厅

摘　要：对江西省建成或基本建成的9268座各类水库的调查与分析结果表明，仍有3488座水库存在不同程度的病险隐患。在分析病险症状及产生原因、安全风险排序、加固效益分析的基础上，结合各设区市社会经济发展水平和病险水库的保护对象等实际情况，经综合分析论证，提出了江西省病险水库除险加固目标任务、加固资金筹措方案及加固保障措施。

关键词：病险水库；安全风险排序；经济评价；资金筹措；保障措施；江西省

江西省共建成或基本建成各类水库9268座，约占全国水库数量的1/9，在各省中排名第二。这些水库作为国民经济的重要基础设施之一，对全省国民经济建设发展起着举足轻重的作用，在防洪、灌溉、供水、发电、保护生态等方面发挥了巨大效益。尽管1998年以来，中央和各级政府加大了投资力度，对大江大河大湖主要堤防、大型病险水库和重点中型病险水库进行除险加固，防洪标准有所提高。但大多数病险水库尚未进行除险加固，防洪标准低、安全隐患多、管理手段落后、预测预报和通讯预警设施差等安全问题非常突出，是防洪安全的心腹之患，给下游城镇、人民生命财产以及主要交通干线等基础设施造成严重的威胁。

根据《江西省国民经济和社会发展第十个五年计划水利发展重点专项规划》目标任务："基本完成病险水库的除险加固，确保在设计标准内不垮坝。"按照江西省委省政府提出的"搞好病险水库加固整修，从根本上解决病险水库的除险加固问题"的要求，为进一步掌握全省水库工程安全状况，统筹规划病险水库除险加固工作，进一步调查摸清现有病险水库基本情况，五年内基本消除现有的病险水库的工作目标，经对全省小（2）型以上水库安全状况进行全面普查，编制了江西省病险水库除险加固规划。

1　病险水库评定

1.1　建设成就

据统计至2002年年底，全省共建成或基本建成各类水库9268座，其中大型水库25座，中型水库224座，小（1）型水库1329座，小（2）型水库7690座。已建成的249座大中型水库中，土坝和土石混合坝共有191座，占总数的77%；浆砌块石、混凝土重力

本文发表于2004年。

坝和拱坝58座，占总数的23%。其中以柘林水库总库容79.2亿m^3为最大，井冈冲水库混凝土砌块石拱坝坝高92m为最高。

全省水库总库容280.16亿m^3，其中兴利库容157.7亿m^3；水库有效灌溉面积达1461万亩，占全省有效灌溉面积2841万亩的54.1%；水库年供水量达119.03亿m^3，占全省水利工程年供水的52.2%；水库发电装机174万kW，年发电量达40亿kW·h，基本形成了蓄、引、提并举，防洪滞洪、灌溉供水、治水办电三结合的水利工程体系，为促进社会经济发展、确保人民群众生产生活安全、改善生态环境等发挥了不可替代的作用。

1.2 病险水库评定原则

根据江西省水库的特点，对全省水库进行安全普查，评定病险水库的基本原则是：

（1）截至2003年4月30日，已完成安全鉴定工作的水库，且经部大坝中心核查或省大坝安全中心核查定为三类坝的水库。

（2）对未进行安全鉴定的大中型和重点小（1）型病险水库，以水库上报材料为基础，由省水利厅技术委员会组织专家进行核定。核定原则为：影响5000人口以上；影响京九、浙赣铁路、高速公路等重要交通干线或重要设施，距离10km以内；且病险情严重、不能按设计正常运用的水库核定为病险水库。

（3）一般小（1）型和小（2）型病险水库，以各设区市水利（务）局核定的为准。

据本次安全普查核定，至2002年12月31日，有3488座水库存在不同程度的病险隐患，占全省水库总数的37.6%；病险水库总库容为80.1亿m^3，占全省水库总库容的28.6%。其中大型11座，占相应水库总数的44%；中型153座，占68.3%；重点小（1）型287座，占21.6%；一般小（1）型488座，占36.7%；小（2）型2549座，占33.1%。

1.3 病险水库安全风险排序

为权衡病险水库除险加固的轻重缓急，更合理地安排使用加固资金，对全省未进行安全鉴定工作的80座大中型病险水库进行病险安全风险排序。安全风险排序的原则是主要考虑各坝的相对安全程度，水库经济效益则置于从属位置，按照安全风险排序方法进行。

根据水库管理单位提供资料：工程有关原设计及历次加固设计报告、工程特性数据（如地理位置、集雨面积、坝高、总库容、下游人口、灌溉面积等）；地质勘察、质检资料；工程施工记录（包括历年除险加固工程）；工程运行管理资料，包括运行过程中的变化（含观测资料），曾经出现过的险情和异常情况，水位和泄洪情况（含出现过的最高水位及出现年份）等。经分析比较安全风险排序采用以现场评分（site rating）法为主，综合评定大坝、溢洪道、输水建筑物、金属结构的风险值。通过对80座大中型病险水库的现有病险情进行的安全风险排序，安全评定结论为：1座大型水库即油罗口水库的现有病险情为高风险性；79座中型水库中，有20座水库的现有病险情为极高风险性，59座水库的现有病险情为高风险性。

2 病险水库除险加固目标及原则

2.1 病险水库加固目标

经对各地自然条件、水库工程现状和社会经济等条件比较分析，按照水库病险安全风险指数高低、地理位置重要性和存在问题，结合当地水资源状况和社会经济发展水平等实际情况，区别轻重缓急，因地制宜地提出全省病险水库除险加固目标。

加固规划总目标是：根据江西省病险水库实际，把除险加固的重点放在大中型病险水库及重点小（1）型病险水库，恢复和治理水库工程的安全性状；以科技为先导，遵循自然和经济规律，采取水库工程保安加固建设、水库防汛信息化建设、配套完善监测与管理设施、强化水库安全管理能力建设等综合措施，分阶段解决水库病险安全问题，促进水库经济效益、生态效益和社会效益的协调统一，努力实现水库工程安全、设施齐全、功能完备、管理高效规范、环境园林化的高标准水库目标。

在2007年底前，全面完成现有3488座病险水库的除险加固任务，分两期进行：近期即在“十五”期间内完成全省现有大型、94座中型及部分重点小（1）型、小（2）型病险水库；中期即2007年前完成剩余的59座中型以及全部小（1）型、小（2）型病险水库的除险加固任务。

2.2 除险加固的基本原则

2.2.1 坚持行政首长负责制的原则

病险水库除险加固的责任在地方各级人民政府，按分级负责的原则，进一步建立健全行政首长负责制，逐座明确落实病险水库除险加固的地方政府行政责任人、技术责任人和岗位责任人。

2.2.2 坚持统筹规划、突出重点、分步实施的原则

将加固治理中涉及工程规模、地理位置、技术复杂程度等因素进行综合考虑、统筹安排。做到兴利与除害相结合、防洪与抗旱并举，处理好全局与局部、长远与当前的关系，优先实施病险风险指数高、危害值大、地理位置重要的水库。

2.2.3 坚持因地制宜、注重实效、高起点高标准的原则

从实际出发、注重实效、量力而行，集中人财物，分片、逐步进行综合治理。从现代化管理的需要出发来设计除险加固工程，做到加固一个、除险一座、摘帽一个、完善一座、美化一座，实现水库工程的安全运行、高效管理、环境美丽的高标准。

2.2.4 坚持科学治理、标本兼治、建设管理的原则

对病险水库除险加固的前期工作、项目申报和审批、建设资金、建设管理、监督管理和建后管护等都应按有关规定认真执行，依法治水。充分利用国内外现有的科技成果和高新技术手段，科学治理。

2.2.5 坚持开发式治理、建立良性运行管理机制的原则

水库除险加固建设与经济发展相结合，在重视工程加固建设的基础上，同时根据水库的自然条件，合理开发利用水土资源，在提高抗御洪旱等自然灾害能力的同时，提高水库资源的可持续利用，提高水库自我造血功能。同时，加快水库管理单位的体制改革，建立水库管理单位良性运行机制。

2.2.6 坚持社会办水利的原则

在各级政府加强对病险水库除险加固的领导、加大病险水库投入的同时，广泛动员和组织社会各种力量参与除险加固建设，建立和完善多层次、多元化、多渠道的投资体系。

2.2.7 实施降等使用和报废管理的原则

对既丧失防洪、灌溉、供水水库基本功能，又不能进一步开发利用、经济效益差、投入与产出失重的水库，作降等使用或报废处理。对于在规划期间内未完成除险加固的水库，实施降等使用和报废管理的办法。

3 病险水库除险加固资金筹措方案

对于江西省3488座不同类型不同病险隐患的水库工程，要得到全面彻底的除险加固，资金量投入是关键，经测算共需投资80.79亿元。按规划目标，2003—2005年需要投入44.27亿元，年均14.76亿元；2006—2007年需要资金36.52亿元，年均18.26亿元。规划结合全省经济发展水平和资金筹措能力等实际情况，考虑加固工程特点，提出配套资金补助标准和资金筹措方案。建议按照分级管理、分级负责原则，实行中央、省、市、县、乡多层次、多渠道筹措资金：一是争取国家对病险水库除险加固专项资金；二是省及省以下各级水利建设基金和预算内基建投资；三是地方财政单列专项资金；四是小农水经费以及农业开发、以工代赈、老建扶贫、水利岁修等资金集中使用；五是水管单位水费及综合经营收入；六是“一事一议”农民投工投劳；七是承包、租赁、拍卖水库经营权收益。

4 效益分析

4.1 效益估算

参照已批复的老营盘水库等48座大中型病险水库除险加固设计书的经济效益分析方法，经推算病险水库除险加固完成后，现有病险水库的总库容能够得到全面恢复，可恢复防洪库容30亿m^3，兴利库容50亿m^3，恢复改善灌溉面积800万亩，年新增供水能力10万m^3，年新增发电量3万kWh，年新增多种经营约6000万元。灌溉效益采用分摊系数法计算，按有无项目对比计算总增产值，在水利和农业两部门分摊，由总增产值乘以灌溉效益分摊系数计算灌溉效益。经调查江西省各灌区所采用的灌溉效益综合值，通过分析计算，可得本项目多年平均灌溉效益达16亿元。再考虑新增的供水效益、发电效益和多种经营等，多年平均效益约为17亿元。

规划实施后，水库的防洪能力得到明显提高，水库的滞洪削峰效果、防洪除涝能力进一步增强，水库下游的近2000万人口防洪安全将得到明显提高，减免下游洪涝灾害损失，保障经济社会持续稳定发展；并将减少用于防洪抢修、修复水毁工程的资金，社会和经济效益极其显著。

水库经除险加固后，对治理水土流失、遏制生态环境恶化有明显的作用，有效地改善农业生产条件，解决城镇、农村的生产和生活用水，以促进社会经济的可持续发展。同时，为水库管理单位增创收、依托水库本身走上良性发展的轨道，提供了强有力的经济保障。

4.2 经济评价

经济评价采用的主要参数按《水利建设项目经济评价规范》要求确定，按分析计算的费用与效益，并根据选取的评价参数、折现计算基准点以及费用、效益发生和结算原则，经进行投入产出平衡分析计算，求得项目的各项国民经济评价指标：经济内部收益率为

24%，经济净现值为45.87亿元，经济效益费用比为1.8，各项经济指标均较好。

为研究投资和效益的测算误差对国民经济评价结果的影响进行敏感性分析，经计算成果可知，当投资和效益单项因素或投资和效益同时向不利方面变化时，经济内部收益率均大于12%，高于社会折现率，经济净现值均大于0，经济费用比均大于1，本项目在经济上是合理可行的，并具有一定的抗风险能力。

5 病险水库除险加固保障措施

5.1 组织及政策保障措施

进一步建立健全行政首长负责制。按照属地管理、分级负责的原则，对现有病险水库逐库落实行政责任人、技术责任人和岗位责任人，明确各责任人的主要职责。大、中、小型水库的行政责任人必须分别由设区市、县、乡镇领导担任并在省内主要新闻媒体上向社会公布，接受社会监督。

建立机构，密切配合，加强监管。省政府成立江西省病险水库除险加固工作领导小组，对病险水库除险加固工作进行指导、督促。各设区市按照分级负责原则成立相应的机构，并明确专门的办事机构和专职管理人员，落实责任制，切实做到组织有序、措施有力、工作扎实。

加强工程项目管理，保证配套资金落实。省水利厅每年将组织有关部门对相关设区市的除险加固工作进行检查、协调及管理等，全面推动病险水库除险加固工作的进程。

加快水库管理单位的体制改革。改革是发展的动力，病险水库的除险加固与改革同步，根据省水利厅提出的江西省水利工程管理体制改革实施方案，建立科学、良性的管理、运行、维护、发展机制，使水利工程能够长期发挥应有的效益。

5.2 资金保证措施

积极争取加固资金的投入。各地积极采取借、贷、赊、融等多形式、多层次、多渠道的办法和措施筹集资金，确保配套资金及时足额到位。按除险加固项目的轻重缓急，控制除险加固的规模，确定加固的优先次序和重点，集中有限资金安排一批、摘帽一批。

各级财政要加大扶持力度，把除险加固建设资金纳入预算，多方筹措病险水库除险加固建设资金：设立病险库加固专项资金；在各级水利基本建设投资中列支专项资金；在安排财政扶贫资金、农业综合开发资金、以工代赈资金等专项资金时，把除险加固建设作为一项内容；采取有力措施，将水利工程产权置换出来的资金投入水利建设；按照“谁投资、谁经营、谁受益”的原则，鼓励社会上的各类投资主体向水库建设投资；积极宣传、鼓励、引导水库及周边群众参加小型水库的加固治理建设，调动和保护农民搞好小农田水利。

加强资金管理。严格实行项目资金专户管理，做到专款专用，保证项目资金用到除险加固工程中，杜绝挪作他用，并接受审计等部门监督检查。

5.3 技术保证措施

认真做好安全鉴定和技术认定。所有小（1）型以上病险水库必须依据规定进行大坝安全鉴定，并经有关部门核查。各县（市、区）抓紧对辖区内小（2）型水库进行全面的安全技术认定工作。

加固各阶段的设计审查坚持“大病大治、小病小治”的原则，实事求是，突出重点，把有限资金用在最需要加固的项目和地方。同时，配备必要的管理设施和大坝安全监测手段，为除险加固后的管理工作创造条件。

规范除险加固的建设管理及除险摘帽工作。认真执行项目法人制、建设监理制和招标投标制，确保除险加固工程质量和进度。加固工程经有关主管部门组织验收合格的，进行病险水库摘帽工作，并按江西省水利厅赣水管〔2002〕59 号文件《江西省水库大坝注册登记办法》要求，进行水库注册登记复查、换证等工作，加强建后管护，做到建管并重。

实行水库降等使用和报废管理办法。对符合有关规定条件的实行水库降等运行管理和报废处置措施，以减轻防洪压力，集中有限的财力治理重点病险水库。由于地方资金落实不到位或其他主观原因，在 2007 年底前没有进行除险加固的病险水库，实行强制性降等使用或报废，以消除防洪安全隐患。

6　结语

当前，水利已摆在基础设施的首位，全社会对水利的认识不断提高，水忧患意识普遍增强，要求改变水环境的愿望越来越迫切。必须抓住这个历史难得的发展机遇期，振奋精神，开拓进取，加快推进病险水库除险加固的进程，及早消灭现有病险工程，构筑适应发展的现代化的水利基础设施防御体系。为实现江西省在中部地区崛起提供防洪安全和水利保障，以水资源的可持续利用保障经济社会的可持续发展，为全面建设小康社会作出贡献。

岩底水库主坝白蚁危害及治理

李建锋[1]，王小春[1]，阮绮水[2]

1. 江西省水利科学研究所；2. 江西省上饶市信州区水务局

摘　要： 介绍岩底水库主坝蚁害现状，通过对“传统法”和“诱杀法”的比较分析，提出了灭蚁施工方案，并认为用诱杀法灭蚁效果最为彻底，不会复发。

关键词： 土坝；蚁害；灭蚁方案；岩底水库

1　概述

上饶市岩底水库于1958年动工兴建，主坝经历年加高，于1986年基本建成。水库总库容1040万m^3，集雨面积12.8万km^2，灌溉面积933hm^2（1.4万亩），其主坝坝顶长215m，最大坝高21.5m，坝顶高程121.50m，正常蓄水位117m，为均质土坝。经调查发现主坝存在黑翅土白蚁和黄翅大白蚁，这二种土栖白蚁是公认的危害堤坝的大敌，坝面白蚁活动痕迹十分明显，按照土栖白蚁危害程度等级进行评估，危害程度已达四类蚁害工程标准（严重），已到了非治理不可的地步，否则将导致堤坝“溃于蝼蚁”的严重后果。

2　蚁害种类

根据兵蚁、工蚁、蛀食痕迹及泥被泥线特征鉴定危害蚁种为黑翅和黄翅两种土栖白蚁，该类土栖白蚁的巢穴喜筑于酸性黏土中，巢深1.5～3m，群体相当庞大，一窝成年蚁巢有1个主巢和10～20个副巢，白蚁数量多达上百万只，对堤坝危害极大，是堤坝内的“定时炸弹”。

3　白蚁危害现状

3.1　背水坡

杂草灌木丛生，主要有茅草、刺庞草、黄荆和土茯岭等，为白蚁生存提供了天然食料和屏障，坝坡上蚁害迹象十分醒目，泥被泥线如蜘蛛网状，随处可见，查到分群孔遗迹5处；尤其是“岩底水库”四个大字区域内白蚁活动十分频繁，挑开新鲜泥被泥线可看到许多正在外出采食的工蚁及护卫兵蚁，最明显的两处泥被泥线长达3～5m，宽0.02～0.03m，白蚁活动常现带长达20m。危害范围从坝脚至坝肩，桩号为0+000～0+215。据

本文发表于2013年。

大坝工管人员反映，“岩底水库”四个大字下方高程约115m，在汛期高水位时，常常出现散浸现象，初步判断与蚁害穿坝有关。汛后曾在背水坡散浸处示范挖出黑翅土白蚁主巢1窝，黄翅大白蚁主巢1窝。

3.2 坝顶

坝顶虽来往人较多，白蚁活动受到一定干扰，但杂草丛处白蚁及泥被泥线随处可见，查到分群孔遗迹3处，危害范围的桩号为。0+000～0+215，启闭机房的门框已被白蚁全部蛀空，留在木框上的白蚁从坝内带出的大量红黏土清晰可见。

3.3 迎水坡

迎水坡茅草丛生，红条石护坡至高程120m，但块石上却很容易查找到白蚁及泥被泥线，查到分群孔遗迹1处，危害范围从坝坡正常水位至坝肩，桩号为0+000～0+215。

3.4 坝两岸及山头

主坝两岸及山头植被良好，主要植被为松树，不少枯树已被白蚁蛀空，在地面枯枝落叶中查到许多正在取食的白蚁及新鲜泥被泥线，查到分群孔10余处，这是白蚁生存和繁殖的大本营，也是大坝上白蚁的主要来源之一。

3.5 分群现象

每年5月暴雨前后，发现有大量有翅繁殖蚁从坝肩及两岸集体飞出，像蝗虫一样在空中飞舞，如乌云一般遮天蔽日，群飞结束坝顶上很容易发现一堆堆白蚁脱落的叶翅，这足以说明坝上白蚁已发展为成年期的蚁群，白蚁危害越来越猖獗。

3.6 水库管理站

水库管理站院内外种有许多绿化树，绿化树遭土栖白蚁蛀食严重，一株苦楝树上有3m多高的白蚁取食痕迹，一栋木质结构仓库因土栖白蚁长期蛀食已于2000年倒塌。

4 危害程度分析

大坝上茅草灌木丛生，且地表枯枝落叶多，其植被生态环境十分有利于潜伏在坝内的白蚁取食、生存和繁殖，坝面上白蚁活动痕迹频繁，实属罕见，坝上发现的白蚁是危害堤坝最厉害的两种土栖白蚁，主巢、副巢、蚁道间路路相通，可形成四通八达的地下交通网，尤其是修筑于坝顶心墙和坝肩下方的蚁巢、蚁道，对大坝威胁最大。由于上下游坝坡均发现分群孔和泥被泥线，说明白蚁的主干道已串通坝顶心墙，形成漏水隐患，一旦汛期水位暴涨居高不下，将可能形成大的漏水通道，造成大坝长期散浸、集中渗漏和管涌，甚至导致跌窝或垮坝的险情。大坝早在1980年就发现了白蚁迹象，曾采用挖巢法、灌毒水法处理过白蚁，但处理后白蚁依然较多且逐年加重，加上历年加高，使蚁巢、蚁道越埋越深，长期困扰着大坝的安全，有效治理蚁害已成为大坝管理的头等大事。

5 灭蚁方案的比较与选择

灭治土栖白蚁方法很多，20世纪50—80年代常用的方法称为“传统法”，水利部于1990年在全国推广使用的堤坝白蚁防治新技术“找标杀法”则被称为“诱杀法”。两种方法比较见表1。

表1　灭蚁方案比较表

方案	传统法	诱杀法
方案描述	（1）追挖蚁巢法； （2）筑毒土层毒土墙法； （3）熏毒烟法； （4）灌毒水毒浆法	（1）查泥被泥线施药法； （2）查分群孔施荡法； （3）设引诱坑（堆、桩）诱蚁施药法
方案比较	（1）直观，施工期短，但灭蚁不彻底，易复发； （2）破坏坝体严重，易造成新的隐患； （3）药物毒性大，操作不安全，对生态环境污染严重	（1）诱杀施药受季节影响，施工期长；但灭蚁彻底，效果好，不会复发； （2）不须追挖巢，对坝体不破坏，安全保险； （3）高效低毒，无污染，操作安全

通过比较，诱杀法具有高效低毒、安全保险、易操作实施等优点，根据江西省20多年治理白蚁实践以及南方兄弟省份水利系统积累的经验，采用诱杀法最为适宜。

6　诱杀法灭蚁的施工方案

根据土栖白蚁的生活习性，江西省已采用诱杀法治理了大中型水库30多座和3338hm^2（5万亩）以上圩堤10余座及10个园林绿化工程，均取得了显著效果，实践证明，该技术是目前土栖白蚁防治效果最理想、最彻底的方法，治理蚁害效果持久，不会复发。具有诱杀效果极佳，用药量少，无污染，不破坏坝体、苗木，投资省，操作简便，易推广等优点。其诱杀机理和诱杀方案叙述如下。

6.1　诱杀药物和诱杀机理

（1）诱杀药物。主要为灭蚁灵原粉和灭蚁毒饵条。

（2）诱杀机理。纯灭蚁灵为结晶体粉末，无味、不溶于水，属慢性胃毒剂，对白蚁无驱避作用，工蚁取食或沾染药物后，不会马上死亡，仍能在群体中生活，群体中其他品级白蚁（如蚁王蚁后、兵蚁、繁殖蚁及幼蚁），由工蚁哺育，白蚁个体间的相互交哺、喂食、吮舐和抚慰等生活习性使灭蚁药在整个白蚁群体中广泛传递，最后导致全巢白蚁中毒死亡。诱杀正是利用了白蚁这些习性来达到灭蚁目的的，一般来说，一个上百万的蚁群只要有0.5%的个体取食了药物，就收达到全巢覆灭的效果。

6.2　诱杀方案

6.2.1　诱杀时间

2001年10月至2002年7月。

6.2.2　诱杀方案及进度

采用以诱杀为主的土栖白蚁防治技术，先进行大面积诱杀施药，然后局部补治，反复灭杀，直到蚁巢内白蚁彻底中毒灭亡或霉烂。白蚁被灭杀后可挖巢检验灭蚁效果，该方法效果直观显著，不会复发。

2001年10月至2002年5月进行诱杀，方法如下：

（1）设引诱坑施药诱杀。在治理范围内布置数排诱杀坑，排距5m，坑距10m，诱杀坑呈梅花形布置，坑内放入白蚁喜食的引诱料，柴上喷药或投毒饵，然后加盖防雨布并复一层0.15m的土踩紧压实，防止雨水渗入坑内，使坑内不透光不漏水，保持白蚁适宜的温湿度，为白蚁提供良好的取食环境，让白蚁自由取食。待诱集到较多白蚁时再集中一次性喷灭蚁灵粉，让白蚁沾染药粉回巢传染。

（2）设引诱堆施药诱杀。将从大坝上割下的茅草灌木，捆成堆均匀摆放在引诱坑之间的空白地方，并与少量黏土混杂，待堆内诱集到白蚁时再施药。

（3）泥被泥线施药诱杀。在白蚁活动旺季（气温达 15～30℃时），地表泥被泥线大量出现，在地表或枯枝落叶、老树兜、枯柴废板料、灌木草丛、乱石砖块、树上树下查找白蚁并施药，让白蚁沾染药粉带毒回巢传染。

（4）分群孔内施药诱杀。在 4—5 月重点查找白蚁修筑的新鲜分群孔，分飞前塞入灭蚁毒饵条，让白蚁取食毒饵传染。

（5）在坝体两岸或背水坡坝脚蚁迹集中处进行示范性挖巢灭蚁，取出蚁巢并在巢位喷施 1%氯丹乳剂后回填夯实。

通过以上方法的施药诱杀，使白蚁不断取食或沾染药粉，反复传染，直到蚁巢内白蚁慢性中毒后彻底死亡，坑（堆）内、坝面、树上树下和浪渣线查不到白蚁活动为止。

2002 年 6—7 月对治理范围内的诱杀效果进行了复查，并通过炭棒菌指示物开挖死巢验证灭蚁效果。若有漏网的白蚁则进行补杀施药，直至全部消灭白蚁为止，并由甲方派员检查验收。

7 坝内死巢处理措施

白蚁群体尽管被消灭了，但遗留在坝内巢穴依然对堤坝防洪威胁很大，尤其是白蚁在坝顶心墙及坝肩下酸性黏土留下的死巢、蚁道、空洞或其他漏水隐患，应结合大坝整治加固工程措施（如锥探灌浆、冲抓套井回填或劈裂灌浆等）一并处理。

8 结语

岩底水库白蚁诱杀分 2～3 期进行，目前已完成了第一期大面积诱杀施药工作，并取得了初步成效。预计通过精心普查、反复施药诱杀和局部补治，可以彻底消灭大坝及其附近的白蚁，确保白蚁不再复发，并通过工程处理措施，使岩底水库达到无蚁害的工程标准。

江西省圩堤（区）现状及其治理的思考

熊冲玮[1]，李建锋[1]，刘　亮[2]

1. 江西省水利科学研究所；2. 江西省水利水电学校

摘　要： 本文简要介绍了江西省圩堤（区）概况、特点及治理情况，针对圩堤治理所带来的新问题，阐述了作者的观点。

关键词： 江西圩堤；治理；思考

1　江西省圩堤（区）概况及特点

江西省位于长江中下游南岸，东邻浙江、福建，南接广东，西连湖南，北毗湖北、安徽。主要属长江、珠江两大流域，其中长江流域占97.6%，珠江流域中2%。此外，尚有285km^{2}属东南沿海的钱塘江及韩江流域诸水系。

境内地势南高北低，边缘群山环绕，中部丘陵起伏，北部平源坦荡，四周渐次向鄱阳湖倾斜，形成南窄北宽以鄱阳湖为底部的盆地状地形，其间水系纵横，河网密布。赣、抚、信、饶、修五大河穿越于广大丘陵与盆地之间，蜿蜒汇入我国最大的淡水湖—鄱阳湖。每遇洪水季节，江湖水位上涨，洪水灾害频繁，特别是鄱阳湖及五河中下游地区，洪水位大都超出地面数米不等，洪水灾害尤为严重，防洪圩堤成为我省防洪工程的重要组成部分。

全省圩堤（区）主要分布于赣、抚、信、饶、修五河干流中下游及鄱阳湖滨湖地区。圩区总面积约7760.14km^2，占全省土地总面积的4.65%，保护耕地67.82万hm^2（1017.23万亩），保护人口1156.18万人。现有66.67hm^2（1000亩）以上圩堤884座，堤线总长6405.04km，其中0.67万hm^2（10万亩）以上圩堤18座，堤线长度108.71km，保护耕地29.02万hm^2（435.28万亩）保护人口397.34万人；0.33万～0.67万hm^2（5万～10万亩）圩堤27座，堤线长度790.10km，保护耕地11.18万hm^2（167.69万亩），保护人口264.48万人；0.07万～0.33万hm^2（1万～5万亩）圩堤117座，堤线长度1581.07km，保护耕地13.88万hm^2（208.24万亩），保护人口216.18万人；66.67～666.67hm^2（1000～10000亩）圩堤722座，堤线长度2949.31km，保护耕地13.73万hm^2（206.02万亩），保护人口278.18万人。圩堤现状防洪能力3～100年一遇不等。

全省圩堤具有堤线长、分布范围广、不少圩堤兴建年代久远、防洪标准偏低、涵闸老

本文发表于2003年。

化等特点。半数以上圩堤是新中国成立前群众修建的，这些圩堤兴建投运以来，大多无任何勘察设计资料，对堤基未采用任何有效措施处理，堤身填筑土料较杂，碾压不密实，故这些圩堤堤身、堤基存在较严重的工程隐患。每遇大洪水，险情不断，并曾多次决堤，造成严重损失。新中国成立后，堤防建设有了很大发展。1954 年、1962 年、1965 年大水后连续几年分别进行了大规模的圩堤培修加固，还围筑了一些新堤。1998 年大洪水后，随着垂直防渗技术（如射水造墙、高压喷射灌浆等）在圩堤除险加固工程中的推广运用，相对封闭抒区的数量也逐年增多。

全省各地圩堤经历次加固加高，工程防洪标准有了较大提高，有些圩堤（如南昌市城防堤）达到了 100 年一遇的防洪标准，但各地圩堤防洪能力参差不齐。据统计，全省尚有 81 座 666.67hm^2（1 万亩）以上圩堤防洪能力还不到 10 年一遇，其中不少圩区涵闸老化失修，排灌水系不健全。每逢汛期，外洪内涝，渍涝成灾，险象环生，严重威胁圩区人民生命财产安全。

2 圩区治理的原则与工程措施

2.1 治理原则

圩区治理必须坚决贯彻执行国务院有关洪患治理的十六字方针“整治河道、退耕还林、退田还湖、平垸行洪”，服从流域总体规划布局，做到上下兼顾，统筹规划，土方和建筑物并举，在确保防洪安全前提下，搞好灌溉，排除渍涝，兼治旱、淤。按照“内外分开、水旱分开、留湖蓄洪、排蓄结合、以排为主、灌排分开”的原则，全面规划，综合治理。

2.2 工程措施

目前，圩区治理的工程措施主要有疏浚内圩外河、联圩并圩、加固加高圩堤、改造病险涵闸、健全排灌水系、退耕还湖等措施。

（1）疏浚圩内圩外河道。结合圩堤加固，大力搞好沟河的清淤疏浚，改善环境，恢复沟河引排功能。

（2）联圩并圩。将零星分散的小圩联并成大圩，缩短圩堤长度，既能减少修筑土方和挖占土地，又能减轻防洪负担，减少圩堤渗漏，增加圩内水面积，提高圩内河网的调蓄能力。

（3）加固加高圩堤。对圩堤险工险段进行加固加高，目前主要采用以下工程措施：加高培厚，混凝土、草皮护坡，坡式、墙式护岸，填塘固基，水平铺盖，堤内排水减压井，垂直防渗帷幕等。

（4）病险涵闸改造。结合圩堤加固，对病险涵闸进行接长加固，拆除重建等处理，使圩区建筑物真正达到能挡、能排的要求。

（5）合理配置动力，健全排灌水系。根据圩内沟河面积大小、要求预降的深度、稻田滞蓄能力、水旱作物品种布局、道路村庄所占面积等，合理配置排涝动力。同时因地制宜地建立排灌水系，平整土地，解决排、灌、引、降、航矛盾。

（6）退耕还湖。圩区治理时，对影响防洪的，一律退耕还湖；对影响排水的进行分圩让路，畅通排水。

3 圩区治理的思考

1998 年全国流域性的特大洪水后，国家加大了堤防建设的力度。堤防建设的重点是对那些达不到防洪标准和汛期出险的堤防进行除险加固。堤防出险主要是堤基渗透破坏、堤岸迎流顶冲产生崩岸、堤基软弱土层产生沉降变形破坏、堤身散浸或脱皮滑坡等。所有上述出险问题中，危险性最大的是堤基渗透破坏，牵动全国的九江大堤溃口就是堤基渗透破坏引起的。工程上对于渗透破坏一般采用铺盖、堤内侧设排水减压井、浅基截渗墙和垂直防渗帷幕等措施。水平铺盖由于占地多及黏土缺乏等原因采用较少，减压井由于易老化，亦很少采用。1998 年洪水后，全省采用较多的是垂直防渗技术（射水造墙、高压喷射灌浆等），达到了较好的堤基防渗除险效果，但采用这一方法所带来的新问题却是值得我们思考的。

3.1 堤基垂直防渗阻断地下水的正常排泄带来的环境地质问题

一般说来，江河湖海是区域性地表水和地下水的最低排泄基准面，在没有进行堤基垂直防渗处理前，地下水仍能按照区域性地下水的总体流向，通过最短路径排向江河湖海。然而，当实施了一定范围内的连续性垂直防渗帷幕后，天然状态下的地下水渗流场被人类活动改变了，地下水的排泄通道被阻断了，新的环境地质问题将随之产生。

如果大范围截断地下水的排泄通道，地下水位必将壅高，使得土地产生沼泽化或盐碱化，进而影响农作物、建筑物、人类生存环境。特别是一些靠山临水的地区（如江西景德镇、临川、上饶等城市），植被覆盖良好，大气降水的相当一部分渗入地下，再排入天然河道，实施了封闭的垂直防渗后，将使这些城市全年在水中浸泡。

3.2 堤基垂直防渗阻断江河湖水正常补给地下水带来的环境地质问题

堤基垂直防渗除阻断地下水的正常排泄外，同样也会阻断江河湖水正常补给地下水，这样也将打破区域地质环境的平衡。地下水主要靠江河湖水集中补给的一些地区，当地工农业用水除了靠江河湖水外，另一主要来源便是开采地下水。由于截断了江河湖水向地下水的正常补给通道，势必形成只采不补的恶性局面，导致地下水位的下降，影响地下水的正常开采和利用，甚至引起建筑物基础应力条件的改变。产生建筑物失稳的严重后果。

再者，在外河来水量相同的条件下，若阻断江河湖水对地下水的补给，必将导致江河水位偏高，按理我们应该加高加固堤防。而实际上，设计时对此未予考虑，防洪标准没有相应提高，造成生态环境的恶性循环。

4 结语

1998 年洪水后，国家加大了堤防建设的投入，全省掀起了一场轰轰烈烈的堤防建设的高潮，达到了较好的治理效果，工程防洪标准也有了较大提高。面对成绩，我们既要总结经验，但同时也不能忽视所带来的问题。我们治理的目的是改善自然环境，而不是使其恶化。因此，对圩区采用垂直防渗措施时，可否考虑在部分堤段采用其他工程措施，避免形成“全封闭”圩区；圩区封闭的相对程度达到多少比较合适，而不至于破坏环境，这些都有待于我们做进一步的分析研究。

参考文献：

[1] 韦港，冀建疆．堤防工程与环境地质问题［J］．水利规划设计，2000，1.
[2] 江西省圩堤基本情况汇编［R］.

江西省10万亩以上重点圩堤蚁害现状及防治对策

王小春，刘晓容，张绍付

江西省水利科学研究所

摘　要： 对江西省18座10万亩以上重点圩堤蚁害的蚁种、原因、危害现状和1998年汛期部分大小圩堤出现蚁害险情进行了调研和分析，提出了今后防治措施，并按圩堤蚁害程度分为一类、二类、三类、四类，对今后圩堤白蚁防治有着重要的指导意义。

关键词： 重点圩堤；蚁害；调查；防治对策

江西省现有大小圩堤4000余座9800多km，其中保护耕地10万亩以上的重点圩堤有18座，均为土堤。圩堤土壤厚实略带酸性，堤身杂草灌木丛生，温湿度适宜，有丰富的水源，是土栖白蚁营巢生存和取食的大本营。近10多年来南方省区圩堤因蚁害出险告急的事故时有发生，如广东省梅县地区大堤1986年曾发生62处决口倒堤，其中就有55处（占89%）为蚁害导致，直接经济损失超5亿元。古人云："千里之堤，溃于蚁穴"，江西省圩堤蚁害现状如体呢？为了全面摸清全省各堤蚁情，提出今后江西省重点圩堤白蚁防治对策，按照江西省水利厅的部署，江西省水利科学研究所于1997年6月开展了江西省10万亩以上重点圩堤白蚁危害的课题调研，现将调研现状及防治对策分述如下。

1　圩堤蚁害调查

1.1　蚁种

调查发现危害圩堤白蚁种类繁多，根据形态特征鉴别主要蚁种有黑翅土白蚁、黄翅大白蚁、家白蚁、黄胸散白蚁和歪嘴土白蚁，其中对圩堤危害最大的有黑翅土白蚁和黄翅大白蚁，号称"无牙老虎"，是水利工程的蛀虫。

1.2　圩堤产生白蚁的原因

1.2.1　堤基蚁患

建堤前，由于未清除堤基内的蚁巢，给大堤留下蚁患，由此使圩堤基础出现早期渗漏。

1.2.2　白蚁蔓延

附近山坡林木、村庄等环境蚁源蔓延到堤上来。如周围环境中枯萎树木、老树兜和杂草灌木丛中经常孳生大量白蚁，它们为了生存和发展，便到处寻找食料和水源，而圩堤有草皮

本文发表于2000年。

护坡和野生杂草灌木，这是土栖白蚁生活所需的食料，同时堤身土壤为略带酸性黏土，具有保温保湿作用，为白蚁提供了良好的生存条件，因此周围环境中蚁源容易蔓延到堤身。

1.2.3 有翅繁殖蚁飞上堤

在每年4—6月白蚁纷飞季节，由于大堤周围夜间灯光明亮，江湖水面月光反射，容易将附近山坡和村庄田野中的有翅繁殖蚁吸引到堤身上，雌雄成双配对后入土营造新蚁巢，这是圩堤白蚁的主要来源。

1.2.4 人为造成

有些单位管理制度不严而招引白蚁。如经常在堤上晒柴草、堆放木材，或在堤上建木宅、造坟墓、盖猪栏，或在堤旁种植许多白蚁喜食的林木，这些都容易招致白蚁为害。

1.3 蚁害现状、程度的等级分类

主要调查10万亩以上的重点圩堤18座，调查截止时间为1998年11月底详见表1。结果表明：有14座圩堤发现不同程度的蚁害，占调查总数的78%，有4座圩堤暂未发现蚁害，占调查总的数的22%。

表1　江西省10万亩以上重点圩堤蚁害初步调查统计及治理方案表

序号	圩堤名称	堤身蚁害情况	蚁种鉴定			蚁害地表迹象统计		堤身蚁害部位险情	危害程度分类	治理方案
			黑翅	黄翅	其他	泥被泥线/片	分群孔/丛			
1	红旗联圩	查出大量蚁害	√			75	57	渗漏29处	四类	诱杀、灌浆、预防
2	长乐联圩	查出大量蚁害	√		√	54	19		三类	诱杀、灌浆、预防
3	蒋巷联圩	查出大量蚁害	√	√		66	20		三类	诱杀、灌浆、预防
4	南新联圩	查出大量蚁害	√	√	√	50	29		三类	诱杀、灌浆、预防
5	廿四联圩	查出大量蚁害	√			76	36	渗漏45处	四类	诱杀、灌浆、预防
6	赣西联圩	查出大量蚁害	√			45	14		三类	诱杀、灌浆、预防
7	饶河联圩	查出少量蚁害		√	√	12			二类	诱杀、预防
8	珠湖联圩	暂未查出蚁害							一类	诱杀环境蚁源，灭繁殖蚁
9	梓埠联圩	查出大量蚁害	√		√	53	174	渗漏58处	四类	诱杀、灌浆、预防
10	康山大堤	暂未查出蚁害							一类	诱杀环境蚁源，灭繁殖蚁
11	信瑞联圩	暂未查出蚁害							一类	诱杀环境蚁源，灭繁殖蚁
12	军山湖联圩	暂未查出蚁害							一类	诱杀环境蚁源，灭繁殖蚁
13	抚东堤	查出大量蚁害	√	√	√	34	168		三类	诱杀、灌浆、预防
14	富大有堤	查出大量蚁害	√	√		198	24	渗漏5处	四类	诱杀、灌浆、预防
15	丰城联圩	查出少量蚁害		√		19			二类	诱杀、预防
16	唱凯大堤	查出大量蚁害	√			46	12	跌窝、管涌8处	四类	诱杀、灌浆、预防
17	赣东大堤	查出大量蚁害	√	√	√	812	386	渗漏、跌窝65处	四类	诱杀、灌浆、预防
18	长江大堤	查出大量蚁害	√	√	√	191	96	漏水、跌窝72处	四类	诱杀、灌浆、预防

注　蚁害调查统计资料截止时间为1998年11月底。

根据蚁害调查情况，我们提出将圩堤蚁害程度分为一类、二类、三类、四类。

一类蚁害堤：是指目前暂未发现蚁害，但周围环境100m内发现蚁源的圩堤。一类蚁害堤有4座，为珠湖、康山、信瑞、军山湖联圩，占调查总数的22%。该类堤应采取必要的预防措施，以免蚁害发生。未查到蚁害的原因可能有：①对白蚁危害的严重性认识不足；②由于白蚁危害较隐蔽或蚁害较轻，仔细查找；③虽然有蚁害，由于未能按省厅制定的普查大纲进行全面普查，而只是作了局部检查。

二类蚁害堤：是指蚁害程度一般，堤身仅发现少量白蚁及泥被泥线，且周围环境50m以内发现蚁源的圩堤。其蚁害有进一步发展蔓延的趋势，需采取必要防治措施，把蚁害消灭在萌芽状态之中。该类堤有2座，即饶河联圩堤和丰城圩，占调查总数的11%。

三类蚁害堤：是指蚁害程度较重，堤身发现大量白蚁、泥被泥线和分群孔（或分飞现象）的圩堤。需及时治理。该类堤有5座，占调查总数的28%。

四类蚁害堤：是指蚁害程度较重，堤身发现大量白蚁、泥被泥线和分群孔（或分飞现象），且已造成渗漏、管涌、跌窝、滑坡和决口等险情的圩堤。由于蚁害险情严重，应立即治理，避免只加固不灭蚁，险情年年发生。该类堤有7座，占调查总数的39%。

以上分类是相对的，如不及时预防和治理，任其蔓延和发展，则会导致蚁害堤的升级。

2 圩堤蚁害严重的原因及危害机理

2.1 蚁害严重的原因

2.1.1 长期不治理

部分圩堤自建堤以来，一直未进行蚁害治理，致使蚁害不断蔓延和发展，成为蚁害严重的堤，如红旗联圩、廿四联圩等。

2.1.2 治理方法不当

(1) 只挖巢不诱杀。由于追挖过程中很容易惊动白蚁，造成不少漏网繁殖蚁（未来蚁王蚁后）和补充型蚁王蚁后迅速迁移，在失去原来蚁王蚁后的激素控制情况下，这些漏网的白蚁容易自立王国，经雌雄配对后营造更多的蚁巢，例如长江大堤、赣东大堤、唱凯大堤、信江大堤过去曾挖出大量蚁巢，此次普查发现蚁害仍然相当严重。

(2) 灌毒浆前不诱杀。因事先没有诱杀白蚁，白蚁本身具有较强的自卫能力，会调集大量的工蚁把浆路或主干蚁道迅速封堵，致使毒浆无法渗入主巢和副巢，往往只能杀死少量支蚁道上的外出白蚁，而大部分白蚁却安然无恙，如长江大堤、赣东大堤、信江大堤也曾灌毒浆，但灭蚁效果却很差。

(3) 喷砒霜（或六六六粉）。由于砒霜和六六六粉为毒性较强的触杀性农药，往往只能把外出白蚁迅速击毙，而无法将药物传染给蚁巢内的白蚁。施药方法不外乎将大量砒霜和六六六粉喷洒在挖巢位置或蚁害严重的部位，往往造成白蚁不敢外出已被全部消灭的假象，实际上圩堤深处巢内白蚁依然存在，待药物被雨水冲刷逐渐减弱或散失，1～2年后又会发现大量白蚁活动。如长江大堤、信江大堤和富大有堤等。

2.2 蚁害导致圩堤出险的机理

黑翅土白蚁和黄翅大白蚁，在圩堤内筑巢深度可达3.0m以上，主巢常常修筑在浸润线的上方，多数蚁巢（含副巢）分布在圩堤上半坡，如堤顶或堤肩的正下方 3.5m 深，或堤腰正下方 1.5～2.5m 深，调查中有 12 座堤已发现较多分群孔（为三类以上蚁害堤），说明蚁巢已发展为成年老巢，其主巢与众多卫星副巢间路路相通，形成四通八达的地下“交通网络”。由于圩堤内外坡均发现泥被泥线和分群孔，说明蚁巢的主干道已穿透堤身形成漏水隐患，随时可能造成险情，尤其是汛期水位持续高涨不下时，可能形成天然的漏水通道，造成圩堤长期散浸、集中渗漏和管涌等险情，甚至导致跌窝、决口。我省的红旗、廿四、梓埠、富大有、唱凯、赣东和长江大堤也曾发生过因蚁害导致圩堤漏水、跌窝等险情。

3 1998 汛期部分出险圩堤的蚁害调查

针对江西省 1998 年遭受百年一遇的特大洪水，造成部分圩堤出现管涌、塌陷、滑坡和决口等险情，我们重点调查了几座出险圩堤的蚁害。

3.1 唱凯大堤

1998 年 6 月 24—26 日堤顶漫水 3d，造成多处堤段出险，钟家、周博巷段 2＋500 堤顶突然下陷 1.5m，堤内坡发生严重脱坡、管涌，管涌水流中冲出大量的白蚁。在抢险中发现堤顶下陷 1.5m 以下有直径为 0.6m 的土栖白蚁巢。艾巷段 71＋600，烟墩村 26＋200～26＋400 段，1998 年也同样出现跌窝、漏水等险情，经开挖至 0.7～1.2m 处，挖出直径 0.25～0.45m 的大小白蚁主副巢，共计 8 窝。汛后距在险段 200～400m 范围内检查发现大量活动迹象。

3.2 梁公堤

检查发现约 3000m 堤段有严重的蚁害，查出泥被泥线 89 片，分群孔 54 丛，1998 年汛期累计发现 66 处因蚁害引起的堤身漏水、跌窝等险情。

3.3 富大有堤

富大有堤 2＋000～8＋900 段曾有私人蚁商在此段挖出蚁巢 3 窝。此次检查仍然发现泥被泥线 198 片，分群孔 24 丛。1998 年汛期累计发现 5 处因蚁害引起的堤身漏水，并冲出大量的白蚁。

3.4 信江大堤

信江大堤 1998 年 6 月遭受特大洪水袭击，造成圩堤决口达 16 处，决口长度累计 690m，直接经济损失达 12 亿多元。经调查发现其中有 10 处决口是白蚁危害导致的，在决口断面上检查到较多的活白蚁、冲毁的主副巢、主支干蚁道、候飞室等残缺蚁迹。据鹰潭 1991 年蚁害普查统计，调查 20 座圩堤总长 84.08km，查出 18 座有蚁害，占普查工程的 90％，蚁患 164 处，危害总长度 8200m。

3.5 金湖联圩

共青城金湖联圩四合圩段（5＋765），自 1995 年就发现内坡 2 处漏水，漏水口较小，呈管涌状，由于当时不曾想到是蚁害引起的，仅采用一般工程除险措施进行了处理，未诱杀白蚁。1998 年汛期的 6 月 25 日，水位高程 21.00m（堤顶高程 23.00m），同样地点发现 2 处管涌，因水位高且流速急，2 处管涌口越冲越大（达 15～20cm），并漏浑水，因发现及时并采取堵漏措施，避免了决口。汛后组织劳力进行开挖检查漏水原因，结果挖出 2

窝黑翅土白蚁巢，主巢深约 3.5m，大约的蚁巢约 $0.8m^3$，可容纳 3～4 个成人，小的蚁巢约 $0.4m^3$，巢内有蚁王蚁后、幼蚁、蚁卵、工蚁、兵蚁达上百万只，主巢周围还有 30 多个副巢、候飞室及空腔，蚁巢主干道（5～15cm 宽）达 40 余条。

3.6 九江城防堤

1998 年 8 月 27 日九江城防堤 0＋813～0＋875 段出现大溃堤。汛后 1998 年 10 月我们对决口段及两端 40～100m 范围的内坡进行了详细检查，并设置 48 个引诱坑，都未能查出白蚁活动迹象，但在决口段的内坡堤脚附近的厂房、民宅、绿化树发现有白蚁危害。堤身未查到白蚁原因：①决口段冲毁严重，并抢险重新填筑；②汛期堤上人来人往，和汛后游客多的干扰，加上堤身表土已被踏实板结，不易查出白蚁活动迹象，设置的引诱坑也无法引出白蚁；③堤身暂无蚁害。

3.7 赛湖联圩

都昌县汉港乡的赛湖联圩东风堤段 1998 年汛期出现 7 处白蚁跌窝，最大跌窝深达 3m 以上，汛后检查发现赛湖联圩尚有 6000m 堤遭受较为严重的蚁害。

3.8 西河东联圩

波阳县西河东圩 33＋050～35＋550 段查到泥被泥线 40 片，分群孔 30 丛。1995—1998 年汛期年年漏水，累计发现 13 处大面积漏水和滑坡，并冲出大量白蚁和巢片。

3.9 郭东圩

永修县郭东圩 5＋600～5＋800 段白蚁危害较重，查到大量的泥被泥线和分群孔。1998 年汛期堤身多处漏水严重，汛后在漏水部位挖出 2 窝大蚁巢，最大一窝蚁巢达 $1m^3$，可容纳 4～5 个成人，巢内有数百万的工蚁、兵蚁、幼蚁及蚁卵。

由于时间、人力所限我们未能对更多出险工程进行现场调查，再加上出险时工程管理人员在紧急抢险过程中无暇仔细观察是否有蚁害，而汛后出险现场已经破坏，难以查找蚁巢，所以 1998 年汛期还有很多出险工程与蚁害有关，有待以后进一步调查。

4 防治对策

建议采用水利部推广的堤坝白蚁防治新技术措施，即“找标杀、找标灌、找杀防”。江西省水利科学研究所近 10 年来，采用该技术措施先后治理了潘桥、金桥、黄金、窑里、龙井、长垅、中村、梁公堤等 20 多座大中型工程蚁害，取得了显著效果。省内外实践证明该技术措施与挖巢灌毒水、喷洒砒霜（或六六六粉）和设毒墙（层）相比，具有投资省，见效快，且不污染环境，不开挖坝体，操作简单，效果显著，是目前国内堤坝白蚁防治最理想技术措施。

（1）找标杀：适用于一类、二类、三类、四类堤害堤。即找蚁迹：寻找白蚁活动的泥被泥线和分群孔；标蚁迹：标记白蚁活动区域，尤其是分群孔和常现区的中心位置；诱杀：在白蚁活动区域内见蚁施药，尤其是分群孔和常现区重点施药诱杀；或布坑诱杀；沿堤轴线设置数排梅花形的诱坑，诱集大量折蚁在坑内取食，然后集中喷粉施药诱杀。灭蚁原理为利用白蚁的生活习性，让外出采食的工蚁沾染或取食慢性胃毒蚁药，带回蚁巢相互传喂、舐吸传染，最终导致全巢白蚁彻底中毒而死亡。

（2）找标灌：适用于诱杀后堤段的找巢灌浆。即找炭棒菌灌浆：在白蚁活动区域内，尤其是分群孔和常现区附近仔细查找炭棒菌（死巢正上方长出的真菌指示物），并做上标记，然后灌纯泥浆填巢、堵蚁路；分群孔图像法定巢灌浆：在没有查到炭棒菌情况下，可在分群和常现区附近用图像法找巢，然后灌浆填巢、堵蚁路。

（3）找杀防：对堤段周围环境中的蚁源进行诱杀根治，起到预防的目的。即找杀蚁源：查找圩堤周围的蚁源，然后重点施药诱杀，彻底消灭环境蚁源；恶化生存条件：铲除圩堤100m内较高的杂划灌木，恶化白蚁取食和生存环境，预防白蚁蔓延上堤，确保治理后圩堤长期达到无蚁害目的。

5 建议

5.1 加强蚁害险情监测

从调查情况来看，江西省18座10万亩以上重点圩堤有14座发现不同程度的蚁害，占调查总数的78%，其中有7座圩堤（唱凯、富大有、红旗、廿四、梓埠、赣东、长江大堤）已出现蚁害漏水险情。此外，1998年大洪水中许多圩堤出现渗漏、滑坡、管涌、跌窝等险情，肯定有不少与蚁害有关的险情被漏查，因此必须加强汛前汛后的蚁害险情检查，并采取必要的预防措施减少险情的发生。

5.2 重视普查和年报

重视普查和防治工作，坚持把蚁害治理列入水利年度工作计划中，每年开展一次蚁害普查，要求各地市将蚁害防治计划与水利年度工作计划一并上报。调查过程中发现越是领导重视普查，普查工作开展得早且检查越仔细的圩堤管理单位，发现的蚁害就越多。而有些圩堤，初查时未发现蚁害，后经我所专业人员实地检查，却发现不少新蚁害（如赣西、南新联圩等）。

5.3 完善组织机构，加强科研

健全和完善我省水利工程蚁害防治组织机构，成立江西省堤坝白蚁防治中心站，落实人员和增加治理及科研经费，业务上由江西省水利厅主管，江西省水利科研所负责技术研究和治理工作。

5.4 先灭蚁后加固

过去江西省重点圩堤加固培修时，事先均未治理蚁害，使蚁患严重，增加了治理难度，尤其对目前兴起的用土工合成化纤材料进行处理的堤防工程，由于材料埋入堤土后容易遭到土栖白蚁的蛀穿和分泌物腐蚀等形式的破坏，进而影响材料本身性能和使用寿命，所以今后加固培修时要先灭蚁后加固，对新建圩堤应防治堤基早期蚁害及环境蚁源。

5.5 治后坚持预防工作

江西省部分水库数10年前进行过治蚁，灭除了蚁害，但治后因长期不重视预防，导致近年来又发现蚁害。由于蚁源广，繁殖蚁飞翔能力强（顺风飞可飞翔500～800m），可能会飞到堤上营巢，造成新的蚁害，所以治后的预防要坚持不懈，其所需经费少，花工不多，工程单位可自力更生搞好，关键是领导重视。

5.6 严禁蚁商擅自挖巢，规范治理

近几年来湖北、安徽等地私人挖蚁商在圩堤上挖巢和灌毒水的次数呈上升趋势，挖巢和灌毒水对水利工程是有弊无利，其灭蚁效果差，且容易破坏圩堤，造成新的险情。所以

今后堤坝白蚁防治要严禁私人蚁商擅自挖灌蚁巢，应由用具备水利和堤坝白蚁防治知识的专业队伍按照水利部推广的“找标杀、找标灌、找杀防”技术措施实施治理，根治蚁害，以确保工程安全。

5.7 制定规划

江西省尚无近期和远期治蚁规划，很有必要制订今后5～10年防治规划，并落实治蚁专项经费，按蚁害轻重缓急有计划统一安排全省的白蚁防治，特别是尽早安排三类、四类蚁害堤的治理。

我省水库溢洪道挖潜改造的思考

王仕筠

江西省水利科学研究所

摘　要： 本文讨论了江西已建水库无闸控制溢洪道进口段的挖潜改造问题。从列举的工程实例可以看出，溢洪道进口型式的选择，对提高水库的效益是非常重要的。

关键词： 挖潜改造；溢洪道进口段；无闸控制；水库效益

1　水库挖潜改造是现实的需要

水利资源的合理开发利用，是可持续发展思想内涵中的重要组成部分。在积极开发新资源的同时，应注意到许多开发条件较好的坝址大部分已经开发，因此在已建成的水利枢纽中，除面临加强管理外，更有依靠科技进步，走技术改造之路的需要。此外，20 世纪 90 年代以来，随着劳务、材料和设备价格的放开，与 80 年代的价格水平比较，90 年代中期水电工程建设的主要材料及设备价格分别上涨了 230%～470%，是造成工程造价大幅度增加的主要原因（见表 1）。过去建水库是国家拨款，现在由于国家经济政策的调整，水库是中央、省、地方集资贷款建设。新建一座中型水库，一般要耗资数千万元，一座小型水库也需几百万元，这对一般的县、市都是一件不容易的事。例如，赣南近期完建的长滩、罗边和仙人陂 3 座中型电站，年应付利息就分别要 400 万元、1300 万元和 1047 万元。根据江西省 90 年代新建部分电站的调查，单位千瓦的造价大概约 0.8 万元左右（表 2），而对条件较好的电站进行技改，可以节省投资（见表 3）。因此，在新建水库的同时，如果我们也注意选择一些潜力大且改造比较简单的水库，利用它们的已有设施和水力资源，走依靠科技进步，走技术改造之路，使这些水库重新焕发活力，往往可以取得投资少、收效快的效果。

2　江西省中、小型水库溢洪道改造的潜力

江西省中型水库为无闸门控制的溢洪道约占 66%左右，小型水库则超过 90%，几乎均采用渠式或宽顶堰。由于流量系数 m 值低，一般 $m=0.300\sim0.360$，故单宽流量小，需要的过流宽度大，因此这是一种最不经济的进口型式，但对改造而言则潜力较大。如将引渠挖深 H_0 值（溢流水头），加建一座堰高 $P=H_0$ 的实用堰，可使 $m=0.440\sim0.460$，这样在不改变正常高水位和 H_0 值情况下，就可增加泄流量 20%～50%，据资料分析，江

本文发表于 2000 年。

西省中、小型水库无闸溢洪道堰顶以上的滞洪库容占总库容的 1/5～2/5，是兴利库容的 1/4～2/3，这对主要任务是蓄水抗旱的中、小水库而言，可以认为库容的利用价值是较低的。此外，我们还可以用指标 L/H_0（L 表示溢流堰长度）来检验我省无闸溢洪道设计的经济性。通常 L/H_0 越大，相对来说堰顶以下的蓄水库容就大，工程效益也高。江西省资料统计 $L/H_0<20$ 的占 75%，其中 $L/H_0<10$ 的达 40%左右，而 $L/H_0>30$ 的仅占 3%，而国外设计采用的各式异型进口如扇形溢洪道、Y 形溢洪道、迷宫堰溢洪道等，其 L/H_0 值都能超过 50，相比之下国内无闸溢洪道的 L/H_0 都不大（见表 4）。总的来看，国内的溢洪道工程在进口设计方面是大有潜力可挖的，如果能对已建工程的进口因地制宜地进行改造，将 L/H_0 提高到 40～60 或者甚至更大一些，经济效益就相当可观了。笔者到过的水库，不少溢洪道具备改扩建的可能性。

表 1　　水电造价增长幅度情况

时间	50、60 年代	70 年代	80 年代	90 年代初期	90 年代中期	90 年代末
造价/(元·kW^{-1})	1035	1190	1540	2720	4280～5500	8000
与 50、60 年代造价比/%		115	149	263	414～500	773

表 2　　江西省近期新建电站单位千瓦造价

水电站名称	南康罗边	上犹仙人陂	上犹龙潭	安远蔡坊	定南长滩	寻乌斗晏	宁都三门滩	于都渔翁埠	泰和南车	遂川安村	井岗山井岗冲
总库容/万 m^3	1590	1915	11560	2654	1155	9630	840	1085	15300	1965	1870
总投资/亿元	0.86	0.73	2.20	0.70	0.60	2.10	0.48	0.65	1.20	1.20	1.40
每 kW 投资/万元	0.82	0.66	0.55	1.00	0.75	0.56	0.75	1.08	0.94	1.00	1.17
每 kWh 投资/元	2.14	1.76	2.49	3.23	1.92	1.60	1.92	3.02	3.30	3.00	3.50

表 3　　吉水螺滩电站技改效益

技改方案	提高正常高水位	增加有效库容	原装机	新增装机	年增电量	每 kW 投资	每 kWh 电投资
溢流坝顶加橡胶坝	2.5m	740 万 m^3	6400kW	5000kW	1355 万 kWh	3660 元	1.35 元

表 4　　进口型式对 H_0、Q、W 的影响

组次	水库名	控制段宽度 W /m	H_0 /m	流量 Q /($m^3\cdot s^{-1}$)	进口型式	按宽度 W 计算的 m 值	$\frac{L}{H_0}$
A	波阳军民	30	3.83	350	无坎宽顶堰	0.351	7.8
	波阳滨田	27	4.36	352	渠式	0.323	6.2
	阿尔及利亚哈扎	30.4	1.90	350	迷宫堰	0.992	45.7
B	高安樟树岭	45	3.17	370	渠式	0.329	14.2
	美国鲍德曼	36.6	1.80	387	迷宫堰	0.990	60.7
C	高安碧山	40	2.25	221	宽顶堰	0.370	17.8
	浙江清溪口	40	1.75	525	U 形进口（实用堰）	0.449	65.2

3 挖潜改造的工程实践

3.1 迷宫堰的应用

迷宫堰的优点是在有限的宽度范围内可布置较长的溢流长度，因此能大幅度增加泄量，或者流量相同时可以减少堰顶水头 H_0，这样在最高洪水位不变条件下，抬高堰顶高程增加兴利库容。但是迷宫堰只有在相对堰高较大的条件下（如 $P/H_0>1.4$）才有较好的效益。因此渠式进口改为迷宫堰一般可能要增加一些溢洪道挖深，但效益可观。这几年迷宫堰在江西省水库改、扩建工程的应用，引起了地、市水利主管部门的关注。已经完建和正在施工的有好几座。

崇仁县石路水库，兴建于 1958 年以灌溉为主的中型水库，原溢洪道是无坎宽顶堰。1996 年改建将溢洪道最大挖深 1m，改建成迷宫堰（见图 1）。在最高洪水位基本不变情况下，正常蓄水位提高 0.5m，增加兴利库容 130 万 m^3，相当原兴利库容的 13.6%，总共投资 10 万元左右。

贵溪忠心水库，原溢洪道过水宽度一般只有 6.9～8m，泄洪能力仅达到设计要求的 18%。扩建曾考虑扩宽和闸控两方案，前者由于地形条件，两侧扩宽工程量太大根本不能实现。后者管理不便（小型水库）。1998 年扩建工程设计改为迷宫堰，工程造价低，运行安全方便。

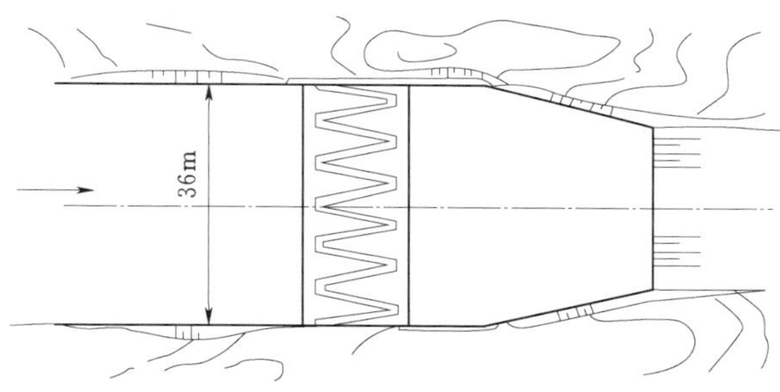

图 1 石路水库迷宫堰

3.2 异型进口应用——贵溪枫林湾水库扩建

原溢洪道进口为宽顶堰，溢流长度 25m，后接 1：3 陡槽段和鼻坎挑流。由于泄洪能力不够，使水库长期控制蓄水。1988 年进行改造，在校核水位（$P=0.1\%$）和堰顶高程（即正常蓄水位）都不变的条件下，通过改变进口段的平面布置，溢流堰增长至 40m，在堰后设过渡段使过堰水流转向为原陡槽流向一致（图 2）。除此之外，溢流堰的堰面采用江西省水利科研所的科研成果简易Ⅳ型低堰，增大 m 值，使泄洪能力满足了要求。工程至今运行情况良好。

3.3 在溢流堰（坝）上加橡胶坝

橡胶坝具有可充气与放气的特性，因此它对减轻洪水和环境保护起到很好的作用，而且坝后沉积的泥沙、碎石等可通过定期放气加以排除。目前全世界约有 600 多座橡胶坝，其中美国有 50 多座，这些坝的高度一般低于 5m，最长的达到 122m。

芷南水电站是吉安枫渡水电厂的二级电站，1966 年开始发电。电站上游 7km 处是枫

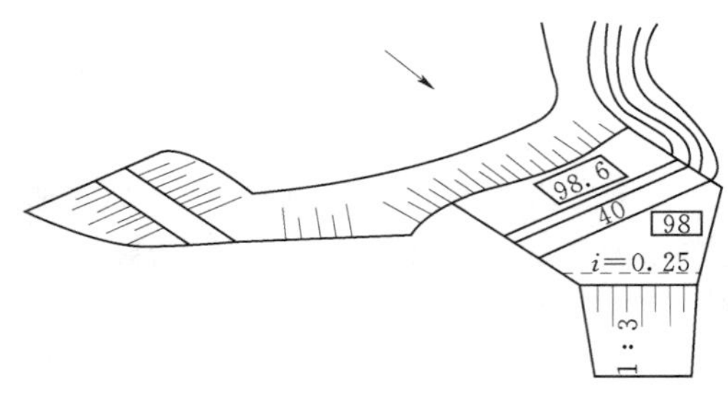

图2 贵溪枫林湾水库

渡一级站。两站正常发电时有 $7m^3/s$ 未能被芰南二级站利用。枫渡电站正常发电时，下游尾水位为136.7m，而芰南二级站的溢流堰顶高程为133.5m，水位差3.2m，原设计一级站的发电尾水位与二级站的正常蓄水位差为1.6m，故其间尚有1.6m水位差值可以利用。考虑到回水和其他不利因素影响，挡水建筑物的高度定为1.5m，在原奥氏溢流堰上安装单锚枕式充气橡胶坝，坝体部分投资31万元，计入淹没补偿费合计总投资60万元左右，施工期只花30d。加坝后，在原发电设备没有进行增容改造的情况下，年增电量176万kWh，以0.25元/kWh计，一年可增收44万元，全部收回橡胶坝的工程投资。如果再把现有2台机组进行增容改造，每台出力可由1250kW增加到1950kW，这样年增电量将超过400万kWh。

4 思考与讨论

1996—1997年期间，笔者曾对省内70个县（市）近百座水库进行过调查，得到的看法是：现有水库的溢洪道中，有相当一部分改、扩建后能取得明显效益。有的水库枢纽建筑物能适应水位抬高后的要求，上游库区淹没不大，甚至没有淹没赔偿问题，挖潜改造条件优越。但是也有的水库规模早已达到设计要求，只由于溢洪道没有完建（如永修云山水库、瑞昌幸福水库和横港水库等）或者早期移民留下的问题一直没有解决（如新余江口水库），致使历时30～40年的水库，至今仍处于低水位运行，效益有限。此外，20世纪80年代以来开展的病、险水库修复工作没有将工程的除险加固与挖潜改造较好地结合起来，水库的挖潜改造是一项很有意义的工作。全国已建的84000多座水库中，如按座数计，99.5%左右是属中、小型水库，它们数量多作用大。

在调查中看到的溢洪道改、扩建方案，基本上保持原渠式或宽顶堰，采用扩宽或加大溢流水深 H_0 即加高大坝或降低正常蓄水位的办法增加泄流量。一般来说，扩宽的开挖量或加坝的填筑量都较大。如果能从改变原进口型式着手，进行方案比较，择优选用，既可节省投资，还可提高效益。

为说明问题，下面讨论几个工程实例。

4.1 宁都团结水库

宁都团结水库属大型，具备良好的扩建条件，如大坝曾按“75.8洪水”保安标准复核，运行正常；设置有净宽150m的非常溢洪道；现正常蓄水位242m，移民高程245.00m；装机2×1250kW，而引水系统的断面尺寸满足3×1250kW的过水要求。由于该县径流电站占的比例大，扩建该水库有利于电网调峰，改善供电质量。改造方案之一，可考虑将正常溢洪道20m溢流前缘的宽顶堰，改为实用堰，使流量系数 m 值提高，就可使堰顶高程抬高1m，闸门工作桥等可不变动。估算由此可增加年发电量250万kWh。

4.2 万载卅把水库

万载卅把水库以发电为主，集雨面积大，径流量大，库容小。大坝也是按“75.8洪水”标准加固的，1992 年被评为优良工程，大坝运行正常。近六年来水库在正常蓄水位 128m 以上运行达 360 多天，特别 1993 年有 1/3 的时间在高水位运行。由于汛期泄洪频繁，仅 1993 年一年就泄洪 7000 万 m^3，水力资源利用率低，且增加下游沿河两岸广大地区的洪水压力和洪灾损失。1969 年兴建水库的设计书中，就肯定了正常蓄水位 130m 的可行性，并作为后期目标。据近年调查，正常高水位 130～132m 新增加，淹没农田 0.2hm^2（3 亩），山地 16hm^2（240 亩），人行道改道了 3km，迁移农户一家。但是正常高抬高 2m（即水位 130m），可年增发电量 250 万 kWh。改造部分建筑物的投入资金 120 万元，3 年收回。

4.3 崇义长河坝水库

崇义长河坝总库容 1260 万 m^3，大坝为浆砌石重力坝，中段 20m 是溢流坝段。该库的改造方案可考虑采用坝顶加橡胶坝抬高正常蓄水位，其效益列入表 5。值得注意的是电站生产的电能是 80%属自供区使用。根据调查，正常高水位提高 2～3m，上游库区仅淹没农田 0.1333hm^2（2 亩），搬迁 1 户农家。橡胶坝总投资约 20 万元。

表 5　　　　长河坝改造效益

情况	正常蓄水位 /m	调节库容 /万 m^3	年均发电量 /kWh
现状	364.37	567	219
改造后	366.37	715	230
	236.00	367.37	795

4.4 宁都老埠水库

宁都老埠水库的大坝尺寸满足正常高水位抬高 1.5m 的要求。1990 年前多数年份水库的运行水位均超过正常高水位 327.5m，最高达 328.8m。1990 年大坝又经大面积坝基帷幕灌浆和坝体套井回填的防渗处理，这几年水库也曾在 328.3～329m 高水位运行。此外，水库是以发电为主，现一、二级电站的装机容量均偏大，适合于抬高正常蓄水位。由于正常水位的抬高，减少了溢洪道弃水的次数，从而也减轻现溢洪道平面布置上的不合理（泄槽 90°转弯）造成的严重冲刷。

改造方案是拆除现溢洪道进口控制段上高度 2.7m 的砌石实用堰，以迷宫堰替代，其他部分建筑物不动。水库主要特征值列入表 6，改造坝体总投资 33.3 万元左右，4 年收回投资。

表 6　　　　宁都老埠水库改造方案特征值

进口型式	堰段宽度 /m	堰顶高程 /m	设计水位 /m	校核水位 /m	大坝顶高程 /m	设计流量/ ($m^3 \cdot s^{-1}$)	校核流量/ ($m^3 \cdot s^{-1}$)	兴利库容 /万 m^3
直线实用堰（现况）	24	327.5	331.27	332.25	335	336	474	1255
迷宫堰（建议）	24	329.0	331.44	332.37	335	357	492	1365

4.5 修水南茶水库

修水南茶水库初具规模，1979 年停建时，库区淹没处理已经按正常高 214m 的原方案完成。大坝心墙填筑至高程 208m，坝壳 202m，坝基高程 185m。大坝右侧有 20m 宽溢流口作为临时溢洪道，其底板高程为 202.00m，相应兴利库容 365 万 m^3。1997 年水库所在的乡政府争取到 100 万元扶贫贷款资金，决定用于续建这座水库，主要目的是增加发电量。乡政府向设计者提出，按 100 万元决定规模。经方案比较，采用迷宫堰看来是较好的（见表 7）。初设结果溢洪道改造 42 万元，大坝加高加固 32 万元，电站改造 21 万元，加上税金和设计费共 106.4 万元，3 年收回投资。

表 7　　修水南茶水库方案比较

进口型式	最高洪水位/m	正常水位/m	兴利库容/万 m^3	新增灌溉面积/hm^2	新增年发电量/万 kWh
宽顶堰	208.8	202.00	420	85.33	76
直线实用堰	208.8	204.16	490	128	110
迷宫堰	208.8	206.00	610	224	200.8

4.6 波阳军民水库

波阳军民水库属大型，总库容为 1.84 亿 m^3，由于大坝除险加固没有完成，水库一直控制蓄水。30m 宽的正常溢洪道（渠式）渠底板高程 80.15m，低于正常蓄水位 2m，$Q=350m^3/s$，$H_0=3.83m$。水库还有一座能泄 $Q=1370m^3/s$ 的非常溢洪道。按洪水频率 $P=0.05\%$的最高洪水位计，现大坝的安全超高达 3.62m，以防浪墙计有 4.62m。一俟大坝除险加固完成，该库潜力较大。该库挖潜增效方案，在溢洪道进口控制段上建一座堰高 $P_2=2.5m$ 的实用堰，堰面也采用简易Ⅳ型堰面，使之满足条件：上游堰高 $P_1>0.52H_0$，下游堰高 $P>0.74H_0$，则 m 值可取 0.450，$H_0=3.3m$，故现最高洪水位 85.98m 不变情况下，溢流堰顶高程为 82.65m 即该方案能将原正常高水位再提高 0.5m。另一个改造方案是迷宫堰，其堰顶高程 87.70m，可获得很好的效益。

4.7 丰城观桥水库

丰城观桥水库每年弃水频繁，当地农民为了能让水库多蓄水灌溉，近年在汛末常用 5cm 左右的木板在溢洪道上挡水。县里拟定的溢洪道改造方案是建 20m 宽的开敞式折线实用堰，堰高 $P=0.87m$，$m=0.367$，$H_0=2.19m$，堰顶高程 49.67m（正常高）。笔者考虑，是否能将溢洪道挖深 0.3m，使之满足采用简易Ⅳ型堰面的前述条件，则 $m=0.450$，H_0 可由 2.19m 降为 1.9m，堰顶高程抬高为 49.96m，增加 0.29m 深度的兴利库容。

4.8 高安碧山水库

当初为解决溢洪道泄洪流量不足（差 $33m^3/s$），而将正常蓄水位由 58.01m 降至 57.47m，减少 0.54m。据了解该县的水库溢洪道进口型式单一，不是渠式就是宽顶堰。其实碧山水库溢洪道，如果改为一般的实用堰将 m 值提高至 0.370，就满足了泄洪要求。当然如果再进一步采用简易Ⅳ型低堰，正常高水位不但不降还可以抬高 0.22m，达到 58.23m。从这个例子看出溢流堰的堰面曲线对 m 值的影响是较大的。说明“优化设计”